AGJ
2587

Learning Resources
Brevard Community College
Cocoa, Florida

D1609089

DATE DUE	
DEC 0 9 1996	
DEC 0 9 1996	

Human Lead Exposure

Editor
Herbert L. Needleman, M.D.
Professor
Department of Psychiatry
Western Psychiatric Institute and Clinic
University of Pittsburgh
Pittsburgh, Pennsylvania

CRC Press
Boca Raton   Ann Arbor   London

Library of Congress Cataloging-in-Publication Data

Human lead exposure / editor, Herbert L. Needleman.
    p. cm.
  ISBN 0-8493-6034-X
  1. Neurotoxicology. 2. Lead—Toxicology. I. Needleman, Herbert L., 1927-
  [DNLM: 1. Environmental Exposure. 2. Lead—adverse effects.
3. Lead Poisoning.   QV 292 H918]
RC347.5.H86   1991
615.9'25688—dc20
DNLM/DLC
for Library of Congress

91-20525
CIP

This book represents information obtained from authentic and highly regarded sources. Reprinted material is quoted with permission, and sources are indicated. A wide variety of references are listed. Every reasonable effort has been made to give reliable data and information, but the author and the publisher cannot assume responsibility for the validity of all materials or for the consequences of their use.

All rights reserved. This book, or any parts thereof, may not be reproduced in any form without written consent from the publisher.

Direct all inquiries to CRC Press, Inc., 2000 Corporate Blvd., N.W., Boca Raton, Florida 33431.

© 1992 by CRC Press, Inc.

International Standard Book Number 0-8493-6034-X

Library of Congress Card Number 91-20525
Printed in the United States   2 3 4 5 6 7 8 9 0

PREFACE

Lead is a metal of considerable utility and versatility. It is also a persistent, ubiquitous danger. Warnings about its threat appear and reappear throughout history from ancient times to the present. An extraordinary amount of new information has been collected in the past few years about the nature of lead toxicity. This book is an attempt to chronicle this explosion of modern knowledge by some of the leading contributors to the field.

Thomas Kuhn's book, *The Structure of Scientific Revolutions,* has radically altered our understanding of how progress in human knowledge evolves. Kuhn portrays the course of scientific development not as a smooth uphill march towards objective truth, but as a struggle between contending visions and designated "paradigms" in his schema. Kuhn shows that any new information that challenges a governing paradigm is first ignored, then contested, and then forced to fit the preexisting paradigm. Each paradigm has its own set of favored tools, and these tools limit the accumulation of information to that which reinforces the paradigm. When the body of challenging data becomes too large to be digested, the old set of primitive beliefs, rules, tools, and language are overthrown, and a new paradigm is enthroned. In toxicology, the fashionable tools of the old paradigm were best exemplified by the LD_{50} which can only perceive and report whether the animal is alive or dead. Similarly in neurology, the Babinski sign is only interested in whether the toe moves up or down.

In the last decade, enormous progress has been made in the study of lead toxicity (certainly enough to define a paradigm shift). This has occurred synchronously with changes in the tools employed. Toxicology moved from measuring binary outcomes (life/death) to graded measures of performance, including behavior. Neurology came to recognize that many behaviors, including cognition, attention, and perception, are sensitive and critical measures of the state of the brain, deserving at least equal status with reflexes or CAT scans as indices of healthy function.

Only 50 years ago, the prevailing belief in pediatrics was that if a lead-poisoned child did not die, there were no durable sequelae. This dogma changed dramatically in 1943 when Randolph Byers showed that 19 of 20 children who had recovered from acute lead poisoning had behavioral or cognitive deficits. Byers asked whether a significant proportion of children with school failure might have undiagnosed lead toxicity, and opened the modern era of lead toxicology.

The question of whether silent doses of lead are associated with meaningful neuropsychological deficit received considerable attention in the past decade and a half, and became one of the most controversial fields in public health. In the early and mid 1970s, initial studies of children used small samples, less than sensitive measures of effect, and unsophisticated statistical analyses. Some showed a lead effect, and some did not. Fuel was added to the controversy because the presence or absence of low level lead effects became critical to decisions regulating lead in the air, gasoline, and housing stock. These decisions involved huge amounts of funds. The stakes, and the temperature of the discourse, were correspondingly high.

Epidemiologists entered the area and sharpened the focus of newer human investigations of lead at low dose. In the later 1970s, studies with larger samples, better measures of covariates, and sophisticated multivariate modeling began to be published. Bench scientists picked up the tools of neurobiology, molecular biology, and behavioral analysis and used them in the study of lead at low dose. These newer studies produced a wealth of new information, showing lead effects in systems previously not known to be vulnerable and at doses lower than previously acknowledged to be toxic. Studies of animal behavior found effects that paralleled to a remarkable degree those observed in children.

A consensus on health effects at lower dose began to emerge. In 1986, the Environmental Protection Agency reviewed the studies of children's IQ function at low dose and asserted that neurotoxicity ensued at blood lead levels between 10 and 15 µg/dl, and perhaps even lower. The pace of change is displayed when one recognizes that in 1960 a blood lead level of 59 µg/dl was considered harmless. The Agency for Toxic Substances and Disease Registry (ATSDR), in their 1988 Report to Congress, also concluded that the toxic level of lead in blood was 10 to 15 µg/dl. The ATSDR report indicated that the new definition of lead toxicity radically increased the estimates of the number of affected children. It seems clear that lead toxicity is the most common preventable disease of American children.

Despite the unprecedented amount of data establishing that this is the most common serious disease of childhood, little attention or redress has been given to the problem. There are at least four discernible reasons for this mismatch. First, for a long time, the disease has been considered solely a problem for poor inner-city minorities. This group has little access to the levers of power or public opinion, and this irony was compounded by some professionals who blamed the mothers for the problem. If only they took better care of the children, some said, the disease would not have occurred. Second, because billions of dollars are at stake, the lead industry weighed in on the argument and persevered in asserting that lead is harmless at doses below those that cause symptoms. Third, academic medicine and pediatrics, because the disease is one of low technology, have tended to ignore lead toxicity. Many pediatric house officers are not trained in the diagnosis and treatment of lead poisoning. Fourth, some government agencies have been less than arduous in meeting their responsibility to redress the problem. The Centers for Disease Control (CDC) is a clear exception to this observation. CDC, within the limits of a truncated budget, has pursued the detection and prevention of lead toxicity with vigor and intelligence.

The contributors to this book examine the field of lead poisoning from a number of perspectives. In the first section, the historical and environmental background of lead is discussed. Jerome Nriagu looks at the pharmaceutical uses to which lead has been put in many ancient cultures and notes that centuries ago many physicians warned that practitioners may be doing more harm than good by prescribing the metal. Jane Lin-Fu, who perhaps more than anyone after Byers has brought the question of low level toxicity to awareness, examines the modern history of lead toxicity. Paul Mushak, the senior author of the ATSDR report, discusses the analysis of lead in various media, with particular emphasis on quality assurance and control. Cliff Davidson and Michael Rabinowitz then survey the sources and pathways of lead to humans.

In the next section, the neurobiology of lead is examined. Ellen Silbergeld presents an approach to neurotoxicity. Donald Fox reviews the recent information on the impact of lead on special senses: vision and hearing. Gary Goldstein uses his studies of lead's effects on protein kinase C to develop an intriguing model of its effects on brain development. Deborah Rice reviews her elegant studies of lead given over the lifespan to nonhuman primates and notes the extraordinary correspondence to observations of deficits in children.

In the next section, data on human exposure is presented. Lead poisoning was first observed in the workplace. Thomas Matte, Philip Landrigan and Edward Baker review the modern studies of occupational lead exposure. Richard Wedeen examines the relationship between lead exposure and renal disease. David Bellinger and Herbert Needleman review the studies of lead at low dose and children's intelligence scores, and discuss some of the methodologic issues encountered. James Burchfiel, Frank Duffy, Peter Barthels and Needleman report on quantitative EEG findings in low level lead exposure, and note that using more sensitive methods of analysis of signal uncovers effects at lower exposures. Joel Schwartz first reviews the data on lead's effects on blood pressure, an area of extreme importance, and then looks at lead's impact on growth and endocrine function. Not all studies of lead and IQ have found effects at the customary $P = .05$ level of statistical significance. Constantine Gatsonis and Needleman review their meta-analysis of 24 modern studies of lead at low dose.

In the final section, the book moves away from the biology of lead exposure to look at economic, legal, and sociological issues. Richard Frank, a health economist, examines a new force in stimulating lead abatement litigation. Samuel Hays, an environmental historian, examines in detail the forces arrayed in the struggle for lead control and comments on the relationship between discipline, training, and affiliation on judgments and role in the lead dialectic.

If this book has a point of view, it is that as better methods of data collection and analysis are employed, effects are found at lower doses. The bottom has not been found. It is this phenomenon, at work in both human and animal studies, that has lead to the downward revision of the definition of lead toxicity. It is now estimated that one child in six in the U.S., rich or poor, black or white, has a level of lead in the neurotoxic range. Being white and economically comfortable do not vaccinate a child against lead exposure. But being poor certainly increases the risk. Over half of the black children in poverty enter first grade with toxic levels of lead in their blood. The societal costs for this exposure can only be imagined.

Yet this epidemic of lead poisoning is completely preventable. Prevention of lead poisoning creates remarkable opportunities for improving housing stock and reducing unemployment. Recognizing the real possibility of eradicating this disease and the benefits that will flow from this action has drawn the Department of Health and Human Services to draft a plan to eradicate — not control — childhood lead poisoning. Their plan sets the time to accomplish this at a decade. If this book, by illuminating the data on toxic mechanisms of lead and the barriers to its removal, can enhance this prospect, it will have achieved the goal of the authors and the editor.

Herbert L. Needleman
Pittsburgh, 1991

THE EDITOR

Herbert L. Needleman, M.D., is Professor of Psychiatry and Pediatrics at the University of Pittsburgh School of Medicine, Pennsylvania. He graduated from Muhlenberg College in 1948 and received his M.D. in 1952 from the University of Pennsylvania School of Medicine. He trained in pediatrics at the Children's Hospital of Philadelphia and in psychiatry at Temple University. Dr. Needleman served as Associate Professor of Pediatrics at Harvard Medical School, prior to joining the medical staff of the Children's Hospital of Pittsburgh as Director of the Behavioral Science Division. He is now Director of Lead Studies at the University of Pittsburgh School of Medicine.

Dr. Needleman is a diplomate of the American Board of Pediatrics and the American Board of Psychiatry and Neurology. He is a member of the American Academy of Pediatrics, the American Academy of Child and Adolescent Psychiatry, the American Pediatric Society, American Association for the Advancement of Science, Sigma Xi, and the Institute of Medicine of the National Academy of Sciences.

For his research in the study of lead exposure he has received the First Scientific Studies Award of the Association of Children with Learning Disabilities, the Sarah Poiley Medal of the New York Academy of Sciences, and the Charles A. Dana Award for Pioneering Achievements in Health and Higher Education.

Dr. Needleman has presented more than 50 invited lectures at international conferences, testified several times before the United States Senate and House of Representatives, and is a consultant for various U.S. government agencies. Dr. Needleman is the author of over 80 papers and over 40 book chapters, and has edited 3 books on environmental health. He is currently studying the effects of lead on higher order social behavior.

CONTRIBUTORS

Edward L. Baker, M.D., M.P.H.
Director
Public Health Practice Program Office
Centers for Disease Control
Atlanta, Georgia

Peter H. Bartels, Ph.D.
Professor
Department of Microbiology and Optical
  Sciences Center
University of Arizona
Tucson, Arizona

David Bellinger, Ph.D.
Assistant Professor
Department of Neurology
Harvard Medical School
Boston, Massachusetts

James L. Burchfiel, Ph.D.
Associate Professor
Department of Neurology
University of Rochester Medical Center
Rochester, New York

Cliff I. Davidson, Ph.D.
Professor
Departments of Civil Engineering and
  Engineering and Public Policy
Carnegie Mellon University
Pittsburgh, Pennsylvania

Frank H. Duffy, M.D.
Associate Professor
Department of Neurology
Children's Hospital
Harvard Medical School
Boston, Massachusetts

Donald A. Fox, Ph.D.
Associate Professor
College of Optometry
Department of Biochemical and
  Biophysical Sciences
University of Houston
Houston, Texas

Richard G. Frank, Ph.D.
Professor
Department of Health Policy and
  Management
Johns Hopkins University
Baltimore, Maryland

Constantine Gatsonis, Ph.D.
Assistant Professor
Department of Biostatistics and
  Department of Health Care Policy
Harvard Medical School
Boston, Massachusetts

Gary W. Goldstein, M.D.
Professor
Departments of Neurology and Pediatrics
Johns Hopkins University School of
  Medicine and the Kennedy Institute
Baltimore, Maryland

Samuel P. Hays, Ph.D.
Professor
Department of History
University of Pittsburgh
Pittsburgh, Pennsylvania

Philip J. Landrigan, M.D.
Professor
Department of Community Medicine
Mount Sinai School of Medicine
New York, New York

Jane S. Lin-Fu, M.D.
Director
Childhood Lead Poisoning Prevention
  Program
U.S. Department of Health and Human
  Services
Health Resources and Services
  Administration
Maternal and Child Health Bureau
Rockville, Maryland

Thomas D. Matte, M.D., M.P.H.
Medical Epidemiologist
Office of the Deputy Director
National Institute for Occupational Safety
 and Health
Centers for Disease Control
Atlanta, Georgia

Paul Mushak, Ph.D.
Consultant and Adjunct Professor
Department of Pathology
University of North Carolina
Chapel Hill, North Carolina

Jerome O. Nriagu, Ph.D., D.Sc.
Research Scientist
National Water Research Institute
Environment Canada
Burlington, Ontario, Canada

Michael B. Rabinowitz, Ph.D.
Instructor
Department of Neurology
Harvard Medical School
Boston, Massachusetts

Deborah C. Rice, Ph.D.
Research Scientist
Health Protection Branch
Toxicology Research Division
Health and Welfare Canada
Ottawa, Ontario, Canada

Joel Schwartz, Ph.D.
Senior Scientist
U.S. Environmental Protection Agency
Washington, D.C.

Ellen K. Silbergeld, Ph.D.
Professor
Department of Toxicology
University of Maryland
Baltimore, Maryland

Richard P. Wedeen, M.D.
Associate to Chief of Staff for Research
 and Development
Veterans Administration Medical Center
East Orange, New Jersey

DEDICATION

Randolph Kunhardt Byers, 1896—1988
Who brought us into the modern era of lead toxicology.

TABLE OF CONTENTS

SECTION I: HISTORICAL AND ENVIRONMENTAL BACKGROUND

Chapter 1
Saturnine Drugs and Medicinal Exposure to Lead: An Historical Outline 3
Jerome O. Nriagu

Chapter 2
Modern History of Lead Poisoning: A Century of Discovery and Rediscovery 23
Jane S. Lin-Fu

Chapter 3
The Monitoring of Human Lead Exposure 45
Paul Mushak

Chapter 4
Lead in the Environment: From Sources to Human Receptors 65
Cliff I. Davidson and Michael Rabinowitz

SECTION II: NEUROBIOLOGICAL FACTORS

Chapter 5
Neurological Perspective on Lead Toxicity 89
Ellen K. Silbergeld

Chapter 6
Visual and Auditory System Alterations Following Developmental or Adult
Lead Exposure: A Critical Review 105
Donald A. Fox

Chapter 7
Developmental Neurobiology of Lead Toxicity 125
Gary W. Goldstein

Chapter 8
Behavioral Impairment Produced by Developmental Lead Exposure: Evidence
from Primate Research 137
Deborah C. Rice

SECTION III: HUMAN LEAD EXPOSURE

Chapter 9
Occupational Lead Exposure 155
Thomas D. Matte, Philip J. Landrigan, and Edward L. Baker

Chapter 10
Lead, the Kidney, and Hypertension 169
Richard P. Wedeen

Chapter 11
Neurodevelopmental Effects of Low-Level Lead Exposure in Children 191
David Bellinger and Herbert L. Needleman

Chapter 12
Low-Level Lead Exposure: Effect on Quantitative Electroencephalography
and Correlation with Neuropsychologic Measures 209
James L. Burchfiel, Frank H. Duffy, Peter H. Bartels, and Herbert L. Needleman

Chapter 13
Lead, Blood Pressure, and Cardiovascular Disease..................................... 223
Joel Schwartz

Chapter 14
Low Level Health Effects of Lead: Growth, Developmental, and Neurological
Disturbances... 233
Joel Schwartz

Chapter 15
Recent Epidemiologic Studies of Low-Level Lead Exposure and the IQ of
Children: A Meta-Analytic Review 243
Constantine A. Gatsonis and Herbert L. Needleman

SECTION IV: SOCIOLOGICAL AND LEGAL ISSUES

Chapter 16
Economic Aspects of the Litigation for Harm Due to Lead Poisoning................. 259
Richard G. Frank

Chapter 17
The Role of Values in Science and Policy: The Case of Lead........................ 267
Samuel P. Hays

Index 287

Section I: Historical and Environmental Background

Chapter 1

SATURNINE DRUGS AND MEDICINAL EXPOSURE TO LEAD: AN HISTORICAL OUTLINE

Jerome O. Nriagu

TABLE OF CONTENTS

I.	Introduction	4
II.	The Saturnine Drugs	5
	A. Metallic Lead	5
	B. Lead Sulfide or Galena	6
	C. Lead Oxides: Litharge and Red Lead	6
	D. White Lead or Ceruse	10
	E. Lead Acetate (Sugar of Lead)	10
	F. Drug Contamination	13
	G. Lead and Diseases of the Reproductive System	14
III.	Lead in Mineral Elixirs	15
IV.	Concluding Remarks	18
References		19

I. INTRODUCTION

Lead and its compounds, being toxic, have excellent antiseptic properties and were among the first drugs of mineral origin to be used. Galena (PbS, or lead glance) the most common ore of lead, was employed as an eye salve in predynastic Egypt perhaps as early as the Badarian period (about 6000 years ago). There is evidence that this mineral was used as eye make-up and salve in the ancient Indian civilizations of Mohenjo-Daro and Harappa (3200 to 2800 B.C.). Such an application for galena is also recorded in the Old Testament and in the inscriptions of the ancient Mesopotamian cultures. It is not clear whether the galena found in the ancient graves of North American Indians had any prophylactic significance.[1] The traditional medicine of most West African cultures includes galena as an eye salve, a practice that probably goes back to antiquity.

Other lead compounds, particularly the oxides, found applications in *materia medica* of the most primitive cultures. Ancient Babylonian medical tablets record the use of lead compounds in plasters, although there are philological problems in identifying the actual lead salts used. The pharmacopoeia of ancient Egypt preserved in the Ebers Papyrus[2] and Hearst Medical Papyrus[3] featured lead compounds as astringents, external cooling agents and Collyria (eye salves). Hindu prescriptions containing lead *(Sisa)* and red lead *(Sindura)* occur in Sanskrit medical compendia such as *Charaka Samahita, Susrata,* and *Vagbahata*.[4-7] Ancient Chinese sources refer to elixir formulations calling for large doses of lead.[8,9]

Therapeutic applications of lead became extremely widespread during the period of the Roman Empire,[10] and were later continued by the Arabic medical profession. Although Paracelsus did not introduce lead into modern pharmacopoeia, he popularized its use as an internal therapeutic agent. Paracelsus and his followers added "salts of lead to the witches brew of plants and animal matter as treatment for virtually all ailments."[11] His influence is reflected in the 1540, 1556, and 1590 editions of *Dispensatorium Valerii* which listed only 28 preparations of metals and metallic compounds. The *Dispensatorium Brandenburgicum* published in 1698 contained about 90 metallic drugs and in 1771, the fifth edition of the *Pharmacopoea Wirtenbergia* had over 185.[12] By 1766, Goulard[13] listed over 100 external therapeutic applications for lead and its compounds alone. Until the turn of this century, chemotherapeutic agents were still based primarily on poisons such as lead, mercury, arsenic, and phenol. The medical profession endeavored to transform these agents into magic bullets through chemical group substitutions that made them less poisonous to the hosts. Saturnine drugs were therefore gradually extended to the treatment of a number of diseases including cancer, hemorrhage, leprosy, dropsy, gout, syphilis, diarrhea, and epilepsy. Woodall's[14] ode to the therapeutic virtues of mercury would seem equally applicable to lead:

> The perfect cure proceeds from thee
> For pox, for gout, for leprosie
> For scabs, for itch, of any sort
> These cures with thee are but a sport

Lead has persisted in the pharmacopoeia of many cultures particular in the developing countries, and folk remedies currently represent an important cause of acute plumbism. Eye cosmetics from the Indian subcontinent (notably the kohls and surmas) still contain up to 80% lead sulfide and some members of the immigrant communities in Britain have recently been poisoned by them.[15] Lead poisoning has also been associated with medicinal products (such as *bajar, bala guti* pills, and *pushyanug churna*) and purported aphrodisiacs prescribed by practitioners of traditional Indian medicine (the Hakims or Vaids). The lead-containing folk remedy of southeast Asia known as "pay-loo-ah", used as a cure for fever and rash, continues to poison the children of Hmong refugees in the U.S.[16] In Mexico, the use of azarcon (lead oxides) for empacho (chronic indigestion) and other gastrointestinal illnesses is accompanied by many cases of acute poisoning;[16] ironically, azarcon is often given when the patient shows signs of lead intoxication! These recent case histories of lead poisoning resulting from folk remedies clearly suggest that large numbers of people were afflicted in previous times when the use of such medicaments was much more prevalent. This report provides an historical outline of the contribution of the medicinal exposure route to the widespread incidence of plumbism.

II. THE SATURNINE DRUGS

There is an arresting dichotomy in the use of saturnine preparations described in the ancient literary sources. Lead salts were prescribed as external remedies and for diseases of women in the Old West. By contrast, Chinese and Indian texts contain numerous recipes for saturnine elixirs which were consumed to attain immortality or longevity. The first section of this review will focus on saturnine drugs particularly of the Western countries while the second part will address the leaden elixers of Asia and the Middle East.

A. METALLIC LEAD

Metallic lead was widely assigned chthonic properties and as such played an important role in ancient magicosacerdotal healing.[10] Lead plates inscribed with invocations, abjurations, and prayers to reinforce the curative effects of remedies or simply to drive off the demonic paroxysm have been found dating to prehistoric times. Lead amulets, talismans, pendants and medallions were worn to protect health. Such applications of lead in occult healing are interesting but probably have little pharmacological significance.

The uses of metallic lead in folklorish medicine of ancient times are succinctly summarized by Pliny:[18]

In medicine, leaden plates are applied to the region of the loins and kidneys for their comparative chilly nature to check the attacks of venereal passions, and the libidinous dreams that cause spontaneous emissions to the extent of constituting a kind of disease. It is recorded that the pleader Calvus used these plates to control himself and to preserve his bodily strength for laborious study. Nero, whom heaven was pleased to make emperor, used to have a plate of lead on his chest when singing songs *fortissimo,* thus showing a method for preserving the voice.

The Galenists usually compounded their lead drugs according to their qualitative composition or homeopathy: hot, cold, wet, and dry in varying degrees. Because of the presumed cooling property of lead, Soranus[19] recommended applying a piece of lead to the navel of a newborn child after the umbilical cord had been shed, to help the wound cicatrize and the umbilicus mold properly into a cavity. Dioscorides[20] recommends using the lead "unmixed" from washed lead as a rub to help the stroke of a sea scorpion or the dragon! Several ancient authors, such as Diocles, Serapion, and Erasistratus, Caelius Aurelianus, and Alexander of Trailles[21] prescribed lead pills for intestinal obstruction. Because of its weight, the lead pill was supposed to dislodge and drive out the obstructing matter.

Galen[23] also noted that a lead plate was widely prescribed for the treatment of nocturnal emissions. On this matter, Aurelianus[23] flatly stated, "Provide the patient with a hard, cold bed, and have him lie on his side when he goes to sleep...Place a long, thin lead plate under his loins or put sponges soaked in cold vinegar water around them." Lewin[24] has even suggested that lead plates were used then in an effort to produce sterility or at least to prevent conception.

For the treatment of ganglion Oribasius[25] and Paulus Aegineta[26] applied "a thick plate of lead, like the vertebrae, and larger than the ganglion, and bind it on, for by its weight this dissolves it in due time." Aetios[27] recommends nearly the same plan of treatment: "bind a piece of lead upon the tumor, and after some days remove it when ganglion will be found much softened; it is then to be squeezed firmly between the thumb and the fingers by which means it will be speedily dissolved.'' To prevent suppuration, a special set of drugs called enhemes were used particularly on flesh wounds.[28] One example contained metallic lead powered together with spodium from Cyprus, silver dross, copper scales, chalcitis, and alum.[29]

The Arabs continued to invoke the refrigerant properties of lead in its therapeutic usage. Al-Biruni (circa 800 to 870 A.D.)[30] noted that "lead and an oil are triturated in a mortar by means of a pestle and the product obtained therefrom is a palliative for hot inflammations because of its refrigerant action, so much so that it is of advantage even in inflammations due to malignant tumors. The foil is used in a poultice for scrofular and glandular wounds and for wounds of the joints and for swellings due thereto. The latter melt and disappear as if they were never present." Many years later, William Salmon, a 17th century physician in London, treated colic due to circumvolution of the intestine with "leaden or golden bullets swallowed."[11] For John Bull's *cholera morbus,* one cartoonist satirized that "steel lozenges and lead pills" were the only remedy.[11]

B. LEAD SULFIDE OR GALENA

Early references to the medicinal application of galena (*plumbum combustum* or *plumbum elotum*) can be found as inscriptions on kohl boxes and in a 13th century B.C. letter in which Pai, a "painter of Amon in the tomb city" asked his son to send him galena for his eye disease.[31] Galena was also used as eye paint to make the eyes larger and brighter, and for protection against eye diseases.

Lead sulfide was used in many drug preparations of the Greco-Roman times. The Hippocratic texts used it in emollient for ulcers.[32] Pliny[18] and Dioscorides[20] furnish a good detail of its medicinal applications:

It is used to make an eye-wash and for women's skins to remove ugly scars and spots and as a hair wash. Its effects is to dry, to soften, to cool, to act as a gentle purge, to fill up cavities caused by ulcers, and to soften tumors; it is used as an ingredient in plasters serving these purposes, and for the emollient plasters mentioned above. Mixed with rue and myrtle and vinegar, it also removes erysipelas, and likewise chilblains if mixed with myrtle and wax.

Similar uses for galena were described by Galen[22] and by Celsus[33] who added that it was used to suppress bleeding. Indeed, analysis of ancient Egyptian kohl samples show that most of them contain galena in an organic matrix (see Nriagu[10]).

Kohls and surmas continue to be used in the Muslim communities to darken the hair and skin around the eyes. Two basic preparations are still in common use (see Table 1). The first is in the form of a powder (surma) applied to the conjuctival surfaces with a needle or stick and the second (Kajal) is in the form of a cream with vegetable fat matrix. A number of cases of lead poisoning has been reported among the Asian communities of Europe, Japan, and the U.S. all attributable to the use of surmas.[34-37] Highly elevated blood lead-levels found in a large number of the children of Asian immigrants in Britain have also been attributed to the use of surma in the family.[35]

C. LEAD OXIDES: LITHARGE AND RED LEAD

The use of metallic rust on wound is noted in the ancient myth of Achilles treating the wound of Telephos with scrapings from the tip of his lance, an interesting treatment which left Pliny[18] wondering as to "whether he did it with a bronze or an iron spearhead."[28] Because of their excellent analgesic and styptic properties, the use of lead oxides on wounds began well before the Trojan War of about 1250 B.C.[38] The Ebers Papyrus, a compendium of ancient Egyptian drugs in use before 1550 B.C. contains about 30 medicaments calling for red lead or "lead earth."[2] The Hearst Medical Papyrus, compiled around 1550 B.C.,[3] also has several confections containing lead oxides.

TABLE 1
Lead Content of Surmas Imported Into the U.K. From India or Pakistan

Name and source	Color of powder	Lead content (%)
MD Hashim Surma, Bunda Road, Karachi	Grey	83
Hashmi Surma, Jowahar Chaharam, Karachi	Grey	80
Nargasi Surma, Hamdard, Pakistan	Grey	77
Surma Moqawi Basar Taj Company, Lahore	Grey	54
Bal Jyoti, Murrari Brothers, Delhi	Grey-black	38
Binger Surma, Hamdard, Pakistan	Black	26
Nag Jyoti, Murrari Brothers, Delhi	White	Trace
Multani, Ayurvedic 36-H Connaught Circusm Delhi	Grey	Trace
Indian Surma	Cream	Trace
Bhimsaini Kajal with Aela, Murrari Brothers, Delhi	Black paste	Trace
Bhimsaini Kajal with Aela	Black paste	Trace

From Aslam, M., Darig, S. S., and Healy, M. A., *Publ. Health (London)*, 93, 274—284, 1979. With permission.

About 20% of the red-lead prescriptions in the Ebers Papyrus go toward the treatment of eye diseases such as xanthelasma, pterygium, blindness and squint.[2] All forms of infectious conjunctivitis were widespread in Egypt hence the common use of leaden collyria.[39] Most of the prescriptions containing red lead are for external applications, mainly as pastils, emollients, and plasters for burns, wounds, ulcers, and alopecia; as poultice to drive away tremblings in the fingers and draw out splinters from the flesh; as ear plug to cure loss of hearing or injected into the ear that discharges foul-smelling matter.[2] Rarely did remedies for internal disorders contain lead, the only instance, in fact, being the use of lead salts to alleviate constipation. The use of medicaments containing fresh lead earth for tongue disease and to strengthen the teeth[2] were also recommended.

The extensive Assyro-Babylonian pharmacopoeia included about 250 medicinal plants, 120 mineral substances, and 180 other drugs not counting those used as solvents and vehicles for the actual medicinal substances.[10] Among them were several lead compounds still to be properly identified. The extensive trade and cultural links[39] suggest that the therapeutical applications of lead compounds by many ancient cultures of the Mediterranean region were similar to those of ancient Egypt, as recorded in the Ebers and Hearst Medical Papyri. The strong influence of Egyptian medicine on the pre-Hippocratic medicine in Bronze-Age Greece has also been noted by several people (see Partington).[32] The possibility that some of the lead used for medicaments in Dynastic Egypt came from Greece (the Laurion mines) has been suggested on the basis of similarity in the lead isotope signature.[40]

The Hippocratic collection has about 30 medicaments containing lead oxides which were recommended mostly for the treatment of *de ulceribus, de fistulis, de mulierum morbis,* and *de natura muliebri*.[32] Galen, Celsus, and their disciples subsequently used saturnine drugs in even greater numbers. Celsus,[33] used litharge as an exedent and *minium* from Sinope as erodent. He used litharge as an antiseptic, to check bleeding, to clean wounds, and he also applied it to putrid flesh, to pustules and to nasal ulcerations. For wounds, Celsus has a list of 34 plasters and ointments, eighteen of which contain heavy doses of lead oxides. Paulus Aegineta[26] provides a very comprehensive list of 90 ancient plasters (compiled mainly from Dioscoridess, Galen, and Oribasius), 38 of which contain lead salts.

Pliny[18] summarizes the preparation and medicinal applications of lead oxide(s) in antiquity, "It (molybdaena, or lead oxide) is used in preparing a particular emollient plaster for soothing and cooling ulcers and in plasters which are not applied with bandages but which they use as a liniment to promote cicatrization on the bodies of delicate persons and on the more tender parts.... Also combined with scum of silver and dross of lead, it is applied warm for fomenting dysentry and constipation." The similarity in the above prophylactic usage of these compounds with those in the much earlier records of the Ebers and Hearst Medical Papyri should be obvious.

Galen[22] also prescribed litharge for intertrigo, but considers the application of litharge on wounds to be a useless treatment. Celsus[44] recommended *minium* as an antiseptic, and for the treatment of nasal polypus and foul genital ulcerations. Lead oxides were used, at times, as an ingredient in eye salves.[33] Archagathos (about 220 B.C.), credited with introducing Greek Medicine in Rome, invented a plaster containing misy, burnt copper, white lead, litharge, and turpentine resin.[32] Herakleides of Tarentum patented many prescriptions containing litharge, alum, vitriol, verdigris, white lead, copper scales, galls, earth wax, cocus, turpentine, and opium.[32] Menekrates of Zeophleta (about first century A.D.) is reputed to have invented the diachylon plaster (very popular in ancient times) which contained litharge, oil, marrow and herb juices.[32] Indeed, excavations of a Roman military hospital near Baden in Switzerland have yielded such bottles of ointments containing lead oxides in various organic matrices.[41]

The litharge moxa was also widely embraced by the Arabic medical profession and later copied into the European materia medica. Al-Kindi (circa 800 to 870 A.D.) used medicaments containing litharge for the treatment of abscesses, scrofula, boils, vertigo alba, lacerations, and dirty wounds, swollen sores in the anus and buttocks, hemorrhoids, and various eye diseases.[42] Al-Biruni (973 to 1051) provided the following details on therapeutic uses of lead oxides (calcined lead) during his time:[30]

Such pastes [containing lead and oil] and calcined materials particularly the washed calcined materials are of use in festering and malignant wounds, tumors and ulcers of the joints. If applied as a poultice to the joints and swollen glands thereof, it melts the latter. The calcined metal, especially the washed variety, is useful for occular wounds and xerophthalmia, and for mammary wounds and polypus. It acts as an anti-aphrodisiac, if a piece of the calcined material is wrapped on the back. It is useful for ulcers of the *membrane virile* and testicular wounds and inflammations.

Similar prescriptions were also given by Al-Razi (born 865 A.D.), Maimonides, (1135 to 1204 A.D.) and much later by Abd al-Razzaq al-Jaza-iri, an early 18th century Algerian (see Levey).[42]

Avincenna (died circa 1073), the great Arabian physician, noted that lead was used as a domestic remedy for *adversus alvi fluxum et ulceru interstinorum* and that it was also a custom to put litharge into water suspected to be unwholesome.[43] Arnold of Villanova (1235 to 1312) used litharge and white lead in a suppository for the treatment of *pruritus vulvae*.[44] Henri de Mondeville (1260 to 1320) employed lead-containing ointments for ulcers, fistules, and eye diseases.[45]

Gale[46] was a typical iatrochemist of the Middle Ages who relied heavily on drugs of mineral origin. In his *Certaine Workes of Chirurgerie,* he listed over 40 drug preparations containing litharge and/or white lead. Among these was a plaster that was especially formulated for "King Henrie the Eyghte to amende the swellying in his legge;" it contained "*lithargi auri, litharge argenti, cerusae, corallo rub., boli armoniaci,* and *Sangui draconis.*"[46] The lead compounds mostly went into balms and plasters for sores in the nose, legs and arms, scabes and ulcers, as well as[46]

...the bittinge of a madde dogge, stinging by a snake, adder, scorpion and the like, iche of the legge and inflammations, excoriation, burning and blistering, comminge of whole humours and cancerous ulcerations, pestilentiale tumours, apostemes, burnynges, ruptures, contusions and ecchymomata, scyrhous of the lyver, spleen, stomache and other parts, reumatike passions, inflammations of the joyntes and the gout, to draw out thynges fired as arrowe beades, dartes, and thornes, etc."

Powders and troches containing lead compounds were recommended by Gale for inflammations, herpes, erisipelas, malignant tumors, and as general antiseptic.

The use of litharge in external medicaments continued to be popular until fairly modern times. The *Pharmacopeia Londinensis,* first published in 1618, included lithargyrum (lead oxide) and cerussa (white lead) primarily for external remedies.[12] *Dispensatorium Hafniense,* the first Danish collection of formulae published in 1658 contained a number of lead recipes mostly in ointments and plasters.[93] The *New Dispensatory of 1753* contained the recipes for a much wider variety of saturnine preparations. The *Pharmacopoea Danica,* published in 1772, listed a number of lead compounds but only one, *trochisci de minio,* was recommended for internal use.[93] A Danish decree issued in 1687 permitted the sale of white lead, litharge, and red lead not only by chemists but also by confectioners, grocers, distillers, druggists, and drysalters.[93] Mercurial salve containing lead was employed by Guy de Chauliac[47] in treating leprosy. Baker[48] noted that litharge was the common basis of most of the plasters in use in his days. He even reported[48] a "most violent and obstinate colic which seemed to have been occasioned by some litharge mixed in with cataplasm and applied to the vagina with a view to allay a troublesome itch."

While Gale was compiling his remedies, the science of pharmacy was undergoing a fundamental change. Before then, medical therapy was governed by the Galen principle of opposites which claimed that if fever resulted from the administration of a drug, then that drug was suited for treating chills; or that if a drug caused purging, it should be used for relieving constipation. During the middle of the 15th century, however, Paracelsus founded a system of therapy based on the doctrine that all diseases should be treated with drugs producing like effects — a drug which induces fever may be suitable for treating fever and a purge beneficial to diarrhea. He curtly noted[49] that "Lead hath in it remedies for those diseases which be caused and bread in the miners leade." Paracelsus eschewed the mixtures of herbs of Galen and the traditional healers in favor of simple but powerful medicaments based primarily on mineral substances. According to him, even the deadliest poison could be transformed into an innocent and powerful remedy after proper chemical "correction" and with due attention to dosage, use at the propitious times, and for an appropriate disease.[49]

When ... Sol and Luna have to be proved and purified, Saturn is added to them, and this has the effect of thoroughly purging them. Nevertheless, it is of that nature that it takes away their [malleability?]. It has the same effect on men, with great pains, as Jupiter and Mars. Being mixed with cold, it cannot act mildly. It has the very greatest powers and virtues, whereby it cures fistulas, cancer, and similar ulcers which come under its own degree and nature. It drives the same kind of disease from man as it expels impurities from Luna [silver]. But if it does not go out altogether at the same time, it brings more harm than good. Consequently, whoever would use it must know what diseases it cures ... and what effects Nature has assigned to it. If this be well considered, it can do no harm.

Paracelsus' influence in the development of metallotherapy in the West has been well documented (see for example, References 50 and 51). His success in using mercury to treat the supposedly incurable syphilis, in particular, did much to popularize the prophylactic benefits of toxic metals which subsequent physicians employed liberally in larger and larger doses.

In a 1700 dissertion called *Triumphus Lithargyriatorum* written by the Norwegian Jorgen Nielsen Seerup, the opinions of many authorities were pieced together in an effort to show that lead preparations could be used internally with perfect safety; he went on to claim that litharge could cure constipation, colicky pain, diseases of the spleen, worm infestations, and some infectious diseases including the plague.[93] Although Goulard[13] noted that his secret extracts of saturn were the most efficacious remedies for large variety of external complaints, he nevertheless cherished the belief lead could be deprived of its noxious quality by certain preparations when administered in small doses. In this *Primitive Physic,* Wesley[52] included the following potent remedy:

Vinegar of lead [actually lead acetate]. Take of litharge, half a pound; strong vinegar, two pints. Infuse them together, in a moderate heat, for three days, frequently shaking the vessel; then filter the liquor for use. This medicine is in little use from a general notion of its being dangerous. There is reason, however, to believe that the preparations of lead with vinegar are possessed of some valuable properties, and that they may be used in many cases with safety and success.

It should be noted that litharge and white lead have only limited solubility in water and must be triturated with an acidic or organic solvent before they can be administered orally. Otherwise, they were swallowed as pills. By contrast, lead acetate is readily soluble in water and thus was the compound of choice for the majority of internal lead remedies (see following section).

D. WHITE LEAD OR CERUSE

Pliny[18] aptly observed that the medicinal properties of white lead (*psimithium,* or sugar of lead) basically "are the same as those of other lead compounds, only it is the mildest of them all, and beside that, it is useful for giving women a fair complexion." It is a very common ingredient in ancient ophthalmic medicines or collyria.[10] About one half of the 55 ancient ophthalmic collyria listed by Paulus Aegineta contain white lead.[26] These ingredients were rubbed together, mixed with a small quantity of rainwater, and then used in the treatment of infections of the eyes, tonsils, and suppurated ears.

Soranus[19] suggests the annointment of any part of an infant bruised in delivery with white lead, while Celsus[33] recommends the application of a white lead ointment to the umbilicus of prominent navels, where surgical measures were not needed. The Bible did not specifically mention lead in any medicaments. Among the remedies mentioned in the Talmud are various powders, potions, plasters, and collyria[53] which probably included lead as an active ingredient. For the treatment of *rushcheta,* a mixture of aloe, gum arabic juice, white lead *(aspidka),* litharge *(martika),* malabathrum berries, and glaucium was recommended by the Talmud; all the ingredients were to be placed in a thin rag in summer or in cotton wool in winter and then applied to the site of the ailment.[53]

The basic uses of white lead established in ancient times were continued until fairly recent times. Ibn Jazlah (died 1100) prescribed it for ear and eye diseases, swellings and ulcer. Al-Razi copied the prescriptions in Dioscorides and also used it for scorpion stings and to reduce swellings. Al-Kindi applied ceruse to a swelling of the buttocks, and employed it in drug for the pulsation of the rectum and its swelling, in a plaster to reduce hemorrhoid pain, and in several eye remedies. Al-Samarqandi (died 1222) made it an important ingredient in enemas. Even today, ceruse is still in use as an astringent in the Middle East, especially in Iran and Iraq.[30,42,54]

Ceruse and other lead salts were subsequently extended to prescriptions for dropsy, epilepsy, hemorrhoids, gout, and cancer and were key ingredients in antisyphilitic and antileprosy ointments (see References 11 and 46). During the middle of the 18th century, ceruse and minium were freely used in troches in the French military hospitals.[50] *Cerussa* was among the few metallic compounds in the 1612 edition of *Dispensatorium Valerii,* and was likewise included in *Pharmacopoeia Londinensis* published in 1618 while *The New Dispensatory* of 1753 provided recipes for a variety of drugs containing this lead salt. In general, the use of ceruse was closely similar to that of litharge (preceding section), and both compounds were often blended together to increase the potency of the remedy.

E. LEAD ACETATE (SUGAR OF LEAD)

The boiling of must (grape juice) in lead pots to produce a sweet tasting syrup containing "the sugar of lead" or lead acetate was well known to the Romans.[10] Presumably, pure lead acetate was first isolated by reacting scraps of lead or lead oxides with soured wine *(vin aigre)* or vinegar. The synthesis probably was achieved in Medieval time for the lead acetate to be entitled to an alchemical symbol and an assortment of names including saccharum, *bleizucker,* salt of Saturn, plumbous acetate, virgin's milk, etc.[55] Noah Biggs, the famous chemical philosopher, noted that by 1651 the reaction was familiar to all the kitchen wenches:[56]

> Vinegar how weak soever, put into a peuter saucer, and suffering to stand a while, by and by begins to put forth its active, acid corrosive spirit; and in the vinegar you shall perceive clearly a certain white mother as it were swimming in the vinegar; and the bottom of the saucer shall be damask'd with white streaks, yea, ... a certain substance like cerusse shall be scraped off ... This by practice may be observ'd, ... and it is so trivial and common a business, that it is known to all kitchen wenches.

It should be noted in passing that in 1612, Jean Beguin successed in producing acetone from lead acetate thereby achieving an important landmark in the history of organic chemistry.[55]

Typically, preparations containing lead acetate were used in treating fever, hemoptysis and hematemesis, menorrhagia, gonorrhea, coughs, diarrhea and dysentery. It would appear that up to the 16th century, the medicinal use of lead acetate was neither regular nor extensive. According to Gale[46] it was not mentioned specifically in the Arabic materia medica, and the European iatrochemists rarely employed it. The Paracelsean doctrine that "saturnus puret febres" was upheld by many physicians. Baker[48] noted that it was common to give patients one scruple of sugar of lead just before the onset of the paroxysm of quartan fever and he also referred to a certain preparation of lead called *butyrum bezoardicum saturninum* which was greatly trusted in malignant fevers. Pitcairn placed the sugar of lead among the *remedia rarescentian nimium sedantia.*[48]

In his *De luis Venereae,* Jean Fernel (1497 to 1558) warned against internal use of lead and reported the incidence of his friend who, on the advice of an empiric, took powdered lead instead of sugar for arthritis and ended up with a severe case of lead poisoning.[50] Guillaume de Baillou (1538 to 1616) found widespread incidents of colic and paralysis among patients who received treatment from quacks for their fever.[50] Petrus Borellus (1620 to 1689) was outraged by the habit of *chymici* and *chymiatri* of his days of prescribing sugar of lead and justified his concern by reference to his own friend who suffered acute lead poisoning by dosing himself with this particular lead compound.[50] The 1698 *Dispensatorium Brandenburgicum* recommended internal use of *saccharum saturni* and *magisterium saturni* (lead sulfate) as well as *tinctura antiphtisica,* a solution of lead acetate and ferrous sulfate.[12]

With the founding of chemical pharmacology and therapeutics by Paracelsus in the 16th century, lead acetate quickly became one of the most common ingredients in mineral therapies for a wide variety of diseases. After recounting the many external applications of white lead, Hermann Boerhaave (1668 to 1738) abhorred the internal uses of *saccharum saturne*:[57]

> I cannot imagine it safe to give it, as some do, internally. Almost all the modern physicians, I know, administer it in intermittent fevers, and other distempers; but with what success they best can tell. With me, it stands in the catalogue of poisons. And tho it may cure the fever, it is apt to leave a worse disorder behind... A famous Italian physician once told me that no medicine was used by the monks with such great success as this, for extinguishing all desires to venery, which kind of appetite it infallibly destroys.

Extensive use of lead acetate in the monastries as an antiaphrodisiac, as alluded to by Boerhaave,[50] may explain the apparent frequency of monastic epidemics of lead poisoning reported in Europe during the Middle Ages and later.

Preparations containing lead acetate found particular favor in the treatment of diarrhea. *Rasa-Jalanidhi* recommended about a dozen lead-containing rasas for diarrhea and associated fevers.[58] Mann[59] reported that the indiscriminate use of lead acetate in treating diarrhea was responsible for the epidemic of "dropsical swellings" among the soldiers during the war of 1812. The *New Dispensatory* of 1753 stressed that the sugar of lead was particularly effective against diarrhea, among other ailments, but warned that it occasioned symptoms of another kind, often more dangerous than those removed by it. Harlan,[60] hired to minister to employees of a white lead plant in Philadelphia who had been afflicted with lead poisoning, "found no remedy so well adapted to the purpose [of diarrhea] as the sugar of lead." Harlan apparently based his calamitous treatment on his previous observation of the remedial action of sacharum Saturni against dysentery. Laidlaw[61] actually used himself as a guinea pig to determine that 10 grains (600 mg) of lead acetate a day for 5 d was the upper limit of human tolerance since he did not develop colic on this regimen until the sixth day. Until recently, pills containing 100 mg of lead acetate and 5 mg of opium were still being prescribed for symptomatic treatment of diarrhea. Intravenous injection of a suspension of such pills in water has resulted in an outbreak of lead poisoning in a group of drug addicts in Glasgow in the 1970s.[62,63]

Lead acetate was also widely prescribed for internal bleeding:[64] "[I] affirm with great certainty that sugar of lead will succeed where nothing else is of effect in haemorrhages." Pierre Pomet (1658 to 1699) recommended 3 to 4 grains (180 to 240 mg) of lead acetate for stopping a flux of the belly or to heal sore throat (see Reference 50). In his *American Modern Practice,* Thatcher[65] observed that "saturnine anodyne pills were particularly effective against uterine hemorrhage." Barker[48] likewise referred to the use of "tinctura saturninae" in the treatment of "haemorrhagia uterina".

Gout is often regarded as one of the supervenient symptoms of lead poisoning.[10,11] The past medical records, however, show many instances where lead was actually prescribed for gouty patients. The following remediation for gout was reported by Riverius in 1657:[66] "Sal Saturni, that is salt of lead, dissolved in subtil spirit of wine, easeth pains wonderfully...An infusion of litharge made in vinegar, the vinegar being a little evaporated till it grow sweetish, doth much good to an hot gout."[66] The beneficial effects of lead acetate on gout were also not lost on Mayerne:[67]

Sugar of lead may be safely taken inwardly, with appropriate conserves, and it doth actually mitigate and sweeten (the humours) as its taste doth witness: but it raketh off, and abeteth venerial desires

. . . (perhaps much to the advantage of the gouty patient).

The repugnant use of lead acetate in the treatment of childhood epilepsy was advocated by Dr. Benjamin Rush,[68] a pioneer physician and a signatory of the Declaration of the American Independence: "I was first led to prescribe sugar of lead in epilepsy by hearing that a man had been cured of it by swallowing part of a table spoonful of white lead by mistake, instead of a table spoonful of loaf sugar." Spence[69] claimed to have confirmed Rush's therapy by reporting that he had successfully cured himself of epilepsy by swallowing lead acetate although he acquired severe colic in the process. The College of Physicians of Philadelphia has preserved Rush's medical bag containing the vial of the extract of Saturn used in his house calls.[11]

In his *Treatise on the Effects and Various Preparations of Lead,* Goulard[13] trumpeted the many therapeutic virtues to be derived from lead and its different preparations. His extract of saturn was made as follows:[13]

Take as many pounds of litharge of gold as 32 oz quarts of white vinegar; put them together into a proper kettle, and let them boil, or rather simmer, for an hour, or an hour and quarter, taking care to stir them during the ebullition with a wooden spatula; take the kettle off the fire, let the whole settle, and then pour of the liquor which swims upon the top, into bottles for use. I shall call this liquor the Extract of Saturn, which is to undergo a further modification, as I shall direct.

Goulard saw no limit to the potential applications of his secret extract which was generally sold as "vegeto-mineral water" made by "putting two tea-spoonfuls or 100 drops of the extract of saturn into a quart of water and four tea-spoonfuls of brandy." Goulard's water was claimed to be efficacious against fluxions of the membrane of the tympanum, and occasional deafness; for the cleaning of wounds and moistening the bandages and pledgets before they are covered with the cerate of Saturn; for washing of old, callous, foul ulcers; for washing ulcerated and occult cancers; against contusions and bruises; against extravasation of blood or echymoses; against trobus proceeding fromphlebotomy; against inflammations of the tendons, aponeurosis and ligaments; against phlegmons and abscesses; against sprains, excoriations and burns; against the king's evil, inflammations caused by gunshot wounds, fistulous sinusses, fistulas whether of the eye or anus; against inflammations or curdlings of milk in the breasts, abscees and ulcers in those parts; against erisipelas, piles, chilbains, anchilosis, whitloes, tetters, the itch, ruptures with strangulation, gangrene; against contraction of the tendons and tumours and inflammations attending luxations and fractures.

Goulard's extract attained singular popularity for well over 100 years as the testimony by Tanquerel des Planches[70] clearly shows: "Every day, in the practice of surgeons at the Paris hospitals, great use is made in fomentations, or simple external application of Goulard's extract...."

As late as 1925, lead acetate was still being administered intravenously for cancer.[71] The *Chemical Formulary* published in the 1940s still contained saturnine remedies, including for instance, lead subacetate in poison ivy lotion.[72]

F. DRUG CONTAMINATION

One can certainly appreciate Pliny's[18] remark on the "marvellous efficacy of human experiment, which has not left even the dregs of substances and the foulest refuse untested in such numerous (medicinal) ways." For example, *sory* or inkstone, a mixture of lead, iron and copper sulfates, was used in a prescription for loosening a carious tooth.[33] Many other by-products of lead pyrometallurgy such as slag of silver, dross of lead, golden ash, flower of silver, lead stone and *spuma plumbi,* and *cadmea spodos,* were also used in ancient medicaments, mostly as constituents of plasters, cataplasms, emollients, and pastils. The efficacy of these secondary materials may be exemplified by those of *scoria argenti,*[18] which is said to be a by-product of the lead smelting process and "has an astringent and cooling effect on the body, and has healing properties as an ingredient in plasters, being extremely effective in causing wounds to close-up, and when injected by means of syringes, together with myrtle-oil, as a remedy for straining of the bowels and dysentery." It was also used as an ingredient in emollient plasters administered on proud flesh of gathering sores, or sores caused by chafing or running ulcers on the head. The natural ores of zinc, copper, bismuth, and other base metals widely used in mineral therapy tend to be polymetallic, and unless they are carefully purified, would transfer their lead to the patient.

There is some evidence that a number of medicaments used internally were contaminated with lead either fraudulently or from the use of leaden (or lead-tainted) mortars, pestles, jars, and dispensatories. The maceration of the medicinal ingredients with lead contaminated wines or boiled grape juice would be another important source of lead contamination. The fraudulent adulteration of ancient medicaments with lead was certainly a matter of much concern as the following indignant condemnation by Pliny[18] suggests: "Nowadays whenever they (the medical practitioners) come on books of prescriptions, wanting to make up some medicines out of them, which means to make a trial of the ingredients in the prescriptions at the expense of their unhappy patients, they rely on the fashionable druggists' shops which spoil everything with fraudulent adulterations, and for a long time they have been buying plasters and eye-salves ready made; and thus is deteriorated rubbish of commodoties and the fraud of the druggists' trade put on show." From the Roman Empire times until the Middle Ages, apothecaries were sometimes called *pigmentarii,* a term denoting dealers in dyes and paints but who also prepared the medicines ordered by the physicians, as remarked by Pliny above. Lead pigments were quite common in Roman times,[10] and some adventitious lead contamination of drugs might have been occasioned by the common use of shop equipment and dispensatories.

G. LEAD AND DISEASES OF THE REPRODUCTIVE SYSTEM

The ancients were very much aware of the effects of lead on various functions of the human reproductive system. *Rasa-Jalanidhi* claimed that phthisis is caused by, among other things, an excessive loss of semen and for its remedy prescribed a number of lead-containing drugs. Among the medicaments found to be effective against spermatorrhea were *Byadhi ba'rana, Pottali rasa, Bhubaneshwara parpati,* and *Shankara raa*. Pliny[18] noted that lead plates fastened to the loins or over the kidney was used to cool the heat of fleshy lust. The rubbing of leaden salves to prevent nocturnal emissions in athletes was described by Agricola who attributed the remedy to Galen.

Al-Biruni[30] recommended lead oxides for ulcers of the *membrum virile* and for testicular wounds and inflammations. Ointments containing mercury and lead were widely used to soothe the syphilitic sores that ravaged Europe during the Middle Ages.[73,74] With Paracelsus came the increased use of lead acetate as a favorite treatment for gonorrhea; Jean Astruc (1684 to 1766), for example, prescribed small doses of lead compounds for chronic cases.[50]

In his *Historia Animalium,* Aristotle[76] notes that some women of his time prevented conception "by anointing that part of the womb on which the seed falls with oil of cedar, or with ointment of lead or with frankincense co-mingled with olive oil." In his classic text, *Gynaecology,* Soranus of Ephesus[19] advocates the following rational contraceptive technique: "Conception is prevented by smearing the mouth of the womb with old (sour) oil or honey or cedar gum or opobalsam, either alone or mixed with ceruse (white lead), or with ointment which is prepared with myrtle oil and ceruse." Lead is a well known spermicide.[77] Soranus[19] then goes on to suggest boiled honey mixtures and winey concoctions for inducing menstruation and for preventing conception or for inducing abortion. The efficacy of such a prescription may be related to the widespread contamination of the wines and hyromel/oxymel with lead.[10] Oribasios[25] states that collyria (which often contain lead), were applied to the uterus, presumably for contraceptive purposes.

Aetios of Amida[27] notes that "smearing the cervix before coitus with honey or cedar resin alone or in combination with lead or liquid ointment with myrtle and lead or liquid alum or galbanum with wine aids contraception." He continues "Or else, (prepare) lead with oil in pessary; put in the vagina before coitus or sooner." Several moralists and satirists (notably St. Jerome, Juvanal, and Martial) spoke about the practice, relatively common among the womanhood of the Roman empire, of drinking potions with the intention of preventing conception or inducing abortion. Some of such contraceptive decoctions presumably contained lead as an active ingredient. The abortifacient properties of lead continued to be exploited by large numbers of women and their healers until modern times. In 1905, it was deemed necessary to warn the British public about the tragic effects of the then increasing use of lead compounds for female irregularities and to induce abortion.[78] As late as 1950s, there were still reports of three women who died from acute lead poisoning after ingesting 15 grams of lead oxide to induce abortion.[79]

The Hippocratic corpus referred to the use of leaden instruments to dilate the cervix. Leaden tubes for intrauterine medication and for insertion into the vagina and rectum to prevent postoperative cicatrical contractions and adhesions are mentioned.[80] Mal Samuel (circa 180 to 254 A.D.), rabbi, physician, astronomer, and one of the most important sages of the Talmud recommended the use of "leaden tubes whose edge is bent inwards" for examining the uterus.[81] The use of lead oxides in the treatment of ulcerous wombs is recommended in the Hippocratic text.[29] Galen[22] and Celsus[33] prescribed lead salts for ulcers, abscesses and indurations of the breast, uses that continued until recent times. For example, Tanquerel des Planches[70] drew attention to lead colic induced by the injection of Goulard's extract into the vagina. During the 1890s, for example, a small epidemic of lead poisoning in infants was reported stemming from lead acetate ointment prescribed by a druggist for the breasts of nursing mothers.[92]

Lead compounds have been used as an aphrodisiac since prehistoric times. *Rasa-Jalanidhi*[58] noted that lead is especially efficacious in diseases affecting the semen, increases longevity and procreative energy if taken regularly for a long time. Lead purified in the prescribed way was recommended as an aphrodisiac.[58] The elixers of the Chinese also included lead compounds for their aphrodisiac properties. Al-Biruni observed that lead compounds were used for this purpose during the Islamic era of the Middle Ages.[30] A number of Western alchemists and iatrochemists experimented with lead as an antiaphrodisiac[67] and Boorhaave[57] intimated that such a use triggered the epidemics of lead poisoning in European monasteries during the Middle Ages. And this particular use for lead has yet to be put to rest. The prevalence of renal disease discovered recently among the Asian immigrants in Britain has been attributed to secretive use of a saturnine aphrodisiac called "rustneg".[82]

III. LEAD IN MINERAL ELIXIRS

Unlike in the Old West, the consumption of mineral elixirs and drugs assumed a religious fervor in ancient China and other ancient cultures of the East. The Chinese use of inorganic and metallic substances in therapy clearly began in remote antiquity. A compendium of chemical substances in the text *Chi Ni Tzu* (presumed to be of fifth century B.C. vintage) listed metallic lead and a number of its compounds which probably had pharmaceutical uses.[9] The literary records, however, suggest that elixir preparations made from lead and other toxic metals began to be consumed in heavy doses or administered on long-continued basis during the Later Han, or about the first or second century A.D.[9] For many centuries after that, the adepts and their disciples continued the hard and painful process of testing the effects of metallic preparations upon the human mind and body. Sivin[83] has found over 1000 titles on elixirs of immortality in the alchemical literature, while Okanishi Tameto[84] indexed 2405 recipes for elixirs (defined by the word *tan* in the title) in only 321 medical compilations. A few examples show that large amounts of lead were often employed in preparing some of the common elixirs.

In *Tshan Thung Chhi* (Kingship of Three), dated to about 142 A.D., Wei Po-Yang, who is widely regarded as the father of alchemy, recommended the use of 15 oz each of lead and mercury in the preparation of the famous "cyclically-transformed elixir".[9] Ko Hung (circa 283 to 343 A.D.), the systematizer of Chinese alchemy, maintained that material immortality could only be attained by the consumption of one of the major elixirs such as *huan tan* (cyclically-transformed elixir) and *chin i* (portable gold elixir). Among the lead-containing elixirs he recommended, one finds *tan hua* (a deadly mixture of mercury, realgar, alum, lead, lead carbonate, salt, and other ingredients); *Thai-Chhing tan* (a mixture of mercury and its salts, lead, gold, vinegar, mineral sulfates, etc); *Yin Tzu tan fa* (cinnabar, lead, and mica suspensions); *Chang Tzu-Ho tan fa* (mercury, copper oxide, lead, red pinacled millet with crushed jujube-dates as vehicle); and *Chou hou tan fa* (cinnabar, mercury, lead, and copper or one of its alloys).[9] The consumption of such elixirs containing large quantities of toxic metals certainly has to be injurious to the health of the adepts in spite of their claim to the contrary.

Sun Ssu-Mo (circa 581 to 672 A.D.) listed 67 different elixirs but gave the recipes for only 32, most of which entailed large quantities of lead, mercury, and/or arsenic as the active ingredients.[83] On the dosage and efficacy of such a "minor cyclically-transformed elixir", Sun Ssu-Mo says: "Every day after eating, take three pills wrapped in jujubes. It cures epilepsy, melancholia, possession by goblins, and so forth. Taken over a long period, it hardens the bones and marrow, aids circulation of the blood, moistens the skin, brings out the color in the face, quiets the soul, and puts one in touch with the immortals."[83] The perceptive remark that the elixir "puts one in touch with the immortals" should not be doubted.

For making the "lead elixir", Sun Ssu-mo recommended 4 lb of lead ripened in fire and 1 lb of quicksilver cleaned by grinding with salt. The elixir was said to be a cure for "disorder due to hot factors, possession by demonic forces, epilepsy, and autumnal intermittent fevers."[83] One of the earliest recipes for the lead-mercury elixir is given in the *Yellow Emperor's Canon of the Nine-Vessel Spiritual Elixers,* written before 142 A.D. For preparing the Black-and-Yellow *(hsuan huang),* the Canon advocates the use of 10 lb of quicksilver and 20 lb of lead among other ingredients.[9] The few examples above from ancient Chinese texts and inscriptions show that the practice of ingesting metallic elixirs was widespread in ancient China. Perseverance in the taking of the elixirs was counseled over and over again in many of the ancient texts,[9] a rather unfortunate advice. There is ample evidence that metallic (elixir) poisoning was common, and the demise of many emperors and high government officials by this malaise is recorded in many ancient tests.[85] It is not inconceivable that such poisonings reached epidemic proportions in some segments of the society who could afford the preparations.

Some Chinese alchemists considered the toxicity syndrome from taking metallic elixirs to be quite normal and propitious. One striking passage to this effect (which describes the classic symptoms of lead poisoning) is the early sixth century A.D. text, *Thai-Chhing Shih Pi Chi* (Records in the Rock Chamber: a Thai-Chiing Scripture):[85]

After taking the elixir, one feels an itch all over the body and the face, rather like having the sensation of insect crawling over one. The body, the face, the hands and legs become swollen. One may experience a feeling of repulsion at the sight of food, and vomiting usually follows after a meal. One feels rather weak in the four limbs. Other symptoms include frequent defaecation, vomiting, headache, and pains in the abdomen. No alarm should be caused by these effects, because they are due to the work of the elixir in dispelling all the inherent disorders in the human body.

Considerable effort was devoted to developing the counter measures against such elixir poisoning. In a Thang or Early Sung tract, Hu (Kang) Tzu for example warms that "the five metals have to be purified from all poisons caused by heating. If the ingredients are used for making elixir without their poison removed, and ingested for any length of time, death will be caused when the rules are not followed."[85] Other ancient texts stressed the need to neutralize the elixir toxicity by proper blending of the *yin* and *yang* ingredients according to the theory of categories. Others recommended potent herbal mixtures that could serve both as an elixir and as an antidote for ordinary elixir poisoning. The necessity of building up one's constitution before embarking on the elixir diet is widely recommended. Some traditional detoxifying treatments were also tried including hot and cold water baths, ingesting mixtures containing scallion, soya-bean sauce, wines, etc.[9]

The Chinese protoscientific influence becomes increasingly manifested in Japan from about the beginning of the 6th century A.D.[9] Because the country was richly endowed with gold and silver, the ancient Japanese generally eschewed the gold transmutation dogma but eagerly accepted the chemotherapeutic (elixir) elements of the Chinese alchemy. The extant historiographies reveal that throughout the Heian period (about 795 to 1185 A.D.), the Imperial court and nobility were consistently devoted to longevity medicines of plant and metallic derivation.[9] The general propensity for saturnine drugs and elixirs in ancient Japan is amply documented by the time-capsule specimens deposited in the Shosoin Treasury in 756 A.D. and preserved there to this day.[86,87] Among the 600 or so items (including 60 substances labeled as "medicines") were 128 small triple-wrapped packages containing about 100 kg of lead oxides. The contents of the packages were further graded into superior, medium, and low quality on the basis of the intensity of the orange color of the powder which indicates the lead oxide concentration.[86] As the labels clearly show, the lead oxides in the packages were used for medicinal purposes and presumably in beauty aids as well.

Chemistry in ancient India likewise evolved chiefly as a handmaid of medicine and, somewhat later on, as an adjunct of the Tantric cult.[87] Saturnine medicaments (as opposed to rejuvenative drugs) undoubtedly were used in ancient India, often times as internal remedies. Lead sulfide was described as an ingredient in collyria *(anjanas)* by Charaka, while lead oxides and lead sulfate were prescribed with vegetable drugs for use as plasters in both the *Sushruta Samhita* and *Charaka Samhita*.[87] The *Sushruta Samhita*[1] described lead and zinc as vermifugal and recommended their internal use. The Bower Manuscript, which dates to about the second half of the 4th century A.D., but contains material from much earlier data, also refers to the use of galena in medicine.[88]

Four schools of disease treatment have been recognized in India which can be traced back to ancient times: (1) by rasa or metallotherapy using mercury, other metals and inorganic minerals; (2) by herbs and vegetable drugs; (3) by charms and other magicosacerdotal means; and (4) by modern surgery.[58] It is generally claimed that the first of these schemes (which was primarily cultivated by the Yogis) was of divine origin, the second and third of human origin and the fourth (nonAryan) of barbaric or demonic fomentation. The literary records on Yogic and Tantric medicine certainly show that saturnine drugs, as opposed to rejuvanative elixirs, were extensively used in India since antiquity, often as internal remedies. In the eclectic compilation called *Rasa-jalanidhi* (or Ocean of Indian Chemistry and Alchemy) by Mookerji,[58] it is noted that lead produces a hot sensation in the system, is soothing, bitter in taste and was used to destroy *vayu, kapha, prameha,* and rheumatism.[58] By constant use, lead is claimed to increase memory, longevity, power of digestion, semen production, and procreative energy if taken regularly and for a long time. "Lead incinerated in the best way prescribed is efficacious in phthisis, diseases due to vay, gulma, anemia, giddiness, worms, colic due to excess of phlegm, spermatorrhea, chronic diarrhea, diseases affecting the rectum, and loss of the power of digestion. It also serves as an aphrodisiac."[58] One of the rasas containing nectarized or sweetened lead was said to cure "eighty different kinds of diseases due to an abnormal excess of vayu, and especially dhanusstanbha (tetanus), those due to an abnormal excess of phlegm, all sorts of urinary diseases, asthma, phthisis, anemia, dropsy, fever accompanied with sensation of coldness, and hydrocele."[58] *Rasa-jalanidihi* continued, "Incinerated lead, if taken with sugar, can cure pittam, vayu, headache, eye disease, diseases that affect semen, delirium, inflammation, loss of appetite, and loss of sexual desire, provided there is nothing objectionable in the diet taken by the patient."[58] It should be mentioned that some of the ailments may be associated with lead poisoning and a prescription for more lead would seem highly ill-advised for the patient.

The recipes for a large number of internal drugs (rasas) containing large doses of lead are also given in *Rasa-jalanidhi*. Expectably, a large number of people suffered lead poisoning as a result of the widespread use of lead prescriptions in Yogic medicine. A large number of allopathic (traditional) medicines from the Indian subcontinent still contain large amounts of lead,[82] and the ingestion of such medicaments has resulted in documented cases of lead poisoning among the Indian immigrant population in Europe.[82,89] The deleterious effects of lead taken internally were certainly known in India. *Rasa-jalanidhi* stressed the point that lead not properly purified and incinerated in the prescribed way "gives rise to leprosy, gulma, loss of appetite, anemia, phthisis, phlegm, troublesome impurities of the blood, fever, stone diseases, colic, and fistula."[58] A remedy for lead poisoning was also given: "One is freed from the evil effects of lead, not properly incinerated, if one takes incinerated gold with haritaki and sugar for three days."[58] The rationale for the continuing use of such toxic metallic drugs, however, was clearly articulated in *Rasa-jalanidhi*:[58]

> A medicine prepared mainly from mercury and minerals is superior to that prepared from herbs in as much as (i) the former can be administered in much smaller doses that the latter; (ii) it does not give rise to aversions in the patients who take it; (iii) it cures diseases more quickly than the latter; (iv) it cures diseases which are considered incurable by medicines prepared from herbs. Medicines prepared from mercury, with or without the addition of other minerals, poisons, etc. are the best of all the medicines known to the world.

Unfortunately, no mention was made of the fact that in many instances the cure would have been worse than the disease.

By analogy with the historical uses of the arsenicals and mercurials, the following are also believed to be some of the important medicinal applications of lead and its compounds in the Ayurbedic system of medicine:[90]

Mode of use or application	Diseases
External in medicated oils	Parasitic infections of the skin, eruptions and itchings
External in dusting powder or in ointment	Chronic skin diseases, piles, leucoderma, baldness, alopecia, leprosy, etc.
External ingredients of oral gurgle	Mouth and throat diseases
Ingredients of medicinal cigars	Imbalance of the first and third humors, migrane, hemicrania, earache, diseases of the gum, drowsiness, etc.
Internal in linctus, pills, medicated ghee, etc.	See above

IV. CONCLUDING REMARKS

The preceding sections clearly suggest that large numbers of patients and elixir adepts would have suffered plumbism as a result of repeated and/or continual exposure to leaded medicaments. It should be noted, however, that before Paracelsus, saturnine preparations were primarily prescribed for external remedies and Pliny[18] rightly tells why:

> And cinnabar and red lead are admitted to be poisons, all the current instructions on the subject of its employment for medicinal purposes are in my opinion decidedly risky, except perhaps that its application to the head or stomach arrests haemorrhage, provided that it does not find access to the vital organs or come in contact with a lesion. In any other way for my own part I would not recommend its employment.

It was inevitable however that there was some absorption of the lead in the external applications, the hand to mouth transfer of such lead being a well-known phenomenon that can result in acute poisoning.[72] The application of lead compounds on excoriated skins over an extended period of time could, however, lead to chronic, if not acute, lead poisoning.[48,72] Furthermore, the annointment of womens' private parts may exert an adverse health effect. A classic case history comes from Sir George Baker[72] who reported a "most violent and obstinate colic, which seemed to have been occasioned by some litharge mixed in cataplasm and applied to the vagina with a view to allay a troublesome itching." Thus, even if medicinal exposure was not primarily responsible for the epidemics of plumbism during the Roman times,[10] the added lead absorption from such a route would certainly have been most inauspicious for anyone with an elevated lead burden.

With the introduction of internal metallotherapy in Europe during the 16th century, the number of persons poisoned by the supposed cure with lead must have increased drastically. The medical records contain numerous advisories against the widespread use of lead in internal therapeutics.[11,50] Nevertheless, the use of saturnine drugs for the treatment of internal bleeding, diarrhea, gonorrhea, epilepsy, and fevers as well as in a host of ointments and troches for various wounds, sores, and ulcers continued unabetted until fairly recent times. Today, lead is still widely employed in the herbal or traditional medicines of Asia, Africa, and South America and medicinal exposure still remains one of the main causes of acute lead poisoning.[16,17,82,89,91]

REFERENCES

1. **Farquhar, R. M. and Fletcher, I. R.**, Lead isotope identification of sources of galena from some prehistoric Indian sites in Ontario, Canada, *Science,* 207, 640, 1980.
2. **Bryan, C. P.**, *The Papyrus Ebers,* Ares Publications, Chicago, 1974.
3. **Leake, C. D.**, *The Old Egyptian Medical Papyri,* University of Kansas Press, Lawrence, 1952.
4. **Mitra, J.**, *History of Indian Medicine from pre-Mauryan to Kusana Period,* Tyotiralok Prakashan, Varansi, India, 1974.
5. **Sharma, V. P.**, *Indian Medicine in Classical Age — Chowkhamba Sanskrit Studies 85,* Chowkhamba Sanskrit Series Office, Varansi, India, 1972.
6. **Jaggi, O. P.**, *Yogic and Tantric Medicine,* Atma Ram & Sons, New Delhi, India, 1973.
7. **Zimmer, H. R.**, *Hindu Medicine,* Johns Hopkins Press, Baltimore, 1948.
8. **Needham, J.**, *Science and Civilization in Ancient China,* Vol. 5 (Part II), Cambridge University Press, 1976.
9. **Needham, J.**, *Science and Civilization in Ancient China,* Vol. 5 (Part III), Cambridge University Press, 1976.
10. **Nriagu, J. O.**, *Lead and Lead Poisoning in Antiquity,* John Wiley & Sons, New York, 1983.
11. **Wedeen, R. P.**, *Poison in the Pot: The Legacy of Lead,* Southern Illinois University Press, Carbondale, Illinois, 1984.
12. **Urdang, G.**, *Pharmacopeia Londinensis of 1618,* Hollister Pharmaceutical Library Publ. No. 2, State Historical Society of Wisconsin, Madison, 1944.
13. **Goulard, T.**, *Treatise on the Effects and Various Preparations of Lead,* Translated by G. Arnaud, P. E. Imsly, London, 1766.
14. **Woodall, J.**, *The Surgeons Mate,* R. Young, London, 1639.
15. **Anon.**, *Lead and Health,* Department of Health and Social Security Working Party on Lead in the Environment, Her Majesty's Stationery Office, London, 1980.
16. **Levitt, C., Paulson, D., Duvall, K., Godes, J., Dean, A. G., Roberts, J., and Egenberger, J.**, Folk remedy-associated lead poisoning in Hmong children, Minnesota, *Morbidity Mortality Wkly Rep.,* 32, 555—556, 1983.
17. **Vashistha, K. K., Agee, B., Fannin, S., James, S., Martinez, A., Tilsen, S., and Barr, D. B.**, Use of lead tetroxide as a folk remedy for gastrointestinal illness, *Morbidity Mortality Wkly Rep.,* 30, 546—547, 1981.
18. **Pliny the Elder**, *Natural History,* Loeb Classical Library Translations, Harvard University Press, Cambridge.
19. **Soranus of Ephesus**, *Gynaecology,* Edited and translated by V. Rose, B. G. Teubneri, Leipzig.
20. **Dioscorides**, *The Greek Herbal of Dioscorides,* Translated by J. Goodyear, 1655, Gunther, R. T., Ed., Oxford University Press, 1933.
21. **Alexander of Trailles**, *Ein Beitrag zur Geschichte de Medicin,* Puschmann, T., Ed., Verlag Adolf H. Hakkert, Amsterdam, 1963.
22. **Galen**, *Opera Omnia,* Translated and edited by C. G. Kuhn, Georg Olms Verlagsbuchhandlung, Heldesheim.
23. **Aurelianus**, *On Acute Diseases and on Chronic Diseases,* Translated by I. E. Drabkin, University of Chicago Press, Chicago.
24. **Lewin, L.**, *Die Gifte in der Weltgeschichte,* Verlag, 1920.
25. **Oribasius**, *Oribasius Latinus,* Morland, H., Ed., A. W. Brogger, Oslo.
26. **Adams, F.**, *The Seven Books of Paulus Aegineta,* Vol. 1, Vol. 2, Vol. 3, translations and commentary, Sydnenham Society, London, 1844—1847.
27. **Aetios of Ameda**, *Libri Medicinales,* Olivieri, A., Ed., B. G. Teubneri, Leipzig.
28. **Majno, G.**, *The Healing Hands,* Harvard University Press, Cambridge, 1975.
29. **Hippocrates**, *Oeuvres Completes d'Hippocrate: Traduction Nouvelle avec le Texte Grec en Regared,* Vol. 1—10, Littre, E., Ed., Bailliere, Paris.
30. **Said, H. M.**, *Al-Biruni's Book on Pharmacy and Materia Medica,* Hamdard National Foundation, Karachi, Pakistan, 1973.
31. **Budge, E. A. W.**, *The Mummy,* Cambridge University Press, Cambridge, 1925.
32. **Partington, J. R.**, *A History of Chemistry,* Vol. 1, Macmillan, London, 1970.
33. **Celsus**, *De Medicina,* Translated by W. G. Spencer, Loeb Classical Library Translations, Harvard University Press, Cambridge.
34. **Snodgrass, G. J. A. I., Ziderman, D. A., Gulati, V., and Richards, J.**, Cosmetic plumbism, *Br. Med. J.,* 4, 230, 1973.
35. **Ali, A. R., Smales, O. R. C., and Aslam, M.**, Surma and lead poisoning, *Br. Med. J.,* 2, 915—916, 1978.

36. **Kato, K.,** Lead meningitis in infants, *Am. J. Dis. Child.,* 44, 569—591, 1932.
37. **Byers, R. K.,** Lead poisoning: review of the literature and report on 45 cases, *Pediatrics,* 23, 583—603, 1959.
38. **Nriagu, J. O.,** Occupational exposure to lead in ancient times, *Sci. Total Environ.,* 31, 105—116, 1983.
39. **Castiglioni, A.,** *Proc. CIBA Symposia, August-September,* CIBA-GEIGY, Ardsley, New York, 1943.
40. **Gale, N. H. and Stos-Gale, Z.,** Lead and silver in the ancient Aegean, *Sci. Amer.,* 244(6), 176—192, 1981.
41. **Kobert, R.,** in *Beitrage aus der Geschichte der Chemie,* Diergart, P., Ed., Franz Deuticke, Leipzig, 1909, 103—119.
42. **Levey, M.,** *The Medical Formulary or Aqrabadhin of Al-Kindi,* University of Wisconsin Press, Madison, 1966.
43. **Gruner, O. C.,** *A Treatise on the Canon of Medicine of Avicenna: Incorporating a Translation of the First Book,* Luzac, London, 1930.
44. **Rennau, T.,** *Die Gynakologie des Arnold von Villanova,* Freiburg, 1912.
45. **Pagel, J. L.,** *Die Chirurgie des Heinrich von Mondeville,* Hirschwald, Berlin, 1892.
46. **Gale, T.,** *Certaine Workes of Chirurgerie,* De Capo Press, Amsterdam, 1971.
47. **Goldwater, L. J.,** *History of Quicksilver,* York Press, Baltimore, 1972.
48. **Baker, Sir George,** *Medical Tracts,* 3rd ed., Payne & Foss, London, 1785.
49. **Waite, A. E.,** *The Hermitic and Alchemical Writings of Aureolus Philippus Theophrastus Bombast, of Hohenheim, called Paracelsus the Great,* Shambhala Press, Berkeley, 1976.
50. **Stevenson, L. G.,** History of Lead Poisoning, thesis, Johns Hopkins University, Baltimore, 1949.
51. **Multhauf, R.,** Medical chemistry and "the Paracelseans", *Bull. Hist. Med.,* 28, 101—125, 1954.
52. **Wesley, J.,** *Primitive Physic, or an Easy and Natural Method of Curing Most Diseases,* R. Hawes, London, 1845.
53. **Preuss, J.,** *Biblical and Talmudic Medicine,* Sanhedrin Press, New York, 1978.
54. **Graziani, J. S.,** *Arabic Medicine in the Eleventh Century as Represented in the Works of Ibn Jazlah,* Hamdard Press, Karachi, Pakistan, 1980.
55. **Anon.,** *Gmelins Handbuch de anorganischen Chemie,* Blei — Teil 1A, Verlag Chemi GmbH, Weinheim, 1973.
56. **Biggs, N.,** *Mataeotechnia Medicinae Praxews. The Vanity of the Craft of Physic,* Giles Clavert, London, 1651.
57. **Boerhaave, H.,** *A New Method of Chemistry,* Translated by P. Shaw and L. Chambers, B. Cowse & W. Innys, London, 1727.
58. **Mookerji, K. B.,** *Rasa-Jalanidhi, or Ocean of Indian Chemistry and Alchemy,* Avani Prakashan, Ahmedabad, India, 1984.
59. **Mann, J.,** Practical observations on colica Pictonum, *New Engl. J. Med. Surg.,* 11, 17—21, 1822.
60. **Harlan, R.,** Observations on colica Pictonum and other affections arising from the deleterious operation of lead on the system, *N. Am. Med. Surg. J.,* 5, 16—23, 1828.
61. **Laidlaw, W.,** Remarks on the internal exhibition of the acetate of lead, chiefly with the view of determining to what extent it may be safely administered in the cure of diseases, especially in uterine haemorrhages, *Boston Med. Surg. J.,* 1, 147—152, 1828/29.
62. **Beattie, A. D., Mullin, P. J., Baxter, R. H., and Moore, M. R.,** Acute lead poisoning: an unusual cause of hepatitis, *Scott Med. J.,* 24, 318—321, 1979.
63. **Beattie, A. D., Briggs, J. D., Canavan, J. S. J., Doyle, D., Mullin, P. J., and Watson, A. A.,** Acute lead poisoning, five cases resulting from self-injection of lead and opium, *Q. J. Med.,* 44, 275—284, 1975.
64. **Hill, J.,** *A History of Materia Medica,* C. Hitch & L. Hawes, London, 1751.
65. **Thatcher, J.,** *American Modern Practice, or, a Simple Method of Prevention and Cure of Diseases,* Ezra Read, Boston, 1817.
66. **Riverius, L.,** *The Practice of the Physic,* Translated by N. Culpepper, Peter Cole, London, 1655.
67. **Mayerne, de T.,** *A Treatise on the Gout,* Translated by T. Sherley, D. Newman, London, 1676.
68. **Rush, B.,** An account of the efficacy of sugar of lead in curing epilepsy, *Philadelphia Med. Mus.,* 1, 60—61, 1805.
69. **Spence, T. R. P.,** An account of the sugar of lead, in the case of epilepsy, *Philadelphia Med. Mus.,* 2, 150—154, 1806.
70. **Tanquerel des Planches, L.,** *Lead Diseases,* Translated and edited by S. L. Dana, Daniel Bixby & Co., Lowell, 1848.
71. **Belknap, E. L.,** Clinical studies on lead absorption in the human. III. Blood pressure observations, *J. Ind. Hyg. Toxicol.,* 18, 380—390, 1936.
72. The Chemical Formulary, H. Bennett, Ed., Van Nostrand, New York, 1930 forward.
73. **Reisman, D.,** *The Story of Medicine in the Middle Ages,* Paul B. Hoeber, New York, 1932.

74. **Riddell, W. R.**, Corradinus Gilinus and his Treatist de Morbo Gallico, 1497, *Med. J. Rec.*, 134, 434—455, 1931.
75. **Boerhaave, H.**, *Boerrhave's Aphorisms: Concerning the Knowledge and Cure of Diseases,* Translated by J. Delacoste, B. Cousse & W. Innys, London, 1715.
76. **Aristotle**, *Historia Animalium,* Translated by A. L. Peck, Harvard University Press, Cambridge.
77. **Himes, N. E.**, *Medical History of Contraception,* Schocken Books, New York, 1970.
78. **Hall, A.**, The increasing use of lead as an abortifacient, *Brit. Med. J.*, 1, 582—587, 1906.
79. **Karlog, O. and Moller, K. O.**, Three cases of acuate lead poisoning, *Acta Pharmacol. Toxicol.*, 15, 1—5, 1958.
80. **Milne, J. S.**, *Surgical Instruments of Greek and Roman Times,* Clarindon Press, Oxford, 1907.
81. **Posner, F.**, *Medicine in the Bible and Talmud,* Yshiva University Press, New York, 1977.
82. **Aslam, M., Darig, S. S., and Healy, M. A.**, Heavy metals in some Asian medicines and cosmetics, *Publ. Health, Lond.*, 93, 274—284, 1979.
83. **Sivin, N.**, *Chinese Alchemy: Preliminary Studies,* Harvard University Press, Cambridge, 1968.
84. **Tameto, O.**, A Study of "Tan" Prescriptions, in *Chinese Alchemy: Preliminary Studies,* Harvard University Press, Cambridge, 1968, 67.
85. **Ho Ping-Yu and Needham, J.**, in *Crafts and Craftsmen in China and the West,* Needham, J., Ed., Harvard University Press, Cambridge, 1959.
86. **Kazunosuke, M.**, *A Sketch of Ancient Ancient Mineral Drugs Based on Chemicals Preserved in the Shosoin Treasury of Nara,* Nihon Kobutsu shum no Kai, Kyoto, 1957.
87. **Ray, P.**, *History of Chemistry in Ancient and Medieval India,* Indian Chemical Society, Calcutta, 1956.
88. **Hoernle, A. R. F.**, *The Bower Manuscript,* Government Printing Office, Calcutta, 1907.
89. **Bose, A., Vashistha, K., and O'Loughlin, B. J.**, Azarcon por empacho — another cause of lead toxicity, *Pediatrics,* 72, 106—108, 1983.
90. **Bagchi, K. N.**, *Poisons and Poisoning — Their History and Romance and Their Detection in Crimes,* University of Calcutta Press, Calcutta, 1969.
91. **Lightfoot, J., Blair, H. J., and Cohen, J. R.**, Lead intoxication in an adult caused by Chinese herbal medication, *J. Am. Med. Assoc.*, 238, 1539, 1977.
92. **Holt, L. E.**, Lead poisoning in infancy, *Am. J. Dis. Child.*, 23, 229—233, 1923.
93. **Grandjean, Ph.**, Lead in Danes, in *Lead,* Griffin, T. B. and Knelson, J. H., Eds., Georg Thieme Publishers, Stuttgart, 1975, 6—25.

Chapter 2

MODERN HISTORY OF LEAD POISONING: A CENTURY OF DISCOVERY AND REDISCOVERY

Jane S. Lin-Fu

TABLE OF CONTENTS

I.	Introduction	24
II.	The Discovery of Lead-Paint Poisoning	25
	A. Peripheral Neuropathy Among Queensland Children: The Discovery of Childhood Lead Poisoning in Australia	25
	1. Footdrop and Wristdrop in Brisbane Children	25
	2. Basal Meningitis and Other Forms of Lead Poisoning	25
	3. Lead Poisoning — A Toxicity of Habitation	26
	4. Ocular Neuritis	26
	5. A Plea for Lead Paint as the Source of Poisoning	26
	6. Lead-Paint Legislation in Queensland	27
	B. Lead Poisoning in Children in the United States	27
	1. 1900s to the 1960s	27
	2. Social Turmoil and the Recognition of Childhood Lead Poisoning	27
III.	Meningitis Among Breastfed Infants: Lead in Toiletries	28
	A. Serous Meningitis Among Nursing Infants in Japan	28
	B. Lead Poisoning in Nursing Infants Caused by Toiletries in Other Countries	28
IV.	Lead Poisoning from Burning Battery Casings: The Depression Disease that has Persisted	28
	A. The Baltimore Outbreak in 1932	28
	B. The Depression Disease Past the Depression Era	29
V.	Colic and Convulsions from Lead-Contaminated Food and Drinks	29
	A. Lead Colic Throughout the Ages	29
	B. Lead Convulsions in Philadelphia From Cakes and Buns	29
	C. Outbreaks During the 20th Century	29
	1. Food and Water	29
	2. Improperly Glazed Earthenwares	30
	3. "Moonshine" or Illicit Alcoholic Drinks	30
VI.	The Lead World, Late 19th to Late 20th Centuries	30
	A. Hazardous Lead Sources in the Late 19th Century	30
	B. The Lead World in 1924	30
	C. Lead in Gasoline — A Source of Mass Environmental Pollution	30
	D. Lead Paint on Dwellings: A Recalcitrant Lead Hazard	31
	E. Lead in Food	32

	F.	Lead in Drinking Water 32
	G.	Lead in Dust and Soil 32
	H.	Lead in Folk Remedies 33
	I.	Miscellaneous Lead Sources 33

VII. Epidemiology of Lead Poisoning in Children 33
    A. Early Misperceptions Concerning Childhood Lead Poisoning 33
    B. Reason for Misperceptions 33
        1. Ignorance About the Extent of Environmental Lead Contamination 33
        2. Erroneous Concept of a "Normal" Blood Lead Level 34
        3. Viewing Lead Poisoning as a Clinical Disease 35
    C. Epidemiology of Childhood Lead Poisoning in 1990 36
        1. Basic Facts in Understanding Childhood Lead Poisoning 36
        2. NHANES II Blood Lead Data (1976 to 1980) 36
        3. ATSDR Report to Congress (1988) 37

VIII. Prevention of Lead Poisoning: A Rediscovery of Old Facts 38
    A. The Toxic Effects of Lead in the Fetus 38
    B. The Preventable Nature of Lead Poisoning in Children 39

IX. What of the Future? 39

X. Conclusion 39

References 40

I. INTRODUCTION

Lead poisoning is an ancient disease uniquely neglected by modern medicine. It is perhaps the only preventable man-made disease allowed to remain pandemic for centuries. Since lead was discovered more than 5,000 years ago, the diverse applications of this useful metal have assured continuous anthropogenic activities and widespread environmental pollution.[1] Today, in technologically advanced nations as well as in developing countries, hazardous exposure to lead has continued to be a common problem, especially among young children. Lead poisoning continues to be one of the most prevalent preventable childhood health problems in the U.S., affecting millions of children.[2]

For centuries lead poisoning was largely viewed as an occupational disease of adults. The older literature makes reference to the illness in newborns and children of lead workers, but these were presented as by-products of adult occupational exposure.[3,4] It was not until the turn of the century that lead poisoning in children was recognized as an important distinct entity,[5,6] and only during the last two decades has the problem drawn attention as a major public health problem.[7,8] As recent research has unveiled lead toxicity at lower and lower levels, it has become apparent that more and more children may have been injured.

In his letter on lead poisoning dated July 31, 1786, Benjamin Franklin concluded the correspondence to his friend as follows:[9]

You will see by it, that the opinion of this mischievous effect from lead, is at least above 60 years old; and you will observe with Concern how long a useful Truth may be known, and exist, before it is generally receiv'd and practis'd on.

Two centuries later, little has changed concerning the ignorance and indifference to the mischievous effects of lead. This chapter reviews the modern history of lead poisoning, particularly as it affects children, and examines how many useful facts or truths have been known and existed in the past century, and how few of these have been generally received and practiced on. Lead poisoning in adults, largely from occupational exposure, is discussed in Chapter 9.

II. THE DISCOVERY OF LEAD-PAINT POISONING

A. PERIPHERAL NEUROPATHY AMONG QUEENSLAND CHILDREN: THE DISCOVERY OF CHILDHOOD LEAD POISONING IN AUSTRALIA

1. Footdrop and Wristdrop in Brisbane Children

In 1890, A. Jefferis Turner, a resident physician at the Children's Hospital in Brisbane, Australia, encountered a puzzling case of peripheral paralysis in a 5-year-old boy.[5] The child had bilateral wristdrop, footdrop, weakness of the legs, and persistent vomiting. The original diagnosis of progressive muscular atrophy proved incorrect when the child recovered from the paralysis. Turner compared the case "hypothetically to chronic lead poisoning." The following year, Turner again encountered footdrop and wristdrop in the outpatient clinics of the hospital. He admitted two children to the hospital under the care of J. Lockhart Gibson who agreed with the tentative diagnosis of lead poisoning. In March 1892, Gibson presented these cases as such before the Medical Society of Queensland where he encountered considerable skepticism over the diagnosis.[5]

Stimulated by the skepticism from the medical community and hospital staff, Gibson zealously pursued the investigation of lead poisoning. Later that year, he reported ten cases of lead poisoning seen at the hospital between December 1891 and July 1892.[6] The authors observed that no other cases of lead poisoning had been recognized in Brisbane children and that the diagnosis appeared to have been frequently overlooked. They further noted: "The chief reason for this failure of diagnosis is no doubt that lead-poisoning was never thought of as a possible cause of the symptoms." All ten children in the series suffered from both wristdrop and footdrop.

2. Basal Meningitis and Other Forms of Lead Poisoning

At the 1892 Intercolonial Medical Congress, A. J. Turner reported four children between 5 and 6 years of age with headaches, vomiting, internal squint and double optic neuritis. Three of the four children slowly recovered. The course of illness suggested to Turner that these cases were not due to "morbid growth" or "tubercular" in nature, and was probably a form of basal meningitis.[10] Gibson concurred with the diagnosis in 1892, but in 1897, convinced Turner that these were actually cases of "lead neuritis." In the same year, Turner published a comprehensive review of the subject in the Australasian Medical Gazette.[5] In this paper, Turner analyzed 76 cases (55 girls and 21 boys) of lead poisoning treated as inpatients at the Brisbane Children's Hospital in the preceding six years. Seven ended fatally.

Turner divided lead poisoning into four clinical groups in his review:

1. Paralytic cases characterized by footdrop and wristdrop — He also reported two cases of paralysis of the diaphragm that ended fatally.

2. Cases characterized by abdominal pain and pain in the limbs — Turner noted that such symptoms were also often present in the paralytic cases, but were misdiagnosed as "bilious attacks". He observed that these children "become unusually irritable, neurotic, spoilt by their parents, and quite unmanageable."
3. Epileptic convulsive cases characterized by various forms of "fits" — Turner noted that many of the paralytic cases have had convulsions earlier. He went so far as to recommend that lead poisoning should be considered in any Brisbane child between 5 and 8 years of age who has convulsions without identifiable cause.
4. Ocular neuritis characterized by combined neuritis of both the optic and oculomotor nerves.

3. Lead Poisoning — A Toxicity of Habitation

By 1897, the remarkable frequency of lead poisoning in Brisbane children was indisputable, but the source of the poisoning remained a mystery. Turner astutely observed that lead poisoning was a "toxicity of habitation", since symptoms recurred after children returned to their homes.[5] Testing of tank water suggested that this may be the source, but it was unclear why adults in the same households were not affected. Turner asserted that the medical expert having pointed out that lead poisoning was a problem, "is entitled...as a citizen, to ask that the matter be investigated, that this large amount of suffering, disablement, blindness, and death be put to an end."[5]

4. Ocular Neuritis

In the issue of the *Australasian Medical Gazette* in which Turner's review appeared, Gibson also published a paper entitled "Ocular Neuritis Simulating Basal Meningitis — Plumbism."[11] Twenty-four cases of ocular neuritis (which included both optic neuritis and oculomotor paralysis) had been observed by Gibson since 1891. These children also had wristdrop, footdrop, colic, blue gum lines, and lead in the urine. These signs and symptoms left no doubt that the children suffered from lead poisoning. Tank water was thought to be the source of lead. In a paper published two years later, Turner stressed that children were misdiagnosed because textbook descriptions of lead poisoning were based on findings in adults.[12]

5. A Plea for Lead Paint as the Source of Poisoning

The cause of childhood lead poisoning remained a mystery until 1904, when Gibson published "A Plea for Painted Railings and Painted Walls of Rooms as the Source of Lead Poisoning Among Queensland Children."[13] He noted in the paper that a government analyst involved in studying the children had suggested earlier that lead paint was the source, "but that the idea was pooh-poohed." Gibson found that a majority, and probably all of the affected children, lived in houses with lead paint in rooms or on verandas, and that both fresh paint and old powdery paint in the form of dust adhered readily to children's hands. He demonstrated that a 1-ft^2 calico cloth rubbed against the painted railings had 0.3 g of lead on it. Gibson also made the important observation that the vast majority of affected children either bit their nails, sucked their fingers, or ate with unwashed hands. He further noted that of 85 children treated for lead poisoning in the hospital from 1889 to 1903, 42 were admitted during the hottest months, i.e., December through February, and 28 were admitted in October, November, March and April. Only 16 were treated during the cooler months. He suggested that outdoor activities during the warm months contributed to a greater exposure.

Gibson concluded, perhaps somewhat naively, that having uncovered the source of lead and the hand-to-mouth activities that introduced lead into children's bodies, it was now possible to prevent both primary attack and recurrence through public education. But cases of childhood lead poisoning continued to occur. In his 1908 Presidential Address before the Intercolonial Medical Congress in Melbourne, Turner reported that since 1891, 262 children had been treated for lead poisoning at the Children's Hospital, and that about 20 children were still being admitted every year.[14] Turner observed sadly that "unfortunately we in Brisbane still from ignorance, no longer excusable, allow our houses to be poisoned traps for children's fingers, and every year furnished its quota of ill-health and suffering, crippling, hopelessness and permanent blindness, and occasionally death, as the natural consequence. This is certainly a matter which calls for legislative interference."

6. Lead-Paint Legislation in Queensland

The 1920 Australasian Medical Congress passed a resolution to seek legislation prohibiting the use of lead paint on verandas and surfaces within reach of children.[15] In 1922, legislation to restrict the use of lead paint was introduced in Queensland. Implementation of the legislation was slow, but cases of lead poisoning began to decline in that area.[16] The Queensland experience had little impact elsewhere in Australia. In 1969, Freeman reported that 90 children with lead poisoning were identified in Sydney between 1948 and 1967.[16]

From the 1920s on, several Australian papers also documented childhood lead poisoning as a cause of chronic nephritis later in life.[15,17,18]

B. LEAD POISONING IN CHILDREN IN THE UNITED STATES
1. 1900s to the 1960s

In the U.S., the early Australian papers were initially viewed with indifference; some considered childhood lead poisoning to be unique to Australia.[19] But from the mid-1920s on, lead poisoning was recognized with increasing frequency in U.S. children. Between 1924 and 1933, 89 children were treated at the Boston Infant's and Children's Hospital for lead poisoning.[20] Increasing numbers of reports of severe lead poisoning among children in the 1930s and 1940s led several large old cities to initiate case finding activities in the 1950s and 1960s.[21-26] Invariably, the number of cases uncovered increased as the activities intensified. A perverted appetite for nonfood items, or pica, was thought to be a major contributing factor. Lead encephalopathy occurred with remarkable frequency. Even after the introduction of chelation therapy, reports published in the 1960s still indicated a fatality rate of 25 to 30%. Despite such disturbing data, childhood lead poisoning attracted little public attention or concern. Many considered lead poisoning to be a problem that affected only poor children in slum dwellings. Others who attempted to deal with the problem were frustrated by the high recurrence rate since there was no effective way to resolve the housing problem.[2,7]

2. Social Turmoil and the Recognition of Childhood Lead Poisoning

With the social turmoil of the 1960s, lead poisoning was recognized as an important pediatric public health problem.[28,29] In 1966, Chicago began the nation's first mass screening program for this disease.[30] Many other cities, including Baltimore, New York, and Philadelphia soon followed.[31] In 1967, the Children's Bureau in the U.S. Department of Health, Education and Welfare (DHEW), now the Department of Health and Human Services, issued a publication entitled "Lead Poisoning in Children" that called attention to lead poisoning as a preventable cause of death, mental retardation, and neurological handicaps.[27] This was the first time a Federal agency had publicly acknowledged lead poisoning as a needless cause of mortality and morbidity in children. Intense public education on childhood lead poisoning ensued. In 1970, the U.S. Surgeon General issued a statement on the problem which recommended screening and early identification of children with evidence of undue lead absorption.[32] This was defined as a confirmed blood lead level of 40 µg/dl or more. Until then, 60 µg/dl had been considered the upper limit of "normal" blood lead levels in children. Public pressure together with official support from the DHEW contributed to the passage of the 1971 Lead-Based Paint Poisoning Prevention Act.[33] Under the Act, in mid 1971 (Fiscal Year 1972), federally assisted mass screening of children for lead poisoning began.[2,7]

Screening data collected in the late 1960s and early 1970s revealed that 20 to 45% of children tested had blood lead levels that exceeded 40 µg/dl.[31] Contrary to earlier findings, many affected children did not reside in inner-city slums or have pica.[34] These unexpected findings caused health workers to rediscover the importance of lead dust as well as hand-to-mouth activities as a source of poisoning in children[13] — facts which Gibson had described in 1904.

During the past 15 to 20 years, due to mass screening, public education, a phased reduction of lead in gasoline, and a decline in food lead content, overt lead poisoning has decreased dramatically.[2] However, the problem is still common in some areas. In 1983, 9.6% of 11,736 children screened in St. Louis, Missouri had confirmed blood lead levels of 30 µg/dl or more.[2] Nor have overt lead poisoning and lead encephalopathy totally disappeared. As late as 1984, a 2-year-old girl underwent brain surgery before the correct diagnosis of lead encephalopathy was made.[2,35] While screening programs expanded in the past 20 years, little progress was made in abating lead paint hazards on housing. The problem of lead paint poisoning, however, is not confined to the U.S. A 1987 study in Belgium indicated that lead paint poisoning in Belgium children may be more common than realized.[36]

III. MENINGITIS AMONG BREASTFED INFANTS: LEAD IN TOILETRIES

A. SEROUS MENINGITIS AMONG NURSING INFANTS IN JAPAN

A puzzling form of serous meningitis among breastfed infants troubled physicians in Japan for over 200 years before its cause was identified in 1923.[37,38] The disease was so common that in 1784, a Japanese monograph on pediatrics included a careful description of this illness. The condition was considered to be a form of benign tuberculous meningitis, infantile beriberi, or intoxication from human milk.[37]

Between 1919 and 1924, this form of meningitis accounted for 3.6% (298 cases) of 8,228 admissions to a pediatric service in Kyoto and for 14.6% (72 cases) of deaths that occurred in this hospital.

In 1923, Hirai reported that lead poisoning was the cause of this illness.[38] Evidence of lead poisoning in these infants included basophilic stippling of red blood cells, discoloration of teeth, lead lines on gums, and hematoporphyrinuria.[39] After this report, 33 papers on meningitis in nursing infants were presented at the 1925 Japanese Pediatric Society's meeting. The source was traced to lead compounds in toiletries used by Japanese nursing women.

B. LEAD POISONING IN NURSING INFANTS CAUSED BY TOILETRIES IN OTHER COUNTRIES

Toiletries and cosmetics from both the old countries and the new world, including hair dyes, continue to contain significant amounts of lead.[1] Cases of lead intoxication from this source have been reported from the U.S., France, Germany, and Moslem countries.[40-43]

IV. LEAD POISONING FROM BURNING BATTERY CASINGS: THE DEPRESSION DISEASE THAT HAS PERSISTED

A. THE BALTIMORE OUTBREAK IN 1932

In 1932, a 7-year-old black girl was admitted in coma to the Harriet Lane Home at Johns Hopkins Hospital. A careful history taken by an astute intern, Miriam E. Brailey, led not only to a correct diagnosis of lead poisoning, but also to the uncovering of an important and persistent cause of dangerous lead exposure — the burning of discarded battery casings for domestic fuel. During the next two months, 40 similar cases were identified in Baltimore.[44]

The 40 patients came from 20 families. All were black and all lived in abject poverty and deprivation. In one instance, seven household members were affected. Five patients had definite lead encephalopathy.

The illness was dubbed as the "Depression Disease," since it was common during those years for junk dealers to provide free battery casings to poor families for kitchen fuel. Following the outbreak of lead poisoning, the Baltimore City Health Department halted the distribution of battery casings for fuel. A 1933 reinvestigation found that Baltimore junk shops were no longer giving away battery casings.[44] But elsewhere, lead poisoning caused by this practice continued to occur in Chicago, Detroit, Long Island, and Philadelphia.[45]

B. THE DEPRESSION DISEASE PAST THE DEPRESSION ERA

Burning battery casings has continued. In 1943, three children in Idaho suffered lead poisoning from this practice, two ended fatally.[46] In 1957, 25 children and 17 adults in an Indian reservation at Great Falls, Montana developed lead poisoning from this cause.[46] An outbreak of lead poisoning from burning battery casings occurred in Rotherham, England in 1956[47] and in Bradford in 1967.[48] In 1964, Angle and McIntire reported a case of congenital lead poisoning associated with burning battery casings.[49] As late as 1981, reports of families with lead poisoning from this cause were published.[50,51] Aside from battery casings, burning of wood coated with lead paint and colored newsprints was also reported to be a source of lead exposure.[52,53]

V. COLIC AND CONVULSIONS FROM LEAD-CONTAMINATED FOOD AND DRINKS

A. LEAD COLIC THROUGHOUT THE AGES

Severe lead poisoning caused by contaminated food and drinks has been recorded throughout the ages.[1] From the 16th to the 18th centuries, several outbreaks of lead colic from such sources occurred in France, Italy, Spain, Sweden, Holland, England, Switzerland, Germany, Australia, and colonial America.[1,54] The Colic of Poitou, due to wine, was the best known among these. The Colic of Devonshire was traced to apple cider, and the dry bellyache or West Indian dry gripe was linked to rum in America.

B. LEAD CONVULSIONS IN PHILADELPHIA FROM CAKES AND BUNS

In 1887, Stewart reported an outbreak of mass lead poisoning in Philadelphia traced to lead chrome yellow dye in cakes and buns.[55] This series included sixty-four cases, five of whom were pregnant women. Four infants born to these women developed convulsions in the first two months of life, and two died. In 1895, Stewart again called attention to the problem of lead food dye in 16 patients with convulsions.

C. OUTBREAKS DURING THE 20TH CENTURY
1. Food and Water

Sporadic reports of poisoning from lead in food have continued to appear in the literature. In 1985, Eisenberg et al.[57] reported severe lead poisoning in several West-Bank Arab families traced to contaminated flour produced at the community mills. The source of contamination was the lead fillings securing the housing of the driveshifts to the millstone.

In 1972, Beattie et al.[58] reported that 11 persons from rural Scotland suffered lead poisoning caused by the domestic water supply contaminated by lead pipes and storage tanks. The increased plumbosolvency of soft water in rural Scotland was an important contributing factor.[58]

2. Improperly Glazed Earthenwares

In 1969, severe lead poisoning in five members of a California family was traced to a Mexican earthenware jug used for storing orange juice.[59] In 1973, several persons living in a commune in Oregon suffered severe lead poisoning caused by consumption of plum wine made in a lead glazed tub at the community.[60] Blood lead levels as high as 98 µg/dl were encountered. In 1987, seven persons from Westchester County, New York were poisoned by a homemade fermented beverage stored in a Mexican ceramic bean jug.[61] There have been many single case reports of severe lead poisoning caused by lead glazed earthenware from different countries.[62-65]

In 1985, Wallace et al.[65] reported that of 67 pieces of Mexican glazed potteries tested, 33 were within the safe limit set by the Food and Drug Administration (FDA). Seventeen released between 7 and 100 ppm of lead, and 17 released over 100 ppm. Lead value as high as 9900 ppm were found. A limited survey of 14 Mexican potteries obtained from Hispanic families in the Seattle area found that 6 exceeded the safe limit.[65]

3. "Moonshine" or Illicit Alcoholic Drinks

Gilfillan[66] suggested that lead used to sweeten Roman wine caused plumbism, and contributed to Rome's downfall. In contemporary U.S., lead poisoning resulting from alcoholic drinks has remained a problem. In 1972, the National Academy of Sciences report estimated that more than 30 million gallons of moonshine were produced each year in the U.S. An analysis of 791 moonshine samples found a lead content of 0.001 to 0.99 mg/l in 54.5%, 1 to 10 mg/l in 25.7%, and over 10 mg/l in 3.3%. The highest lead content found was 86.4 mg/l.[67] In 1984, Lacy and Winternitz[68] reported their review of 141 patients who were admitted with acute or chronic alcoholism to the Druid City Hospital in West Alabama. Fourteen patients, or 9.2%, gave a history of moonshine consumption. Five of these patients suffered from lead poisoning; five were not evaluated and three had no evidence of poisoning.

VI. THE LEAD WORLD, LATE 19TH TO LATE 20TH CENTURIES

A. HAZARDOUS LEAD SOURCES IN THE LATE 19TH CENTURY

At the turn of the century, drinking water contaminated by lead pipes or stored in lead receptacles and lead contaminated food appeared to be the most common nonoccupational hazardous lead sources. Among 30 children with lead poisoning reviewed in the 1891 Cyclopedia of the Diseases of Children, 8 were caused by buns and cake colored with lead chromate dye, 5 by drinking water, 4 by toiletries on nursing mothers, and 2 cases each by fresh paint and lead contaminated cider.[40]

B. THE LEAD WORLD IN 1924

Lead poisoning in children was first recognized in the U.S. in the 1920s. In 1924, Ruddock[69] observed that "A child lives in a lead world." Aside from lead paint on windowsills, porch railings, toys, cribs, and other furniture, Ruddock cited lead food and water receptacles, printers' type used as toys, lead objects accidentally swallowed, lead cosmetics, lead ointments used to treat poison oak in children or on nursing mothers' breasts, and lead dye in baked goods and candies as common hazardous sources. He also pointed out the importance of pica.

C. LEAD IN GASOLINE — A SOURCE OF MASS ENVIRONMENTAL POLLUTION

The 1980 National Academy of Sciences report entitled "Lead in the Human Environment", states that each year, various end uses of lead release about 600,000 t of lead into the environment.[70] Automobile emissions were the single largest source of lead and accounted for about 90% of the total lead emissions. Through this process, lead has been added not only to the air, but also to the water supplies and food crops.

Leaded gasoline was first placed in selected markets in the U.S. in 1923. In 1924, Dupont and General Motors created the Ethyl Corporation to market this new product. The addition of tetraethyl lead, a highly toxic compound, to boost the octane of gasoline met with considerable concern from scientists and public health experts. Five fatalities and 35 cases of severe lead poisoning in exposed workers in 1924 led to a temporary halt in the sales of leaded gasoline. Marketing resumed shortly after a blue ribbon committee, convened by the U.S. Surgeon General, concluded that there were no grounds to prohibit the use of leaded gasoline. Despite strenuous opposition from scientists such as Alice Hamilton and Yandell Henderson, the use of leaded gasoline increased dramatically.[71] In 1947, Kay and Reznikoff[72] compared the lead contents of street dust in New York City in 1924 and 1934, and reported an average increase of 50% in lead content. This paper attracted little attention, and the public remained largely oblivious to the intense lead pollution that was taking place. By 1975, the U.S. consumption of lead in gasoline reached 167,400 t.[73]

In the mid 1970s, it became increasingly clear that automobile emissions from combustion of leaded gasoline were a major source of widespread environmental contamination. Industries vigorously challenged the direct relevance of such emissions to human health. The impact of this environmental pollution became indisputable in the early 1980s when the Second U.S. National Health and Nutrition Examination Survey (NHANES II) data became available.[74] This survey demonstrated that between 1976 and 1980, the decline of the mean blood lead level of the U.S. population closely paralleled the decrease in U.S. consumption of lead in gasoline. As the lead consumption in gasoline declined from 53,000 t per quarter in 1976 to 24,000 t in 1980, the mean blood lead level of the U.S. population dropped 37% from 14.6 µg/dl to 9.2 µg/dl.

The NHANES II data were instrumental in forcing the Environmental Protection Agency (EPA) to face the issue of lead pollution from gasoline more directly. Although the EPA had begun to regulate lead in gasoline in 1973, it was only in 1982 that the Agency proposed an aggressive phased reduction of lead in gasoline.[73]

D. LEAD PAINT ON DWELLINGS: A RECALCITRANT LEAD HAZARD

Although exposure to leaded gasoline has declined dramatically in the last decade, lead paint on housing has remained virtually untouched. This is particularly troubling because the 1972 Lead-Based Paint Poisoning Prevention Act (PL 91-695) directed the U.S. Department of Housing and Urban Development (DHUD) to study methods for removing lead-based paint from housing.[33] The 1973 amendment to the Act (PL 93-151) further directed the DHUD to establish procedures for eliminating lead-based paint in federally assisted housing. To prevent future application of lead paint on houses, the amendment directed the Consumer Product Safety Commission (CPSC) to determine a safe lead level in paint. In 1977, the CPSC established 0.06% as the maximum allowable lead level in paint. A year later, under the 1976 amendment (PL 94-317), CPSC banned the use of paint exceeding this level on residential structures, toys, furniture, and cooking and eating utensils.[33] The CPSC regulations prevented the use of lead paint on housing constructed after 1978, but provided no solution to paint on the existing housing stock. During the two decades after passage of the Lead-Based Paint Poisoning Prevention Act, little or no progress has been made in developing new and more effective lead paint hazard abatement techniques. Moreover, recent studies have indicated that current abatement procedures have failed to effectively remove lead hazards in the form of lead dust from homes.[75]

In the 1988 Report to Congress on the Nature and Extent of Lead Poisoning in Children in the United States,[35] the Agency for Toxic Substances and Disease Registry (ATSDR) estimated that 42 to 47 million houses with lead paint exist in the U.S. and that 12 to 13.6 million children under 7 years are exposed to this lead source. Exposure may be through ingestion of paint chips and/or swallowing of lead in house dust and garden soil during normal hand-to-mouth activities. Lead poisoning caused by dust and fumes created in renovation of expensive homes have affected largely well-educated middle class young professionals and their families.[2]

Lead paint on toys, especially those that are imported, remains a problem. In 1985, the CPSC issued a warning against certain baby crib exercisers that had lead paint on them.[76] In 1986, CPSC again issued warnings against toys made with lead paint and imported from Taiwan.[77]

E. LEAD IN FOOD

Contamination of the food chain is inevitable in a heavily polluted environment. Until the mid 1970s, food was a major source of human lead intake. Food crops are contaminated when grown in leaded soil. Residues from lead pesticides present an additional problem.[78] But since the late 1970s, use of lead pesticides in the U.S. has declined sharply. The decline in leaded gasoline has also reduced the lead contamination of food crops.

Lead solder in food cans was another major source of food lead contamination until the last decade.[79] Up to the early 1940s, the solder commonly used in cans consists of 63% lead and 37% tin. During World War II, a tin shortage stimulated the development of a new solder with only 2% tin and 98% lead. The National Academy of Sciences and the National Research Council studied the safety of this new product and concluded that the lead solder presented no hazard in food cans and packaging.[80] However, in the early 1970s, several studies revealed an excessively high lead content in evaporated milk and baby food and juices.[81,82] These findings prompted the Food and Drug Administration (FDA) to take steps to reduce the lead content of canned goods. During the past 15 years or so, the use of glass jars for baby food and juices and improved food canning technology have decreased significantly the lead content of canned goods.[83]

F. LEAD IN DRINKING WATER

The main sources of water contamination in the early part of this century were lead water pipes in homes and lead water storage tanks.[1] In the mid 1980s, lead solder used in joining copper pipes gained attention as a source of lead in drinking water. In 1987, specimens from many water coolers were also found to have elevated lead. The EPA proposed in 1988 to lower the Maximum Contamination Limit for lead in drinking water to 0.005 mg/l. The EPA estimated that 2 million people are served by distributed water that exceeded this limit.[84]

In areas where the water is soft, lead presents a special problem because of the plumbosolvency of such water. Despite the general assumption that mineral deposits in lead pipes collected over many years provide a protective coating, a 1979 study in Vermont, where the water is soft, found that specimens from lead pipes had significantly higher lead content than those from copper or galvanized iron pipes. Water from the latter sources also had elevated lead levels. About one third of the houses surveyed had water lead content that exceeded 0.05 mg/l, the safe limit set by the EPA then. One specimen had a lead level as high as 0.6 mg/l.[85]

G. LEAD IN DUST AND SOIL

Lead in dust and soil is a pervasive source of lead exposure for children. This became apparent in the late 1960s when childhood screening programs discovered a high prevalence of elevated blood lead levels in children that could not be explained by ingestion of paint chips.[8,34] In addition to leaded gasoline other major sources of lead in dust and soil include stationary emission sites such as smelters and manufacturing plants, and chalking and/or flaking lead paint.[66] In the mid 1970s, it was discovered that lead workers transport lead dust home on their clothes, shoes, and hair and expose their children.[86] One study found that 27 of 29 preschool children of battery workers had blood lead levels that exceeded 30 µg/dl. The lead content of dust in the closets and carpets in these children's rooms ranged from 1,700 ppm to 84,050 ppm.[87]

H. LEAD IN FOLK REMEDIES

Lead has been used in folk medicine since ancient times (see also Chapter 1).[1] These ancient folk remedies are still used in developing countries and by immigrants in their new homes. Lead poisoning has been traced to many folk remedies including aphrodisiacs, baby tonics, Chinese herb medicine, bone meals, teething powder, and ointments.[88] Two Mexican remedies (Azarcon and Greta) have very high lead concentrations. These remedies are commonly used in Mexico and by Mexican Americans to treat "empacho", a folk illness characterized by gastrointestinal symptoms. A Chinese folk medicinal powder (Pay-loo-ah) used for reducing high fever was recently found to be the cause of lead poisoning in a Hmong refugee child.[92] Because of the continuous inflow of immigrants and refugees to the U.S., folk remedies are a continuing source of lead poisoning.

I. MISCELLANEOUS LEAD SOURCES

Other lead sources for children include colored ink in newsprints and children's books, food wrappers, toothpastes, cosmetics, batteries, curtain weights, art and craft supplies such as paint or pottery glaze, and solder.[88-91] Surma or Kohl, an eye cosmetic commonly used in the Indian subcontinent and many Arab nations, is an important cause of lead poisoning in some areas.[43] Gasoline sniffing among children and adolescents has caused severe lead poisoning, including encephalopathy.[92,93] Lead shot in wounds has been reported repeatedly as an unsuspected source of lead poisoning,[94] and hazardous exposure in firing range presents a problem to both law-enforcing officers and others.[95,96] Recently, lead contamination of intravenous methamphetamine used by drug abusers was identified as yet another source of poisoning.[97]

VII. EPIDEMIOLOGY OF LEAD POISONING IN CHILDREN

A. EARLY MISPERCEPTIONS CONCERNING CHILDHOOD LEAD POISONING

Up to the early 1970s, childhood lead poisoning was thought to be inextricably linked to old dilapidated housing, and therefore limited to young children living in old inner city slums.[7,22-27] Since then, screening data have refuted this misperception. While young children in dilapidated housing in urban areas are obviously at a higher risk, excess exposure to lead affects children of all socioeconomic backgrounds.

B. REASON FOR MISPERCEPTIONS

The persistent perception that childhood lead poisoning is confined to the poor stems from several factors which are described in the subsequent sections.

1. Ignorance About the Extent of Environmental Lead Contamination

Despite long knowledge of lead toxicity, the dispersion of millions of tons of this toxic metal into the human environment during this century took place virtually unnoticed.

In 1965, Clair Patterson estimated the existing average lead burden in man to be about 100 times larger than the natural burden.[98] He pointed out that in our severely contaminated environment, what is typical should not be considered natural or safe. He noted that the accepted "normal" blood lead level of the U.S. population of 25 µg/dl left no margin for safety. Patterson concluded that "the average resident in the United States is being subjected to severe chronic lead insult." Four years later, Murozumi, Chow, and Patterson reported that the lead concentration in the Greenland ice sheet of 1750, the beginning of Industrial Revolution, was 25 times greater than natural levels that existed in 800 B.C.[99] The lead concentration in ice sheets tripled during the second half of the eighteenth century, and again tripled between 1935 and 1965. This study concluded that the lead concentrations in Greenland snow in 1965 were 400 times the natural level. This study, and Patterson's paper provided information critical to a general awakening to the extent of environmental lead pollution.

From the late 1960s to the early 1970s, many cities found blood lead levels of 40 μg/dl or more in 20 to 45% of the children tested.[31] Most affected children had no pica, and many lived outside of the so-called "lead belts" and were not exposed to paint chips.

These surprising findings suggested that lead must be pervasive in the human environment, and raised several questions: What was the source of lead in these children? How did lead enter the children's bodies? Do "asymptomatic" children with elevated blood lead levels sustain some subtle damage by lead? How should lead poisoning be defined?

By the mid 1970s, many studies revealed a high lead content in street dust and soil, particularly near heavily traveled roadways. A 1971 study of 77 midwestern U.S. cities found lead content of 1636 μg/g in dust from residential areas and 2413 μg/g in dust from commercial sites.[66] Another study in Los Angeles revealed a soil lead content of 3357 μg/g.[66] Concurrently, studies of lead content of dust in children's homes and on children's hands showed dust to be an important source of lead in children.[34,100] It became apparent that normal hand-to-mouth activities, and not pica, was the most common method of lead ingestion in children. Although paint on houses is the major source of lead in house dust and garden soil immediately adjacent to houses, studies in the 1970s clearly implicated leaded gasoline as a more pervasive source of lead dust in the human environment.[70,73]

2. Erroneous Concept of a "Normal" Blood Lead Level

Until late 1960s, 80 μg/dl was generally assumed to be the upper limit of "normal" blood lead levels in adults and 60 μg/dl to be the limit in children. These levels were in fact blood lead levels at which overt clinical symptoms of poisoning become common.[31]

In 1965, Chisolm suggested that the upper limit for children should be lowered to 40 μg/dl,[101] but his suggestion drew little attention. In 1969, as a member of an ad hoc committee that was preparing the Surgeon General's Statement on the subject, this author again proposed to use 40 μg/dl as the upper limit for children, and further suggested that children with levels beyond this limit be investigated for hazardous lead exposure. The proposal encountered fierce opposition in the committee because some cities were then finding such blood lead levels in 20 to 45% of the children screened.[31] Opponents to the 40 μg/dl limit feared that this new standard would frustrate public health officials with an overwhelming caseload. Only after this author agreed to respond to all the letters of complaint that might be written about the new limit did the committee chairman reluctantly consent to adopt this value in the 1970 Surgeon General's Statement.[32]

Anticipating strong public opposition and a deluge of letters, this author amplified and supported the Surgeon General's Statement with a 1972 paper entitled "Undue absorption of lead among children — a new look at an old problem".[31] Published in the *New England Journal of Medicine,* the paper pointed out that "Many papers that attempt to define a normal level seem to imply that values not diagnostic of lead poisoning are normal. In fact, most papers equate the lowest blood lead level diagnostic of lead poisoning with the upper limit of normal." The paper challenged this erroneous concept and emphasized that early identification of children with evidence of undue lead absorption is a critical step in the prevention of overt lead poisoning. It further noted that

1. A critical question that cannot be answered with certainty and has been responsible for the confusion in the diagnosis of lead poisoning is whether lead causes any permanent damage in man in the absence of clinical evidence of toxicity.
2. A closely related question also unanswered is whether lead can damage the central nervous system of young children in the absence of overt signs and symptoms referable to that system.

The new blood lead limit of 40 µg/dl was accepted without opposition and the Surgeon General's office did not receive a single letter of complaint. But in 1971, the American Academy of Pediatrics took a retrogressive step when it recommended that a child's environment be investigated only after a confirmed blood lead level of 50 µg/dl.[102] In 1975, the Centers for Disease Control lowered the limit to 30 µg/dl, and in 1985, further lowered the limit to 25 µg/dl. Recent research suggests that this limit should be further lowered, perhaps to less than 10 µg/dl.[103-111]

3. Viewing Lead Poisoning as a Clinical Disease

Historically, lead poisoning gained recognition as a clinical illness in adults characterized by colic and peripheral nerve paralysis.[1,3,4] When childhood lead poisoning first attracted the attention of Australian health workers, it was in the form of a severe illness characterized by peripheral nerve paralysis, colic, and convulsions.[5,6] In 1933, McKhann[20] still considered lead encephalopathy to be the usual form of lead poisoning in children. Even during the second half of this century, severe lead poisoning was so common that 182 children were admitted with lead encephalopathy to Chicago's Cook County Hospital between 1959 and 1963.[25,29] The erroneous perception of lead poisoning solely as a severe clinical disease is reflected in the 1959 Diagnostic Criteria for Lead Poisoning in Children issued by the National Clearinghouse for Poison Control Centers of the U.S. Public Health Service.[112] Among clinical criteria listed were signs and symptoms of severe lead poisoning such as persistent vomiting, drowsiness, incoordination, paralysis, convulsions, and coma. The criteria for lead absorption called for a blood lead level of 0.08 mg/100 g of whole blood or more. The document did concede that blood lead levels of 0.06 to 0.08 mg/100 g are indicative of abnormal absorption of lead, but stated that such values "are often not a degree of absorption which is capable of inducing symptoms of intoxication."

Lead poisoning continued to be viewed strictly as a clinical disease until the late 1960s when most children with elevated blood lead levels (then defined as 40 µg/dl or more) uncovered through screening programs were found to be "asymptomatic." With this finding, from the 1970s on, research on lead toxicity shifted from clinical studies to investigation of subtle psychoneurological effects in young children. But as early as 1943, Byers and Lord reported that children thought to have recovered from lead poisoning later demonstrated behavioral and learning disorders that contributed to poor school performance.[21] The significance of this paper, like the findings of Gibson, was rediscovered only in the 1970s.

In 1979, Needleman et al.[113] set another milestone in our understanding of the toxic effects of lead in children. Using dentine lead content as marker of past exposure, these investigators reported that children with high dentine lead levels scored significantly less well in the Wechsler Intelligence Scale for Children than those with low levels. None of the children were ever reported to have lead poisoning clinically. The study further found that the frequency of nonadaptive classroom behavior increased in a dose-related fashion to dentine lead level. These authors concluded that "Lead exposure at dose below those producing symptoms severe enough to be diagnosed clinically, appears to be associated with neuropsychological deficits that may interfere with classroom performance." These findings have since been replicated by many other investigators.

During the 1980s, many papers that reported toxic effects of lead at lower and lower blood lead levels were published.[103-111,113] Recent prospective studies have revealed a correlation of the pregnant women's blood lead levels with preterm delivery and reduced birth weight and delayed development in infants.[108,109,111] Bellinger et al.[105,108] found a 6 point difference in the Mental Developmental Index (MDI) in the Bayley Scale at 6 months between the high and low in-utero exposure groups. The mean blood lead level of the high exposure group was only 14.6 µg/dl and that of the low exposure group was 1.8 µg/dl.[105] A follow-up study of these infants at 12, 18, and 24 months showed a persistent difference of 4 to 8 points in the MDI of the two study groups.[108] Others have reported similar findings.[109,112] Schwartz et al.[106] reported a correlation between children's blood lead levels and stature with no threshold. In another study, Schwartz et al.[107] found that blood lead was positively correlated with hearing loss at 500, 1000, 2000, and 4000 Hz.[107] Again, these investigators found no threshold for this effect.

In addition to these clinical findings, "subclinical" toxic effects of lead reported include inhibition of many enzymes,[114-116] disturbance of vitamin D metabolism,[117] reduced peripheral nerve conduction velocity[118] and electroencephalographic changes.[103]

C. EPIDEMIOLOGY OF CHILDHOOD LEAD POISONING IN 1990
1. Basic Facts in Understanding Childhood Lead Poisoning

Lead plays no physiologic role in the human body,[66] and a natural or normal blood lead level should be close to zero. In 1965, Patterson stated that the natural blood lead level is 0.0025 ppm or 0.25 µg/dl.[98] An average blood lead level of 0.87 µg/dl was reported among unacculturated Indians living near the mouth of the Orinoco River.[119] Current literature suggests that there may be no threshold for the toxic effects of lead on human metabolism and neurophysiology.[103-107] At a minimum, a blood lead level of 10 to 15 µg/dl is reported to be associated with undesirable toxicity in young children and fetus.

Although psychoneurological damages associated with relatively "low" levels of lead exposure are relatively "minor" in comparison to mental retardation and paralysis associated with severe lead poisoning, these "minor" effects are irreversible. Recently Needleman et al.[120] published an 11 year follow-up study of children with relatively "minor" neurobehavioral dysfunctions associated with low level lead exposure in early childhood. They reported that these children demonstrated a markedly higher risk of dropping out of high school and of reading disabilities. Affected children also had lower class standing, increased absenteeism, poor eye-hand coordination and other deficits that contributed to school failure later in life.[120] Although some recent studies on low level lead toxicity have reported negative findings, a recent meta-analysis of 24 modern studies of childhood lead exposure in relation to IQ strongly supported the hypothesis that lead, even at very low doses impairs children's intellectual performance.[121] Because of these recent findings the epidemiology of lead poisoning in children should not be limited to overt clinical cases but should include all children with elevated blood lead levels.

2. NHANES II Blood Lead Data (1976 to 1980)

The Second National Health and Nutrition Examination Survey (NHANES II) between 1976 and 1980[122,123] found that 2% of all persons 6 months through 74 years and 4% of children 6 months through 5 years had blood lead levels that exceeded 30 µg/dl; 15% of the total population and 24.5% of children had levels that exceeded 20 µg/dl. Only 22.1% of the total population and 12.2% of children had blood lead levels below 10 µg/dl (see Table 1).

Among black children 6 months through 5 years, 52.2% had blood lead levels of 20 µg/dl or higher while 12.2% had values of 30 µg/dl or more; only 2.5% had values less than 10 µg/dl. This compares with 18.1%, 2% and 14.5% respectively in white children. 7.2% of urban children, as compared to 2.1% of rural children had blood lead levels of 30

TABLE 1
NHANES II Blood Lead Levels in Persons 6 Months to 74 Years — Percent Distribution by Age and Race
(1976 — 1980)

CHARACTERISTICS	BLOOD LEAD LEVELS (µg/dl)				
	< 10	10—19	20—29	30—39	40 >
All races					
All ages	22.1	62.9	13.0	1.6	0.4
6 Months — 5 years	12.2	62.3	20.5	3.5	0.5
Black					
All ages	13.3	63.7	20.0	2.3	0.7
6 Months — 5 years	2.5	45.3	40.0	10.2	2.0
White					
All ages	23.3	62.8	12.2	1.5	0.2
6 Months — 5 years	14.5	67.4	16.1	1.8	0.2

TABLE 2
NHANES II Blood Lead Data — Percent of Children 6 Months to 5 Years with PbB Above 30 µg/dl
(1976 — 1980)

Demographic variables	All races	White	Black
All children 6 months — 5 years	4.0	2.0	12.2
Annual family income			
Under $6,000	10.9	5.9	18.5
$6,000—14,999	4.2	2.2	12.1
$15,000 or more	1.2	0.7	2.8
Place of residence			
Urban >1 million	7.2	4.0	15.2
Central city	11.6	4.5	18.6
Noncentral city	3.7	3.8	3.3
Urban <1 million	3.5	1.6	10.2
Rural	2.1	1.2	10.3

µg/dl or more. An alarming 18.6% of black children from central cities had such levels. The problem also affected the poor disproportionately: 1.2% of children from families with annual income of $15,000 or more, but 10.9% of those from families with incomes of under $6,000 had levels exceeding 30 µg/dl (see Tables 2 and 3). These data indicated that increased lead absorption affects all children, but is more common in black, poor, and urban children.

3. ATSDR Report to Congress (1988)

The Agency for Toxic Substances and Disease Registry (ATSDR) 1988 Report to Congress on the Nature and Extent of Lead Poisoning in Children[35] estimated: 12 million children under 7 years of age are exposed to some lead paint at potentially toxic levels; 5.9 million live in the oldest housing units with highest lead content in paint; 1.8 to 2 million children live in deteriorating old houses; and 1.2 million children have blood lead levels that exceed 15 µg/dl from exposure to lead paint. The EPA estimated that 24,000 children under 6 years have blood lead levels of 15 µg/dl or more due to lead in drinking water; 230,000 have levels between 15 and 30 µg/dl.

The above reports clearly indicate that lead affects children of all races, socioeconomic classes and neighborhoods, but those living in poorly maintained older homes are at increased

TABLE 3
NHANES II Geometric Mean Blood Lead Level in Persons 6 Months to 74 Years — by Race and Age (1976 — 1980)

Characteristics	Geometric mean PbB (µg/dl)
All races	
6 Months — 5 years	14.9
6—17 Years	11.7
18—74 Years	
Men	15.9
Women	11.0
Black	
6 Months — 5 years	19.6
6—17 Years	14.0
18—74 Years	
Men	18.1
Women	12.0
White	
6 Months — 5 years	14.0
6—17 Years	11.3
18—74 Years	
Men	15.6
Women	10.9

risk. Swallowing lead dust and soil during normal play activities is the most common route of lead intake, but children are seldom exposed to a single lead source.[35,70,73] Because of hand-to-mouth activities preschool children, particularly those under 3 years of age, are at greatest risk.

The NHANES II data indicated that only 12.2% of children 6 months through 5 years had blood lead levels below 10 µg/dl.[125] If one arbitrarily accepts this level as "safe" for young children, one might speculate that up to about a decade ago the vast majority of U.S. preschool children had sustained some damage caused by lead. However, the recent phased reduction of lead in gasoline has resulted in a dramatic decline in the blood lead levels of the U.S. population.[73,74] In 1990, the proportion of young children with blood lead levels of 10 µg/dl was likely to be lower than what was found in NHANES II.

VIII. PREVENTION OF LEAD POISONING: A REDISCOVERY OF OLD FACTS

Lead poisoning is a disease that has been carefully studied and recorded for centuries, but the modern history of lead poisoning is a sad tale of inexcusable disregard for knowledge that has long existed, and a rediscovery of such knowledge after generations of human beings, particularly fetuses and young children, have been permanently damaged.

A. THE TOXIC EFFECTS OF LEAD IN THE FETUS

The toxic effects of lead in the fetus were recognized not long after the Industrial Revolution. Many female lead workers were found to suffer from reproductive failures, including sterility, abortion, premature delivery, and stillbirths. These findings eventually led to the exclusion of women from lead industries.[3,4] Following this ruling, it was generally assumed that the problem of in-utero exposure to lead was resolved.

During the first several decades of the 20th century, the subject of in-utero exposure to lead drew relatively little attention, except for occasional reports of congenital lead poisoning from various sources.[89] But from the mid 1980s on, many researchers began to report diverse toxic effects of lead on the fetus. These included preterm delivery, reduced birth weight, and developmental and cognitive deficits in the first years of life. The effects of lead on the unborn were evident at a maternal blood lead level of 10 to 15 µg/dl or lower.[105,108,110,111]

According to the 1988 ATSDR Report, 4,460,600 U.S. women of childbearing age and 403,200 pregnant women have blood lead levels of 10 µg/dl or more.[35]

B. THE PREVENTABLE NATURE OF LEAD POISONING IN CHILDREN

In 1904, Gibson clearly identified lead paint and dust as the major sources of lead poisoning and pointed out the causal role of hand-to-mouth activities.[13] Both he and Turner stressed the preventable nature of childhood lead poisoning.[5,13] But Gibson's important contributions, particularly his observations on the importance of lead dust, remained virtually unnoticed until their rediscovery in the 1970s. As Turner noted in 1908, millions of houses are still "poison traps for children's fingers, and every year furnished its quota of ill-health and suffering, crippling and hopelessness."[14]

IX. WHAT OF THE FUTURE?

In the last 10 to 15 years, reduction of lead in gasoline and decrease in the lead content of canned food have been two major achievements in controlling lead exposure. The recent EPA proposal to lower the limit for lead in drinking water is also a step in the right direction. In contrast, lead paint on millions of housing has remained virtually untouched. Research is urgently needed to find new and more effective lead paint abatement techniques that would assure the removal of not only lead paint, but also lead dust from homes. Since the cost of lead hazard abatement has been a major deterrent to preventing lead poisoning in children, innovative ways to finance these procedures must also be made available.

In 1988, the U.S. consumption of lead totaled 1,230,732 t.[124] A systematic examination of the current applications of lead may identify yet undiscovered lead hazards in consumer goods or in the general environment. All new applications also need to be evaluated for their potential impact on human health.

Public education on the sources and toxicity of lead needs to be intensified. Screening children for early evidence of undue lead exposure must continue until the environment is made safe for human habitation again.

X. CONCLUSION

Lead is an extremely useful metal that has contributed enormously to human civilization. However, modern history of lead poisoning has shown that the desire for technological advances and the financial profits from such progress has blinded generations of human beings to the devastating effects of lead. The 20th century is a period that can justifiably boast of many remarkable achievements in preventive medicine. It is also an era that draws to a close with prevention of childhood lead poisoning a goal yet to be achieved. Centuries from now, will historians write of the 20th century as a period of remarkable scientific and technological progress, but during which widespread regresssion of the human mind and body due to lead was permitted to proceed unchallenged, despite extensive knowledge about the toxic effects of lead? Or will this century be recorded as the turning point in the history of lead poisoning during which man has finally acknowledged the devastation caused by massive environmental lead pollution, and harnessed this toxic metal of antiquity?

REFERENCES

1. **Nriagu, J. O.**, *Lead and Lead Poisoning in Antiquity,* John Wiley & Sons, New York, 1983.
2. **Lin-Fu, J. S.**, Childhood lead poisoning in the United States: a national perspective, in Proc. Natl. Conf. Childhood Lead Poisoning: Current Perspectives, Indianapolis, Indiana, National Center for Education in Maternal and Child Health, Washington, D.C., 1987.
3. **Hamilton, A. and Hardy, H.**, *Industrial Toxicology,* Paul B. Hoeber, New York, 1934.
4. **Aub., J. C., Fairhall, L. T., Minot, A. S., and Reznikoff, P.**, Lead poisoning, *Medicine,* 4, 1, 1925.
5. **Turner, A. J.**, Lead poisoning among Queensland children, *Aust. Med. Gaz.,* 16, 475, 1897.
6. **Gibson, J. L., Love, W., Hardie, D., Bancroft, P., Turner, A. J.**, Notes on lead poisoning as observed among children in Brisbane, *Proc. Intercolonial Med. Congr. of Australasia,* p. 76, 1892.
7. **Lin-Fu, J. S.**, The evolution of childhood lead poisoning as a public health problem, in *Lead Absorption in Children, Management, Clinical and Environmental Aspects,* Chisolm, J. J., Jr. and O'Hara, P. M., Eds., Urban and Schwarzenburg, Baltimore, 1982.
8. **Lin-Fu, J. S.**, Lead exposure among children — a reassessment (editorial), *N. Engl. J. Med.,* 300, 731, 1979.
9. **McCord, C.**, Lead and lead poisoning in early America, Benjamin Franklin, and lead poisoning, *Indust. Med. Surg.,* 22, 394, 1953.
10. **Turner, A. J.**, A form of cerebral disease characterized by definite symptoms, probably a localized basic meningitis, *Trans. Intercolonial Med. Congr.* of Australasia, Sydney, p. 98, 1892.
11. **Gibson, J. L.**, Ocular neuritis simulating basal meningitis, *Australasian Med. Gaz.,* 16, 479, 1897.
12. **Turner, A. J.**, How to recognize lead-poisoning in children, *Australasian Med. Gaz.,* p. 425, October 20, 1899.
13. **Gibson, J. L.**, A plea for painted railings and painted walls of rooms as the source of lead poisoning among Queensland children, *Australasian Med. Gaz.,* 23, 149, 1904.
14. **Turner, A. J.**, Lead poisoning in children, *Trans. Intercolonial Med. Congr. Australasia,* Melbourne, p. 3, 1908.
15. Historical account of occurrence and causation of lead poisoning among Queensland children, *Med. J. Aust.,* 148, 148, 1922.
16. **Freeman, R.**, Chronic lead-poisoning in children: a review of 90 children diagnosed in Sydney, 1948-1967, *Australasian Pediatr. J.,* 5, 27, 1969.
17. **Nye, L. J. L.**, An investigation of the extraordinary incidence of chronic nephritis in young people in Queensland, *Med. J. Aust.,* 2, 145, 1929.
18. **Henderson, D. A.**, A follow-up of cases of plumbism in children, *Australasian Ann. Med.,* 3, 219, 1954.
19. **Thomas, H. M. and Blackfan, K. D.**, Recurrent meningitis due to lead in a child of five years, *Am. J. Dis. Child.,* 8, 377, 1914.
20. **McKhann, C. F.**, Lead poisoning in children, *JAMA,* 101, 1131, 1933.
21. **Byers, R. K. and Lord, E. E.**, Late effects of lead-poisoning on mental development, *Am. J. Dis. Child.,* 66, 471, 1943.
22. **McLaughlin, M. D.**, Lead-poisoning in children in New York City, 1950-1956, *N.Y. State J. Med.,* 50, 3711, 1956.
23. **Jacobziner, H.**, Lead poisoning in childhood: epidemiology, manifestations and prevention, *Clin. Pediatr.,* 5, 277, 1966.
24. **Ingalls, T. H., Tiboni, E. M., and Werrin, M.**, Lead-poisoning in Philadelphia, 1955-1960, *Arch. Environ. Health,* 3, 575, 1961.
25. **Christian, J. R., Celewyez, B. S., and Andelman, S.**, A three-year study of lead-poisoning in Chicago, *Am. J. Public Health,* 54, 1241, 1964.
26. **Chisolm, J. J., Jr. and Harris, H. E.**, The exposure of children to lead, *Pediatrics,* 18, 943, 1956.
27. **Lin-Fu, J. S.**, Lead Poisoning in Children, U.S. Department of Health, Education, and Welfare, Children's Bureau, Washington, D.C., 1967.
28. **Griggs, R. C., Sunshine, I., Newill, V. A., et al.**, Environmental factors in childhood lead-poisoning, *JAMA,* 187, 703, 1964.
29. **Greengard, J.**, Lead-poisoning in childhood: signs, symptoms, current therapy, clinical expressions, *Clin. Pediatr.,* 5, 267, 1966.
30. **Blanksma, L. A., Sachs, H. K., Murray, E. F., et al.**, Incidence of high blood-lead levels in Chicago children, *Pediatrics,* 44, 661, 1969.
31. **Lin-Fu, J. S.**, Undue absorption of lead among children — a new look at an old problem, *N. Engl. J. Med.,* 2856, 702, 1972.
32. Surgeon General, U.S. Public Health Service: Medical aspects of childhood lead poisoning, *Pediatrics,* 48, 464, 1971.

33. **Lin-Fu, J. S.**, The lead-base paint poisoning prevention act (P.L. 91-695): ten years later, in *Childhood Lead Poisoning Prevention and Control, a Public Health Approach to an Environmental Disease*, Cherry, F. F., Ed., Louisiana Department of Health and Human Resources, 1981.
34. **Lin-Fu, J. S.**, Vulnerability of children to lead exposure and toxicity, *N. Engl. J. Med.*, 289, 1229, 1973.
35. **U. S. Agency for Toxic Substances and Disease Registry**, The nature and extent of lead poisoning in children in the United States: a report to Congress, Atlanta, Georgia, 30333, 1988.
36. **Limbos, L., Sand, A., and Clara, R.**, Childhood plumbism due to lead paint in Belgium, *Eur. J. Pediatr.*, 146, 537, 1987.
37. **Kato, K.**, Lead meningitis in infants. Resume of Japanese contributions on the diagnosis of lead poisoning in nurslings, *Am. J. Dis. Child.*, 44, 569, 1933.
38. **Hirai, I.**, Chronic lead-poisoning as a cause of the so-called meningitis, *J. Pediatr. (Tokyo)*, 290, 960, 1923.
39. **Hirai, I.**, The so-called meningitis, *J. Pediatr. (Tokyo)*, 304, 1334, 1925.
40. **Keating, J. M., Ed.**, *Cyclopedia of the Diseases of Children*, J. B. Lippincott, Philadelphia, 1891.
41. **Holt, L. E.**, Lead-poisoning in infancy, *Am. J. Dis. Child.*, 25, 299, 1923.
42. **Byers, R. K.**, Lead-poisoning. Review of the literature and report of 45 cases, *Pediatrics*, 23, 583, 1959.
43. **Shaltout, A., Yaish, S. A., Fernando, N.**, Lead encephalopathy in infants in Kuwait — a study of 20 infants with particular reference to clinical presentation and source of lead poisoning, *Ann. Trop. Pediatr.*, 1, 209, 1981.
44. **Williams, H., Schultze, W. H., Rothchild, H. B., et al.**, Lead-poisoning from the burning of battery casings, *JAMA*, 100, 1485, 1933.
45. **Levinson, A. and Harris, L. H.**, Lead encephalopathy in children, *J. Pediatr.*, 8, 2, 1936.
46. **Ensign, P. R.**, personal communication, 1969.
47. **Travers, E., Rendel-Short, J., and Harvey, C. C.**, The Rotherham lead-poisoning outbreak, *Lancet*, 2, 113, 1956.
48. **Turners, W., Bamford, F. N., and Dodge, J. S.**, Lead-poisoning at Brandford, *Br. Med. J.*, 3, 56, 1967.
49. **Angle, C. R. and McIntire, M. S.**, Lead-poisoning during pregnancy, *Am. J. Dis. Child.*, 108, 436, 1964.
50. **Dolcourt, J. L., Finch, C., Coleman, G. D., et al.**, Hazards of lead exposure in the home from recycling automobile storage batteries, *Pediatrics*, 63, 225, 1981.
51. **DeCastro, F. J., Lazzara, J., and Rolff, U. T.**, Increased lead burden and the energy crisis, *Pediatrics*, 55, 573, 1975.
52. **Duel, W.**, Burning colored newsprint in fireplace or grill can be a health hazard, *JAMA*, 23, 144, 1978.
53. **Hankin, L., Heichel, G. H., and Botsford, R. A.**, Lead-poisoning from colored printing ink: a risk for magazine chewers, *Clin. Pediatr.*, 12, 645, 1973.
54. **Eisenger, J.**, Lead and wine, Eberhard Gackel and the Colica Pictonum, *Med. Hist.*, 26, 279, 1982.
55. **Stewart, D. D.**, A clinical analysis of sixty-four scases of poisoning by lead chromate (chrome yhellow) used as a cake dye, *Med. News*, 51, 754, 1887.
56. **Stewart, D. D.**, Lead convulsions, *Am. J. Med. Sci.*, 109, 288, 1895.
57. **Eisenberg, A., Avni, A., Grauer, F., et al.**, Identification of community flour mills as the source of lead poisoning in West Bank Arabs, *Arch. Intern. Med.*, 145, 1848, 1985.
58. **Beattie, A. D., Dagg, J. H., Goldberg, A., et al.**, Lead-poisoning in rural Scotland, *Brit. Med. J.*, 2, 488, 1972.
59. **Krinitz, B. and Hering, R. K.**, Toxic metals in earthenware, FDA Papers, U.S. Government Printing Office, 1971-435-654153, April, 1971.
60. **Osterud, H. T.**, Plumbism at the Green Parrot Goat Farm, *Clin. Toxicol.*, 6, 1, 1973.
61. **U.S. Centers for Disease Control**, Lead poisoning following ingestion of homemade beverage stored in a ceramic jug, New York, *MMWR*, 38, 379, 1989.
62. **Klein, M., Namer, R., Harper, E., et al.**, Earthenware container as a source of fatal lead-poisoning, *N. Engl. J. Med.*, 283, 669, 1970.
63. **Clark, K. G. A.**, Lead-glazed earthenware, *Lancet*, 2, 662, 1972.
64. **Miller, C.**, The pottery and plumbism puzzle, *Med. J. Aust.*, 2, 442, 1982.
65. **Wallace, D. M., Kalman, D. A., and Bird, T. D.**, Hazardous lead release from glazed dinnerware: a cautionary note, *Sci. Total Environ.*, 44, 289, 1985.
66. **Gilfillan, S. C.**, Lead-poisoning and the fall of Rome, *J. Occupat. Med.*, 7, 53, 1965.
67. National Academy of Sciences, Lead, Airborne Lead in Perspective, Washington, D.C., 1972.
68. **Lacy, R. and Winternitz, W.**, Moonshine consumption in west Alabama, *Ala. J. Med. Sci.*, 21, 364, 1984.
69. **Ruddock, J. C.**, Lead-poisoning in children, *JAMA*, 82, 1682, 1924.
70. National Academy of Sciences, Lead in the Human Environment, Washington, D. C., 1980.
71. **Rosner, D. and Markowitz, G.**, A gift of God? The public health controversy over leaded gasoline during the 1920s, *Am. J. Public Health*, 75, 344, 1985.

72. **Kaye, S. and Reznikoff, P.,** A comparative study of the lead content of street dirts in New York City in 1924 and 1934, *J. Indust. Hygiene Toxicol.,* 29, 178, 1947.
73. U.S. Environmental Protection Agency, Air Quality Criteria for Lead, Environmental Criteria and Assessment Office, Office of Research and Development, Research Triangle Park, NC, EPA 600/8-83-028, 1986.
74. **Annest, J. L., Pirkle, J. L., Makuc, D., et al.,** Chronological trend in blood-lead levels between 1976-1980, *N. Engl. J. Med.,* 308, 1373, 1983.
75. **Farfel, M. R. and Chisolm, J. J., Jr.,** Comparison of traditional and alternative residential lead paint removal methods, Proc. 6th Int. Conf., Heavy Metals in the Environment, New Orleans, LA, 1987, CEP Consultants Ltd., Edinburgh, U.K. 76; **Fischbein, A., Edlund, P. E., and Weisman, I.,** Lead poisoning in homeowners, *Environ. Res.,* 27, 237, 1982.
76. U.S. Consumer Product Safety Commission, Danara baby crib exercisers recalled because of lead hazards, News from CPSC, Washington, D.C., 1985.
77. U.S. Consumer Product Safety Commission, Certain Voltran Lion toys to be recalled and exchanged because of a potential lead paint hazard, News from CPSC, Washington, D.C., 1986.
78. **Elias, R. W.,** Lead exposure in the human environment, in *Dietary and Environmental Lead: Human Health Effects,* Mahaffey, K. R., Ed., Elsevier, New York, 1985.
79. **Boyd, S. D., Wasserman, G. S., Green, V. A., et al.,** Lead arsenate ingestion in eight children, *Clin. Toxicol.,* 18, 489, 1981.
80. U.S. Dept. of Health, Education and Welfare, Food and Drug Administration, Lead in food supply, *Federal Register,* 44, 51177, 1979.
81. **Lamm, S. H. and Rosen, J. F.,** Lead contamination in milk fed to infants, 1972-1973, *Pediatrics,* 53, 137, 1974.
82. **Mitchell, D. G. and Aldous, K. M.,** Lead content of foodstuffs, *Environ. Health Perspect. Exp.,* 7, 59, 1979.
83. **Jelinek, C. F.,** Level of lead in the United States food supply, *J. Assoc. Anal. Chem.,* 65, 942, 1982.
84. U.S. Environmental Protection Agency, Office of Water, Drinking water regulations; maximum contamination level goals and national primary drinking water regulations for lead and copper, Proposed Rules, Federal Register, August 18, 1988, 53, 31516.
85. **Morse, D. S., Watson, W. N., Housworth, J., et al.,** Exposure of children to lead in drinking water, *Am. J. Public Health,* 69, 711, 1979.
86. **Baker, E., Folland, D., and Taylor, T.,** Lead-poisoning in children of lead workers: home contamination with industrial dust, *N. Engl. J. Med.,* 296, 260, 1977.
87. **Dolcourt, J. L., Hamrick, H. J., O'Tuamo, L. A., et al.,** Increased lead burden in children of battery workers: asymptomatic exposure resulting from contaminated work clothing, *Pediatrics,* 62, 563, 1978.
88. **Lin-Fu, J. S.,** Historical perspective on health effects of lead, in *Dietary and Environmental Lead: Human Health Effects,* Mahaffey, K. R., Ed., Elsevier, New York, 1985.
89. **Baer, R. D., de Alba, J. G., Cueto, L. M., et al.,** Lead as a Mexican folk remedy: implications for the United States, in Proc. Natl. Conference Childhood Lead Poisoning: Current Perspectives, Indianapolis, Indiana, National Center for Education in Maternal Child Health, Washington, D.C., 1987.
90. **Lin-Fu, J. S.,** Lead poisoning and undue lead exposure in children: history and current status, in *Low Level Lead Exposure: the Clinical Implication of Current Research,* Needleman, H. L., Ed., Raven Press, New York, 1980.
91. **Levitt, C., Godes, J., Eberhardt, T. M., et al.,** Sources of lead poisoning, *JAMA,* 25, 3127, 1984.
92. **Coulehan, J. L., Hirsch, W., Brillman, J., et al.,** Gasoline sniffing and lead toxicity in Navajo adolescents, *Pediatrics,* 73, 113, 1983.
93. U.S. Centers for Disease Control, Gasoline sniffing and lead toxicity among siblings, Virginia, *MMWR,* 34, 449, 1985.
94. **Selbst, S. M., Henretig, F., Fees, M. A., et al.,** Lead poisoning in a child with a gunshot wound, *Pediatrics,* 77, 413, 1986.
95. **Valway, S. E., Martyny, J. W., Miller, J. R., et al.,** Lead absorption in indoor firing range users, *Am. J. Public Health,* 79, 1029, 1989.
96. **Fischbein, A., Rice, C., Sakozi, L., et al.,** Exposure to lead in firing ranges, *JAMA,* 241, 1141, 1979.
97. **Allcott, J. V., Barnhart, R. A., and Mooney, L. A.,** Acute lead poisoning in two users of illicit methamphetamine, *JAMA,* 258, 510, 1987.
98. **Patterson, C. C.,** Contaminated and natural lead environment of man, *Arch. Environ. Health,* 11, 344, 1965.
99. **Murozumi, M. O., Chow, T. J., and Patterson, C.,** Chemical concentrations of pollutant lead aerosols, terrestrial dusts, and sea salts in Greenland and Antarctic snow strata, *Geochim. Cosmochim. Acta,* 33, 1247, 1969.
100. **Sayre, J. W., Charney, E., and Vostal, J.,** House and hand dust as a potential source of childhood lead exposure, *Am. J. Dis. Child.,* 127, 167, 1974.
101. **Chisolm, J. J., Jr.,** Chronic lead intoxication in children, *Develop. Med. Child. Neurology,* 7, 527, 1965.

102. American Academy of Pediatrics, Committee on Environmental Hazards and Subcommittee on Accidental Poisoning — Committee on Accident Prevention, Acute and chronic childhood lead poisoning, *Pediatrics,* 47, 950, 1971.
103. **Otto, D., Benignus, V., Muller, K., et al.,** Effects of low to moderate lead exposure in young children: two year followup study, *Neurobehav. Toxicol. Teratol.,* 4, 733, 1982.
104. **Piomelli, S., Seamen, C., Zullow, D., et al.,** Threshold for lead damage to heme synthesis in urban children, *Proc. Natl. Acad. Sci. U.S.A.,* 79, 3335, 1982.
105. **Bellinger, D. and Needleman, H.,** Prenatal and early postnatal exposure to lead: developmental effects, correlates and implications, *Int. J. Mental Health,* 14, 78, 1985.
106. **Schwartz, J., Angle, C., and Pitcher, H.,** Relationship between childhood blood-lead level and stature, *Pediatrics,* 77, 218, 1986.
107. **Schwartz, J. and Otto, D.,** Blood-lead, hearing thresholds, and neurobehavioral development in children and youth, *Arch. Environ. Health,* 42, 153, 1987.
108. **Bellinger, D., Leviton, A., Waternaux, C., et al.,** Longitudinal analysis of prenatal and postnatal lead exposure and early cognitive development, *N. Engl. J. Med.,* 316, 1037, 1987.
109. **Dietrich, K. N., Kraft, K. M., Bornschein, R. I., et al.,** Low level fetal exposure effects on neurobehavioral development in early infancy, *Pediatrics,* 80, 721, 1987.
110. **Fulton, M., Raab, G., Thomson, G., et al.,** Influence of blood lead on the ability and attainment of children in Edinburgh, *Lancet,* 1, 1221, 1987.
111. **McMichael, A. J., Baghurst, P. A., Wigg, N. R., et al.,** Port Pirie cohort study: environmental exposure to lead and children's ability at the age of four years, *N. Engl. J. Med.,* 319, 468, 1988.
112. U.S. Department of Health, Education, and Welfare, Public Health Service, National Clearinghouse for Poison Control Centers, Lead poisoning in children, diagnostic criteria, Washington, D.C., 1959.
113. **Needleman, H. L., Gunnoe, C. E., Leviton, A., et al.,** Deficits in psychologic and classrooms performance of children with elevated blood lead levels, *N. Engl. J. Med.,* 300, 689, 1979.
114. **Hernberg, S. and Nikkanen, G.,** Enzyme inhibition by lead under national urban conditions, *Lancet,* 1, 63, 1970.
115. **Piomalli, S.,** The effects of low-level lead exposure on heme metabolism, in *Low Level Lead Exposure, the Clinical Implications of Current Research,* Needleman, H. L., Ed., Raven Press, New York, 1980.
116. **Angle, C. R., McIntire, M. S., Swanson, M. S., et al.,** Erythrocyte nucleotides in children—increased blood lead and cytidine triphosphate, *Pediatr. Res.,* 16, 331, 1982.
117. **Rosen, J. F., Cheaney, R. W., Hamstra, A., et al.,** Reduction in 1,25-dihydroxy vitamin D in children with increased lead absorption, *N. Engl. J. Med.,* 302, 1128, 1980.
118. **Landrigan, P. J., Baker, E. L., Feldman, R. G., et al.,** Increased lead absorption with anemia and slower nerve conduction in children near a lead smelter, *J. Pediatr.,* 89, 904, 1976.
119. **Hecker, L., Allen, H. E., and Dinman, D. D.,** Heavy metal levels in acculturated and unacculturated populations, *Arch. Environ. Health,* 29, 181, 1974.
120. **Needleman, H. L., Schell, A., Bellinger, D., et al.,** The long-term effects of exposure to low doses of lead in children: an 11-year follow-up report, *N. Engl. J. Med.,* 322, 83, 1990.
121. **Needleman, H. L. and Gatsonis, E. A.,** Low-level lead exposure and the IQ of children. A meta-analysis of modern studies, *JAMA,* 263, 673, 1990.
122. **Mahaffey, K. R., Annest, J. L., Roberts, J., et al.,** National estimates of blood lead levels: United States, 1976—1980, *N. Engl. J. Med.,* 307, 573, 1982.
123. National Center for Health Statistics, Blood lead levels for persons ages 6 months — 74 years: United States, 1976—1980, DHHS Publication No. (PHS) 84-1683, Ser. 11, No. 223, U.S. Government Printing Office, Washington, D.C., 1984.
124. **Woodbury, W. D.,** Lead, in *Mineral Yearbook 1988,* U.S. Department of the Interior, Bureau of Mines, Washington, D.C., 1988.

Chapter 3

THE MONITORING OF HUMAN LEAD EXPOSURE

Paul Mushak

TABLE OF CONTENTS

I.	Introduction	46
	A. Why Monitor Lead Exposure?	46
	B. Types of Human Lead Exposure Monitoring	46
II.	General Methodological Considerations	47
	A. Exposure Survey Designs	47
	B. Field Aspects: Sampling and Sources of Variance	47
	C. General Laboratory Aspects	47
	D. Quality Assurance/Quality Control (QA/QC) in Lead Exposure Monitoring	48
III.	Environmental Monitoring of Lead Exposure	48
	A. Lead in Air	48
	1. Sample Collection	49
	2. Laboratory Analysis	49
	B. Lead in Food	49
	1. Sampling Procedures	50
	2. Laboratory Analysis	50
	C. Lead in Media Important for Childhood Exposure	50
	1. Lead in Paint	50
	a. Sampling Procedures	50
	b. Laboratory and *In Situ* Lead Measurement	50
	2. Lead in Dust and Soil	51
	a. Sampling Procedures	51
	b. Laboratory Measurements of Lead	51
	3. Lead in Tap Water	51
	a. Sampling Procedures	52
	b. Laboratory Measurements of Lead	52
IV.	Biological Monitoring of Lead Exposure	52
	A. Biokinetic/Metabolic Aspects	52
	1. Lead in Blood	53
	2. Lead in Urine	53
	3. Lead in Teeth	53
	4. Chelatable Lead	53
	5. Lead in Bone	54
	6. Lead in Hair	54
	B. Lead in Whole Blood and Blood Components	54
	1. Sampling Procedures	54
	2. Laboratory Analysis	55
	3. The QA/QC Framework	55
	4. Lead in Plasma	55

      C.   Lead in Urine/Chelatable Lead 56
            1.   Sampling Procedures 56
            2.   Laboratory Analysis 56
            3.   QA/QC Procedures 56
      D.   Lead in Teeth 56
            1.   Sampling Procedures 57
            2.   Laboratory Analysis 57
            3.   QA/QC Procedures 57
      E.   Lead in Bone 57
            1.   Sampling Procedures 57
            2.   Laboratory Analysis 57
            3.   QA/QC Procedures 58
      F.   Lead in Hair 58
      G.   Biological Effects as Lead Exposure Indicators 58
            1.   Inhibition of δ-ALA-D Activity 58
            2.   Elevation of δ-ALA in Urine (ALA-U) 59
            3.   Elevation of Erythrocyte Zinc Protoporphyrin (ZPP) 59
            4.   Other Heme-Related Effect Indicators of Lead Exposure 60

V.   Future Directions 60

References 61

I. INTRODUCTION

This chapter describes available approaches and techniques for quantitative assessment of lead exposure in human populations at risk for lead's effects. Discussion topics include the types of lead exposure monitoring, the physiological underpinnings of biological monitoring and the relative utility of monitoring approaches in public health protection.

A. WHY MONITOR LEAD EXPOSURE?

A general principle of clinical and experimental toxicology is that the amount of a toxicant delivered to site(s) of action determines the severity of the adverse health effect(s). Besides the topic's toxicological dimension, there are also public health policy and regulatory aspects. When a lead policy decision or regulatory action is implemented, its results can only be determined by careful and routine human exposure assessment. There are also economic and societal issues. Are lead workers adequately protected with existing exposure controls? Are certain monitoring approaches socially unacceptable, e.g., preemployment screening for genetically hypersensitive lead workers?

B. TYPES OF HUMAN LEAD EXPOSURE MONITORING

There are two general approaches to assessment of human lead exposure: environmental (external, ambient) monitoring and biological (internal, systemic, integrated) monitoring. These approaches have been the subject of considerable attention, including the 1986 Rochester conference, Biological Monitoring of Toxic Metals,[1] the U.S. Environmental Protection Agency's lead criteria document,[2] and in the review by Mushak.[3]

Environmental monitoring of lead consists of quantitative measurement of lead levels in those environmental media which also serve as exposure routes for humans — air, food, drinking water, dust/soil, and leaded paint. Aggregate external exposure is then assessed by combining such data with media intake rates to give absolute intakes. This form of monitoring is especially valuable in identifying which sources are most important for exposure.

Biological monitoring describes the quantitative assessment of lead in biological media from exposed individuals, the total body lead burden, and the toxicologically active lead burden.

There are other related monitoring methods. Environmental measures are often preceded by emission and source distribution monitoring. Similarly, biological monitoring is often extended to cover biological effect assessment. Lead effect assessment includes such end points as elevated erythrocyte protoporphyrin in response to elevated body lead.

This chapter primarily concerns biological and biological-effect monitoring, since this mode of monitoring is increasingly being used to assess health risk and success of regulatory actions. There is still much value in environmental monitoring, and certain aspects are presented.

II. GENERAL METHODOLOGICAL CONSIDERATIONS

Regardless of the type of monitoring, there are general rules governing practice and effectiveness.

A. EXPOSURE SURVEY DESIGNS

The level of integration of monitoring into the statistical/epidemiological design of the overall population study is important, and design questions of relevance include: (1) statistical adequacy of the sample size; (2) clear role for the specific monitoring in the hypothesis testing; (3) appropriate use or selection of statistical analyses; and (4) feasibility of sampling without effect on statistical reliability.

B. FIELD ASPECTS: SAMPLING AND SOURCES OF VARIANCE

Biological and environmental monitoring of lead exposure entails sample collection and transport as the first stage. The quality of this field work affects the overall assessment. Points of concern include: (1) loss of the lead from samples and contamination; (2) the stability of the monitoring matrix; and (3) which environmental medium to sample and how to sample for exposure relevance.

C. GENERAL LABORATORY ASPECTS

Sample treatment, analytic instrumentation, quantitation of a measurement, and data handling are the general laboratory functions in the monitoring of lead exposure.

In the laboratory, sample storage and handling require care to avoid contamination with lead, to avoid lead loss, and to minimize sample matrix changes. Sample processing in the laboratory is a function of the medium and test protocol specifics. Laboratory ware must be scrupulously cleaned of contaminating lead, laboratory reagents must be lead-free, and every precaution should be taken to avoid lead loss at all stages of analysis.

Lead measurement in various environmental and biological media has technically matured to the point where such analyses can be accurately and precisely carried out in competent laboratories. To some extent, the same can be said about the various biological effect indicators. For example, the Occupational Safety and Health Administration (OSHA) requires a minimum 8/9 or 89% adequate performance for continued certification to do occupational lead exposure testing,[4] a high level of achievable proficiency.

Hundreds of papers have been published on lead measurement in diverse environmental and biological media. However, only a few methodologies have survived the rigorous requirements of monitoring: accuracy, precision, moderate cost and wide applicability. For various media, acceptable methodology will vary with the nature of the matrix and the overall testing protocol.

In most cases, applicable testing involves either definitive or reference methodology. A definitive method[2,7] is one with features that allow analyses to be carried out at highest levels of accuracy and precision, i.e., the "gold standard" for lead measurement in that medium. Reference methods[2,5] are those which are shown to have requisite accuracy by comparison to a definitive method or are otherwise acceptable through informed scientific consensus.

For lead analysis in various media, isotope dilution mass spectrometry (IDMS) is the definitive method.[2,5-7] Reference methods for lead in biological media include atomic absorption spectrometry (AAS) in its various configurations,[2,8-10] and the electrochemical procedure of anodic stripping voltammetry (ASV).[2,11] A potential reference method is that of inductively coupled plasma-mass spectrometry (ICP-MS).[12-14] Other, variably useful methods do exist, but they must be carefully examined for their capabilities. Reference methods for environmental media such as tap water, air particulate on filters, food, dust, and soil include AAS[2,15], ASV[211], X-ray fluorescence analysis (XRF),[2,15,16] and inductively coupled plasma-atomic emission spectrometry (ICP-AES).[2,17] Further comments on methods are presented in the various media-specific subsections below.

D. QUALITY ASSURANCE/QUALITY CONTROL (QA/QC) IN LEAD EXPOSURE MONITORING

Generally reliable instrumental methods are already available for lead in diverse media, and one is primarily concerned with having a framework for assuring the quality of the analyses.

As defined by the World Health Organization (WHO),[18] quality assurance encompasses all steps taken to ensure reliability of analytical data, i.e., a general framework for assuring reliability. Quality control has more to do with the routine implementation of this framework. For lead, this would mean using external and internal QA/QC with reference samples for testing precision and accuracy of lead measurement in whole blood.

General statements for QA/QC in toxic metal exposures[1] include: (1) a likelihood that many older papers on lead monitoring are of mixed or low quality; (2) no analytical method is 100% reliable at all times; (3) laboratory personnel training and both external and internal quality control checks are essential, including Standard Reference Material (SRM) for analysis of media-specific lead; (4) regional and national interlaboratory proficiency surveys for lead should be established and routinely used; (5) statistical criteria, as described in Clarkson et al.,[1] should drive decisions about retention or rejection of analyses (i.e., avoidance of Type I and Type II errors); (6) devices such as a regression analysis of reported vs. reference values allow decisions of rejection or acceptance of methodological performance; and (7) authors of manuscripts should include enough QA/QC and statistical analysis data for judgments on the reliability of the results.

III. ENVIRONMENTAL MONITORING OF LEAD EXPOSURE

Assessment of lead in those media by which risk populations are apt to be exposed, especially young children and fetuses, is discussed here. These media are air, food, and those relevant to childhood exposure: paint, dust/soil and tap water. Table 1 provides a summary of these techniques.

A. LEAD IN AIR

Humans are exposed to airborne lead either by direct inhalation or secondary to air lead fallout onto soil and vegetation and as dusts.[2,19,20] With the phasedown of lead content of

TABLE 1
Environmental Monitoring of Human Lead Exposure

Media/Sampling	Method[a]	Comments	Ref.
Ambient air: lead particulate collected on filters; other types in Ref. 2	Flame/flameless AAS; XRF; ICP-AES	Stationary sources still important; sensitivity to ≤1 ng Pb/m^3	2,15,22
Food: complex sampling to duplicate population diets: market basket and total diet food list	Flame/flameless AAS; ASV; various QC SRMs available	Pb levels low; methods adequate but sampling is complex	26—29,31
Paint: *in situ* surface testing or discrete paint layer collection	Portable XRF surface scanning at site; lab samples decomposed and measured by flameless AAS	Limit of XRF unit ca. 0.7 mg Pb/cm^2, needs calibration; lab samples pose few detection problems	2,31
Dust/Soil: site sampling complex, not standardized; bioavailability differs with matrix	Flame/flameless AAS; XRF; QC materials available	Pb levels high in dusts and certain soils; methods are quite adequate	2,34,35
Tap water: first-flush, full-flush, random grab sampling gives different results	Flame/flameless AAS, ASV; extraction/preconcentration steps often required; sediment filtered and also tested	Pb levels are low but methods adequate; high levels with standing in leaded plumbing	2,26,31

[a] See text for abbreviations.

gasoline in the U.S. and elsewhere, the rate of lead being added to the atmosphere and other likely fallout pathways has declined.[2,21] However, emissions from stationary sources continue to a variable degree. In the U.S., much of the available monitoring data is procured through the U.S. Environmental Protection Agency's National Filter Analysis Network (NFAN).[2] State and local units also monitor air lead in testing compliance with Federal ambient air lead standards.

Measurement of lead in air consists of discrete sampling and laboratory analysis steps.

1. Sample Collection

Lead in ambient air is primarily incorporated into particulate matter. This material must first be collected in any of a variety of collectors, including high-volume or personal filter units, impactors, impingers, or scrubbers.[2,15,22]

2. Laboratory Analysis

In the laboratory, collected lead is either leached into test solution or liberated by dry/wet ashing steps. The method of choice for analysis of ambient air samples is AAS. This is also the official Federal method for assessment of compliance with air lead standards.[23] Other methods include the XRF technique and ICP-AES.

Limits of lead detection vary, but the AAS method is capable of quantitation of air lead levels below 1 ng/m^3. With air analysis, increase of sampling time helps to assure that adequate amounts of lead are collected when air concentrations are low.

B. LEAD IN FOOD

Lead enters the human diet through both food production and food processing. While declines in food lead, especially for infants, have been occurring since 1978, there may still be a sizable number of children exposed to undesirable levels of lead in diet.[2]

1. Sampling Procedures

Few sources of lead exposure pose as difficult a task for representative exposure sampling as does food. Human dietary behavior is both complex and highly variable. Any dietary sampling scheme chosen to represent general food lead exposure will always be arbitrary.

In the U.S., nationwide monitoring of lead in food is done by the U.S. Food and Drug Administration (FDA). Monitoring in food has included sampling via the Market Basket Survey,[24] which depicts a typical profile of dietary lead intake, and for information from a newer data base, from the Total Diet Food List[25] based on 100,000 daily diets from 50,000 participants. Up to 3500 categories of food are condensed to 102 groupings for 8 age/sex groups.

2. Laboratory Analysis

Van Loon[26] has critically examined measurement methods for lead in different food groups.

The basic methodologies employed by the U.S. FDA and other laboratory facilities include flame and flameless AAS[27,28] and ASV,[29] with multi-element measurement now being done with ICP-AES.[30]

QA/QC assessment of food lead analysis includes use of SRMs (Standard Reference Material) from the U.S. Bureau of Standards. These include NBS wheat flour, NBS rice flour (SRM 1568), NBS spinach (SRM 1570), NBS orchard leaves (SRM 1571), and NBS bovine liver (SRM 1566).

C. LEAD IN MEDIA IMPORTANT FOR CHILDHOOD EXPOSURE

Although all ages of the U.S. nonoccupational population sustain variable exposure to the same lead-contaminated media, certain sources are of particular concern for children. This is due to either certain child behavior patterns, e.g., eating paint chips or ingesting soil/dust on hands, or certain media use patterns which amplify exposure, e.g., drinking school tap water which has been in contact with leaded plumbing over extended periods.

1. Lead in Paint

Leaded paint is still a major factor in the environmental monitoring of lead exposure in large numbers of young, urban U.S. children. This source is associated with the most severe lead poisoning and represents a huge national depository of persistent pollution. These aspects are detailed by the U.S. Public Health Service to Congress in its recent report on childhood lead poisoning in America.[31]

a. Sampling Procedures

One can either collect leaded paint material for laboratory analysis or carry out *in situ* testing of leaded paint surfaces in residences and public buildings.

Leaded paint surfaces are commonly tested *in situ*, by XRF which probes and registers lead content of all layers of paint. Following Federal court action in 1983, it is officially recognized that leaded paint surfaces need only be present to pose the potential for exposure, i.e., exposure risk is not only present with residential deterioration.[31] Discrete layers of paint can also be isolated and separately analyzed in the laboratory.

b. Laboratory and In Situ Lead Measurement

In situ testing of lead in paint is carried out by scanning suspect surfaces with a portable XRF instrument, equipped with a radioactive irradiation source and calibrated as frequently as possible for accuracy. In the hands of competent personnel, these commercially available units are rated to be adequate for levels of lead in older paints. Often there are reliability problems with the low end of the lead concentration range, below 0.7 to 1.0 mg lead per square centimeter surface (mg/cm^2).[31] In many of the older paints, lead content of up to 50% dry weight is measured.

In situ X-ray fluorescence measurement does not differentiate among the layers of paint. This generally would parallel the exposure reality of a child eating a paint chip or chewing/gnawing on a windowsill. If a surface painting history is known or a layer of paint is suspect, then paint samples can possibly be examined in the laboratory. First, samples are chemically decomposed, usually with strong oxidizing acids,[2] to solubilize the lead. Since lead content of older samples can be very high, serial sample dilution is often required.

Laboratory methods include flame or flameless atomic absorption spectrometry.[2] The ICP-AES technique is acceptable for lead plus other elements, e.g., titanium in pigments of later paint layers.

Generally, levels of lead in paint are high enough that samples can be accurately and precisely measured by normal internal quantitation and standards. Matrix effects become greatly diminished with serial dilutions of samples with high lead content.

2. Lead in Dust and Soil

Monitoring of lead in dust and soil poses many analytical problems. Many of these are encountered at the sampling step and others arise in choice of laboratory techniques.

Factors of physical collection and representativeness must be considered when sampling these media, especially soils. The factors include: (1) the area of the exposure site and depth of soil to be tested; (2) use of total lead or bioavailable lead to quantify exposure; and (3) utility of ancillary tests, e.g., dust or soil lead on the hands of exposed children.

a. Sampling Procedures

Sampling should be directed at those soil and surface dust areas which the child is likely to encounter or is actually observed to encounter during play. Careful, serial vertical core or surface scrape testings should be done.

The question of lead bioavailability in soil and dust has long been a vexing problem in assessing the actual degree of toxicity risk. Often, the geological matrix of the soil is both low in lead and can resist biochemical degradation.

One important pathway of urban child exposure is via interior and exterior dusts, generated in diverse ways.[2,31-33] Interior dust is collected by a combination of vacuuming and wet-wipe techniques. Exterior dusts are collected by outside surface scrapings.[33] Quantitation is done by either mass concentration or by lead level per unit area (m^2).

b. Laboratory Measurements of Lead

Soil/dust-sample processing in the laboratory includes drying, particle size fractionating and treatment with diverse lead-free reagents to give some measure of relative biochemical availability of lead in the samples.[34] Total lead analysis requires full solubilization with hydrofluoric acid.

Laboratory methods for dust and soil which have proven most flexible and reliable over the years are flame/flameless AAS,[35] XRF[26,35,36] and, for multiple elements, ICP-AES.[26,35]

QA/QC practices in soil and dust analysis often employ standardized soil samples of known lead content. In the report of Freiberg et al.,[35] five soil SRMs from the CANMET series[37] and the International Atomic Energy Agency[38] were employed.

3. Lead in Tap Water

Drinking water lead is a potentially significant source of exposure for young children owing to both the pervasiveness of this lead source and water use patterns. In the latter case, the way in which kindergarten and elementary school water fountains are inconsistently used, e.g., after vacations, weekends, and other long standing periods, allows lead to leach into standing water from leaded plumbing.[31]

In homes and public buildings, lead invariably enters tap water via leaching from leaded plumbing (soldered joints, lead lines/connectors, brass fixtures). Water from the supply plant and in water mains is often quite low in lead content and is generally safe for consumers.

a. Sampling Procedures

If tap water sampling is to accurately reflect child lead exposure, especially in schools and child care facilities, then sample collection should parallel water use. One therefore should include collection of first-draw or first-flush samples, i.e., the first water out of the plumbing after standing overnight, weekends, vacations, etc. Fully-flushed collections reflect supply water entering the home/building and will significantly understate the true risk.

Toxic levels of lead in water are still at trace or subtrace levels. For example, the current standard is 50 ppb lead (50 µg/l), and soon may be reduced. Scrupulous care is required to avoid contamination of the collecting vessel. Loss of lead from water to the container's walls is prevented by adding dilute, lead-free acid.

b. Laboratory Measurements of Lead

Tap water is generally an uncomplicated analytical matrix. However, lead levels are often so low that some type of preconcentrating of the element from large sample volumes must be done. Preconcentrating is done either by solvent extraction into an organic phase via a binding agent or by collection on an ion-exchange chromatographic resin.[26] Well water samples may contain heavy sediment, and are first filtered to yield a two-part concentration of lead.

Various laboratory methods have been found to be reliable for water lead analysis. Flameless AAS is the most popular of these.[2,26] The electrochemical method of ASV is also widely used. With flameless AAS, care must be taken to avoid lead loss by volatilization during analysis. For lead and other elements, ICP-AES can be used.

QA/QC steps in drinking water analysis include participation in interlaboratory proficiency testing. As a reference material, several water standards are available from the U.S. Environmental Protection Agency.

IV. BIOLOGICAL MONITORING OF LEAD EXPOSURE

Since biological monitoring deals with lead already in the body, it is an index of integrated exposure from all sources. It reflects the various intake and absorption steps for each medium and each collecting body compartment, e.g., the lung and the gut.

The most commonly used physiological media for lead monitoring are whole blood, teeth, urine, hair, and, in certain cases, bone. These have generally involved *in vitro* testing. There now is growing interest in the use of *in vivo* lead measurement, generally in bone of lead workers or young children. Biological effect indicators closely correlated with lead dose are also discussed in this section.

A. BIOKINETIC/METABOLIC ASPECTS

Much of this chapter deals with the "how" of monitoring. However, the "why" of monitoring, i.e., the physiological basis of such types of exposure assessment must also be understood. Lead in various physiological media from human subjects is related to both external exposure and to various effects in some graded fashion. These relationships reflect lead biokinetics within and between body compartments, based on such factors as the element's biochemical behavior and the metabolic status of the exposed subject. Detailed discussions of the biokinetic underpinning of biological monitoring of lead have been presented.[2,3,39]

1. Lead in Blood

Blood lead is a rather short-term measure of exposure, reflecting relatively short shifts in lead exposure or transitory events, e.g., acute intoxication. Blood lead begins to rise quickly with an increase in lead intake/uptake, as quickly as several hours after ingestion or inhalation. An equilibrium or steady-stage condition is re-established in *increased* exposure by 6 to 8 weeks.[2,39,40]

Over a longer time frame, e.g., the occupational lifetime of a lead worker, blood lead begins to reflect both ongoing lead exposure and resorption of lead from bone. In older adults, the bone lead contribution can be significant, and in the case of retired lead workers, this endogenous contribution is dominant.[2,3,39] Related to endogenous lead behavior are the biokinetics of blood lead in lead workers who retire or change occupations. Measured as a biological half-time, the decay in blood lead with reduced exposure is a function of such parameters as the body lead burdens. One recent study reported that there are two decay times for blood lead.[41] There is a fast compartment, half-time of about 30 d, reflecting soft tissue "washout" and a much slower compartment, with half-time of 5.1 years, reflecting resorption of lead from a bone compartment. With low-level, fixed environmental lead exposure in adults, blood lead is quite stable.[42] In the developing child, however, blood lead can be highly labile, reflecting both developmental changes and basic changes in the nature of lead exposure. In infants, especially, blood lead changes significantly up to about 2 years of age.[43] Over a broad exposure range, blood lead shows a curvilinear relationship. This suggests that at higher exposures, this medium is less reflective of target tissue lead monitoring,[2,3] or that metabolic changes occur to handle lead differently.

2. Lead in Urine

Lead in urine is a highly variable index of ongoing lead exposure, affected by such physiological factors as kidney function, circadian variation, etc. At low levels of lead exposure, biological variation within subjects and between subjects is such as to limit the monitoring usefulness of this medium.[2,39]

Urinary lead rises with increasing lead exposure and the resulting relationship becomes curvilinear upward at high exposures.[44] Urinary lead levels respond rapidly to either rises or declines in body lead.

The level of lead mobilized into urine, i.e., plumburesis, in response to either provocative chelation (the lead mobilization test) or chelation therapy of lead-poisoned children or lead workers, is the most clinically useful aspect of urine monitoring. Lead levels rise quite rapidly after chelant administration, usually within hours.[2,3,39,45]

3. Lead in Teeth

Deciduous tooth lead in children represents cumulative uptake from birth to time of exfoliation. Such accumulation varies with tooth region and dentition type.[2,3,39] Lead is deposited in highest amount in circumpulpal dentine, with much lower levels in primary dentine and enamel. Once lead is sequestered in the mineral matrix of teeth, it does not appear to ever undergo remobilization into blood and is thus a *total* exposure index.

4. Chelatable Lead

In certain childhood or occupational lead exposure situations, a lead mobilization test is done to probe mobile lead stores[2,3,39,45,46] prior to initiation of any further management.

The lead mobilization test is claimed to probe the toxicologically active fraction of mobile lead, given that this index is highly correlated with biological effect indices such as early effects in heme biosynthesis intermediates.[2,39,45] Chelatable lead represents both soft tissue lead and one or more kinetic compartments within bone.[2,3,39,45,46] In one multiclinic study[45] this test was better at revealing possibly toxic body lead stores than blood lead at lower levels (<30 μg/dl). However, Weinberger et al.[47] have noted that data analysis of 248 lead mobilization tests for children in one program showed little positive response below a PbB of 40 μg/dl.

5. Lead in Bone

The skeleton represents the major repository of lead in the human body, containing 90% or more of total lead content. This compartment was historically considered to represent sequestered (immobilized) lead; that is, once absorbed lead was lodged in bone, it ceased to be a toxic threat. Data from the areas of bone physiology and lead toxicology show that bone is a living tissue and that it accumulates lead within at least two, and probably three, compartments.[2,3,39,46] One of these compartments represents a rapid component, resembling that of soft tissue. A second apparently corresponds to lead in trabecular (spongy) bone, while lead with the longest turnover time is in cortical bone, the third compartment.

As noted above, chelation mobilization probes the most labile fraction of body lead. This includes soft tissue and one or more subcompartments of bone fractions.[2,3,39,45,46,48] The fraction of cortical bone lead labile to chelation is not clear.[39,41,46,48] Rosen et al.[46] report that bone lead measured *in vivo*, using XRF of children's tibia (mainly cortical bone), is highly predictive of the outcome of the lead mobilization tests in lead-intoxicated children. In lead workers, Schütz et al.[48] showed a close correlation between chelatable lead and lead in trabecular bone, but little correlation between chelant mobilization and cortical lead measured *in vivo* (workers' finger bones).

Christoffersson et al.[41] showed that lead loss from cortical bone in the fingers of retired lead workers, measured *in vivo*, indicated an average half-time of bone lead of 7 years. "Inertness" of cortical bone lead, at least in occupational exposure, is less than that estimated from different metabolic models, including estimates based on bone strontium-90 mobility. These workers also noted good concordance between finger-bone lead declines and the slower blood lead component. Some sizable fraction of cortical lead therefore may be remobilizing into blood of these retired lead workers.

Given these data, all bone lead compartments should be viewed as potential sources of "endogenous" lead exposure. This bone lead burden is of particular concern in nonoccupationally exposed groups such as postmenopausal women having osteoporosis.

6. Lead in Hair

Lead is deposited in hair.[2,39,49] This occurs in a time-integrated fashion reflecting hair growth concurrent with active lead exposure.[49] Lead deposition into hair is physiologically complex and highly variable between subjects. Such variables as sex and hair color complicate interpretation of monitoring data.

B. LEAD IN WHOLE BLOOD AND BLOOD COMPONENTS

Lead in whole blood is currently the most widely used indicator of systemic lead exposure in humans. Whole blood lead analysis has produced an extensive literature, including a still growing information base for implementation of QA/QC measures. Only a brief overview can be provided here.

1. Sampling Procedures

The medium of most value is whole blood obtained by venous puncture.[2,39,50] Finger puncture (capillary blood) poses problems, involving both a high contamination risk and lead dilution occurring with serum leakage when the puncture is squeezed.[2] One exception appears to be capillary blood from heel puncture of newborns and infants, where one study found little contamination.[51]

Even in cases of severe lead poisoning, lead appears in whole blood at the trace or subtrace level. Extreme care is therefore required to minimize contamination and sample lead loss.[2,39] Collection precautions include the use of lead-free blood tubes, needles, anticoagulant, and a thoroughly cleaned puncture site.

Based on current testing and evaluations, whole blood samples can be preserved for long periods of time, up to several years, by sealing and freezing at $-20°C$.[52]

2. Laboratory Analysis

The technique of IDMS is the definitive method for lead in whole blood.[53] As normally employed, the procedure is time-consuming and expensive, and it requires high levels of operator expertise. The method is rarely used for routine testing and primarily serves as a confirming or validating method for samples analyzed by other means. One recent variation on the basic technique is that of ICP-MS. It may be somewhat more practical for monitoring applications.[12] Routine measurements of lead in whole blood are usually done by the various types of AAS or the technique of ASV.[2,39]

Both flame and flameless variations of AAS have been used, and there is increasing popularity of the flameless (furnace, electrothermal) form. The Delves Cup microanalysis variant is the most popular flame procedure.[9] The latter is specific and sensitive to lead levels as low as 1 $\mu g/dl$.

By using matrix modifiers, the L'vov platform and spectral background correction steps, the flameless method is a reliable form of AAS analysis, with the added advantage of speed and adaptibility to automated operation. Miller et al.[54] used this combination of techniques for blood lead, reporting a sensitivity of 1.4 $\mu g/dl$ and a precision of 2 to 5%.

Electrochemical methods, chiefly ASV, are extensively employed. For best results, the blood matrix should ideally be chemically decomposed. Sensitivity and precision are as good as that of flame/flameless AAS. Care must be exercised in certain cases, such as with elevated blood copper levels during pregnancy.[55,56] Roda et al.[56] have noted that a modification of the most popular, commercially available configuration of ASV is required to avoid systematic instrumentation error, especially at lower levels (<40 $\mu g/dl$). This is accomplished through manual peak-height measurement in lieu of electronic peak recording.

3. The QA/QC Framework

Various QA/QC schemes have been established for analytical reliability of blood lead in laboratories in the U.S. and elsewhere. U.S. programs include the external proficiency testing program of the U.S. Centers for Disease Control (CDC).[2,52] This program has two components, one which provides proficiency testing for occupational surveillance laboratory certification and the other furnishing a service to pediatric screening clinics/laboratories. A key part of this effort is use of bovine blood samples having variable lead content added *in vivo*.

Such states as California, New York, and Pennsylvania also offer more limited proficiency testing samples.

One can also employ a variety of certified reference materials, the most useful of which are those provided by the National Bureau of Standards. Blood at two levels of lead, SRM 955, is available. Use of SRMs also has its problems. Since SRM levels for a particular batch on hand can become known over time, it is essential that their use be unknown to laboratory staff, avoiding operator bias.

Various QA/QC programs are used outside of the U.S., and these are discussed by Friberg.[57]

4. Lead in Plasma

Plasma lead theoretically reflects the fraction of mobile lead being directly delivered to tissues, i.e., it is the more accurate reflection of delivered dose. In practice, this advantage is greatly diminished by such methodological complications as a high contamination risk from external sources and erythrocyte lead. The plasma pool is only 0.1 to 0.5% of total whole blood content under equilibrium conditions.

Ong and co-workers[58] have shown that plasma lead in lead workers was only moderately correlated with certain biological effect indicators, while lead in both whole blood and erythrocytes was highly correlated.

C. LEAD IN URINE/CHELATABLE LEAD

Urinary lead analyses are most reliable and probably most useful when employed during lead mobilization (provocative chelation) or chelation therapy procedures. Lead levels rise greatly, making analysis easier.

1. Sampling Procedures

Lead in urine is present at very low, often ultra-trace, levels and the usual precautions have to be exercised to avoid contamination. Loss of lead can occur by either adherence to the vessel wall or by adherence to precipitated material in the urine sample.

In standard practice, acid-washed polyethylene collection jars to which dilute nitric acid has been added are used. The volume of the container depends on whether 24-hour, 8-hour, or spot sampling is desired. Plumburesis associated with the lead mobilization test involves either 24-hour or 8-hour sample collection.[2,3,39,45] Collection during 5-d chelation therapy regimens is often an inpatient procedure and is invariably done on a 24-hour basis. Spot sampling may only be feasible in screening of lead workers. This requires correction for some kidney index, commonly the creatinine content. Lead mobilization testing is increasingly done for 8 h instead of the traditional 24-hour period.[2,3,39,45]

2. Laboratory Analysis

Urine is a complex matrix biochemically and has both organic and inorganic constituents. It is common practice to either chemically decompose the samples or to isolate lead by chelation/extraction.[2] This is especially true with either lead mobilization or chelation therapy, where large excesses of chelant in urine might affect lead analysis procedures. Acid wet ashing is often done to counter this problem.

Urine lead analysis in the laboratory involves both AAS and ASV methods.[2,39] In the flame AAS approach, lead is usually first partitioned into an organic phase, using a dilute solution of the strong chelant ammonium pyrrolidinecarbodithioate in methyl-isobutyl ketone.[59]

Flameless AAS is also commonly employed. In the procedure of Paschal and Kimberly,[60] chemically modified samples are delivered to the furnace of a flameless AA unit equipped for background correction. Limit of detection is 2.5 μg Pb/l urine and the Relative Standard Deviation is 9 to 14% across the range of urine lead levels likely to be encountered.

In the 8-hour lead mobilization test, EDTA is given at a rate of 500 mg/m^2 body surface and the result is quantitated as μg Pb/mg chelant. A ratio of ≥ 0.6 is positive.[45]

3. QA/QC Procedures

Periodic use of interlaboratory comparison programs is advisable for laboratories doing urinary lead measurement and some of these are described by Friberg[57] and Griffin.[61] For internal laboratory use, a urine lead standard, SRM 2670, is available. It is obtained in freeze-dried form from the U.S. National Bureau of Standards.

D. LEAD IN TEETH

In vitro monitoring of cumulative lead exposure in young children uses shed teeth. While extracted teeth from older subjects can be used in some circumstances, these often have problems such as extensive tooth decay, precluding reliable analysis. Alternatively, *in vivo* measurements of tooth lead have been carried out in children.[2,39]

1. Sampling Procedures

Lead varies with the main regions of the tooth and uniform regional analysis of all samples is required. The same type of tooth should be used across subjects, since this is also a variable.[2,3,39,62,63] Secondary (circumpulpal) dentine is particularly useful as shown by Needleman et al.[62] and Grandjean et al.[63]

Different techniques are employed for isolating regions of the tooth. For example, Needleman and co-workers[62] prepared sagittal sections and isolated the secondary dentine by careful cutting. Sample surfaces should be mildly abraded or treated with a reagent to remove surface lead contamination.

2. Laboratory Analysis

For analysis, tooth samples are either dry ashed at elevated temperature or wet ashed in nitric or nitric/perchloric acids.[64]

Procedures for *in vitro* lead analysis include flame and flameless AAS.[2,39,64] In the flame variation of AAS, lead is commonly separated and concentrated, using a chelating agent and organic solvent. Flameless AAS can be employed, taking into account the matrix effect of calcium and phosphorus. Since lead levels in teeth are relatively high, some dilution to minimize mineral interference is possible.

Anodic stripping voltammetry and other electrochemical methods for lead have been routinely employed.[2,64] In the approach of Needleman and co-workers,[62] small portions of secondary dentine were dissolved in low-lead perchloric acid and analyzed in a commercially available ASV system.

The *in vivo* measurement of tooth lead in children (using XRF) has been described by Bloch et al.[65] In this method, upper incisors were irradiated with a collimated beam of X-rays from a ^{57}Co source.

3. QA/QC Procedures

There are no standard reference materials specifically for lead in teeth. One can, however, substitute bone powder, the most common being animal bone, H-5, available through the International Atomic Energy Agency. This material was used for QC assessment by Keating et al.[66]

E. LEAD IN BONE
1. Sampling Procedures

Bone samples are most commonly acquired for *in vitro* analysis by autopsy.[2,39] In adults, bone biopsy has been carried out. This usually involves the iliac crest[67] or the spinal process of a vertebra.[68]

Precautions governing bone sampling are essentially the same as those for teeth, e.g., protection from surface contamination. Contamination is most apt to occur during muscle and connective tissue removal.

Tissue decomposition involves dry ashing or acid digestion methods, inasmuch as the organic central core, if retained, requires mineralization.

2. Laboratory Analysis

In vitro methods for bone analysis include flame and flameless AAS.[2,39] At present, flameless AAS is the more popular form.

Prior to analysis, dry-ashed and/or acid-decomposed samples must frequently be diluted, to both minimize calcium/phosphorus matrix effects and accommodate the relatively high levels of lead in this medium. In the flameless AAS approach of Drasch et al.,[69] lead in 50 to 200 mg ashed fragments is diluted to 10 ml with acid and deionized water. In the method of Schütz and co-workers,[68] biopsy samples (16 mg) ashed by a tandem wet-dry technique were diluted to 10 ml with acid plus deionized water.

Somervaille et al.[70] compared flameless AAS with XRF using dry ashed metatarsal and tibial bone samples from amputated limbs. Over a bone lead range of 6.5 to 83.3 µg/g, ash weight, mean levels were quite comparable.

At present, interest is increasingly moving toward *in vivo* XRF of lead in bone. *In vivo* studies differ methodologically, vs. *in vitro* approaches, as to instrumentation specifics, and quantitation methods. To date, published studies generally probe lead in either finger in lead workers or tibial bone in children.

Rosen and co-workers[46] described the *in vivo* XRF measurement of lead in the tibial bone of children via generation of low-energy X-rays, a 90° configuration for irradiation and measurement of the resulting L_α lead line. Using an irradiation dose with little health risk for the subjects, a detection limit of 7 µg/g bone was obtained. Precision of the analysis, day-to-day, was 5.1% (95% confidence interval).

Wielopolski et al.[71] examined tibial lead levels in a group of workers with ongoing exposure. A radioactive source, ^{109}Cd, was used to induce lead L X-rays in the bone scan area. At this level of irradiation, mainly surface layer lead is being detected. The detection limit for this system is 20 µg/g wet weight of bone. Similarly, Somervaille et al.[72] used a ^{109}Cd irradiation source and standard instrumentation geometry to examine tibial lead in nonworker and lead worker adults. A detection limit of 18 µg/g wet weight was reported.

3. QA/QC Procedures

As noted earlier, the only standard reference substance for *in vitro* bone testing is animal bone powder, H-5.

QA/QC methods for *in vivo* XRF methods are generally established on a laboratory-by-laboratory basis. This is especially important for equipment standardization that takes account of such parameters as skin thickness overlying the irradiation area and the preparation of calibration samples (phantoms) for quantitation of the XRF signals.

F. LEAD IN HAIR

Lead in hair would appear to be attractive as a monitoring medium. It is easy to collect (noninvasive) and is indefinitely stable. These advantages are strongly offset by the fact that hair surfaces invariably become heavily contaminated by external contact with lead sources. Such contamination is both difficult to remove and difficult to differentiate from endogenous lead content when vigorous cleaning methods are employed. At present, hair lead is viewed as being too problematic to offer a monitoring technique that accurately reflects systemic exposure.[2,39]

G. BIOLOGICAL EFFECTS AS LEAD EXPOSURE INDICATORS

Biological effect indicators of lead exposure primarily reflect effects on various steps in the heme biosynthesis pathway. These are: (1) inhibition of the activity of the enzyme δ-aminolevulinic acid dehydratase (δ-ALA-D), (2) accumulation in urine of an intermediate in heme biosynthesis, δ-ALA, (3) accumulation of the penultimate species in heme biosynthesis, "free" or zinc protoporphyrin; and (4) elevation of coproporphyrin in urine. Several other intermediates in this pathway have found use in monitoring and these are briefly discussed also.

1. Inhibition of δ-ALA-D Activity

Lead in blood is logarithmically and inversely related to the enzyme activity of δ-ALA-D. The enzyme (5-aminolevulinate hydrolase, porphobilinogen synthetase, E.C. 4.2.1.24), an allosteric, zinc-utilizing enzyme with sulfhydryl groups, converts two units of δ-aminolevulinic acid to porphobilinogen.

During blood collection, zinc contamination must be avoided since this element offsets inhibitory effects of lead. Chelating anticoagulants such as EDTA are proscribed, since they will bind lead and affect the enzyme's activity. This enzyme's activity also varies genetically. Laboratories often carry out a full-activation test to examine background activity level in the erythrocyte.

After sampling, analysis should be carried out within 24 h if possible.

The standard method is a δ-ALA-D spectrophotometric technique, using a standardized assay called the European Standardized Method for ALA-D. Porphobilinogen is generated from the substrate (vide supra), the reaction is then terminated with mercury (II) in trichloroacetic acid and the product is analyzed spectrophotometrically at 555 nm. Activity is reported as μM δ-ALA/min/l erythrocytes.

This enzyme's activity is extremely sensitive to lead, such that measurable inhibition occurs over the entire range of "low" lead levels. This indicator is perhaps too sensitive. It may be less useful in probing systemic toxicity than either the measurement of precursor (δ-ALA) accumulation or the popular erythrocyte protoporphyrin analysis.

2. Elevation of δ-ALA in Urine (ALA-U)

Lead's inhibition of δ-ALA-D activity leads to accumulation of δ-ALA in both plasma and urine of exposed subjects. Generally, the level of the precursor in urine is measured.

This precursor is moderately stable in urine when samples are acidified (acetic acid) and refrigerated. Levels of ALA-U are adjusted for urine density or expressed per unit creatinine.[2,39]

The basic laboratory methodology is spectrophotometric. In the most reliable method (Marver et al.[73]), aminoacetone (an interfering substance), is first separated from δ-ALA after both are reacted to form a pyrrole intermediate. This step employs an ion exchange resin, Dowex-1. The separated δ-ALA derivative is measured spectrophotometrically at 553 nm. Most δ-ALA-U methods are sensitive to a level of about 3 μmol/l urine.

One method of δ-ALA analysis, useful for both urinary and plasma δ-ALA is the gas chromatographic approach of MacGee and co-workers.[74] The analyte is isolated, transformed to a pyrrole and derivatized to form a more volatile gas chromatograph species.

Some controversy exists over what level of ALA-U is associated with onset of early toxic responses in humans. The matter is discussed in Chapter 12 of the EPA criteria report.[2] It has been argued that significant elevation in ALA-U only occurs at PbB of 40 μg/dl or above. Others have demonstrated a lower threshold.

3. Elevation of Erythrocyte Zinc Protoporphyrin (ZPP)

With elevated lead exposure, the last step in heme biosynthesis, i.e., insertion of iron into protoporphyrin IX to form heme, is suppressed; protoporphyrin begins to accumulate in the human erythrocyte, and this accumulation is logarithmically and positively correlated with blood level. There is a well documented threshold for the effect at 15 to 20 μg/dl.[2,39,51,75]

In collecting blood samples for ZPP analysis, care should be taken to minimize light exposure and delays in analysis. Sample hematocrit is required for adjustment of laboratory EP values to a normal hematocrit. There are two general approaches to ZPP or "free" protoporphyrin analysis, the field instrument and the laboratory or "wet" methods. Both exploit the fact that protoporphyrin IX shows an intense fluorescence and can be quantitated fluorometrically.

In the more reliable laboratory or "wet chemical" techniques, all of which are micro methods to maximize screening efficiency, small volumes of blood are treated in different chemical ways to allow measurement of the porphyrin as quickly and simply as possible.

In the Piomelli and Davidow approach,[76] "free" porphyrin (i.e., lacking a coordinated zinc atom) is liberated from erythrocytes and eventually partitioned into dilute hydrochloric acid. It is measured at 615 nm with excitation at 405 nm. This procedure also permits analysis of blood on filter paper. Quantitation is by use of pure protoporphyrin IX standard. Chisolm and Brown[77] described a procedure in which the acid partitioning step is varied depending on porphyrin level. Quantitation and calibration involves both the precursor porphyrin and coproporphyrin.

Some laboratory approaches do not entail multiple extraction steps. Lamola et al.[78] described a method where small volumes of whole blood (ca. 20 μl) were diluted with a detergent (dimethyldodecylamine oxide) in phosphate buffer. Fluorescence arising from the zinc protoporphyrin is recorded at 594 nm with excitation at 424 nm.

For field or screening clinic use, commercially available instrumentation, i.e., the hematofluorometer, is commonly used. The various commercial units and their configurations have been discussed in Chapter 9 of Reference 2. Generally, these units are designed to measure oxygenated (capillary) blood. Venous blood must be oxygenated prior to analysis. Some metabolic interferences include the presence of bilirubin and carboxyhemoglobin. Instrumentally, use of a high quality, nonfluorescing cover glass is also required.

In competent hands, the various hematofluorometers appear to perform with adequate precision, although there may be some biases in their accuracy, as noted by Balamut et al.[79] They show that there may be a problem with false negatives in screening. Mitchell and Doran[80] compared the units with a laboratory reference method and found that hematofluorometers showed a negative bias of about 20% at ZPP levels of 50 μg/dl whole blood and would miss about 33% of screening subjects at this level.

Laboratory proficiency testing is conducted by the U.S. Centers for Disease Control. Monthly mailings of three samples each are provided and results are compared to a target performance range of ±15% of the reference laboratory means.

4. Other Heme-Related Effect Indicators of Lead Exposure

A traditional method of lead screening has been the measurement of urinary coproporphyrin (CP-U) in children and lead workers. This has now been largely supplanted by the ZPP analysis. This indicator still retains the advantage of indicating active lead intoxication when elevated.[81]

The standard method for CP-U measurement remains the fluorometric procedure of Schwartz et al.[82] Buffered urine samples are first treated with iodine to convert any precursor of the porphyrin to CP. They are then extracted with ethyl acetate and the porphyrin is re-extracted into dilute hydrochloric acid. Commercially available CP is used as the quantitating standard.

Effects of lead on other steps of the heme pathway are known, but these effects have not been widely exploited as early effect indicators of lead exposure.

V. FUTURE DIRECTIONS

Biological monitoring of lead exposure has experienced great progress in the last several decades. Improvements in quality of monitoring includes elements of both methodology and statistical design. We also understand better the biological basis of the indicators.

Since lead exposure occurs widely and continuously in human populations, i.e., *in utero*, through early development, and beyond, lead monitoring should ideally be expanded to all sectors of the child population. As recommended by the Federal government,[31] pregnant women should also be included in lead screening. Since it is recognized that levels of lead exposure considered "safe" by the biomedical and public health communities continue to be adjusted downward, capabilities of any monitoring design and methodology must be increasingly applicable to continuingly lower levels of exposure. Finally, there still appears to be need for better measures of target tissue lead burdens.

Thresholds for a number of early, subtle effects on both the fetus and the infant/toddler are now accepted to occur at 10 to 15 µg/dl PbB or even somewhat below.[31]

These observations, including official recognition of lower thresholds by EPA[2] and others noted in the report to Congress,[31] suggest that blood lead screening in the future will require precise, accurate methodology for direct measurement of PbB levels below 10 to 15 µg/dl. This is because ZPP levels are not predictive of blood lead levels in this range. Other effect measures may also be needed. With these higher demands on analytical performance, QA/QC frameworks will have to be made more rigorous.

In vivo measures of accumulating lead in young children's bones are important and we can expect that this area will continue to develop. There are already early, promising results from Rosen et al.[46] for children and data from various cited groups for *in vivo* results from lead workers.[2,39,41,48,70,71]

REFERENCES

1. **Clarkson, T. W., Friberg, L., Nordberg, G. F., and Sager, P. R., Eds.**, *Biological Monitoring of Toxic Metals*, Rochester Series on Environmental Toxicity, Plenum Press, New York, 1988.
2. U.S. Environmental Protection Agency, Air Quality Criteria for Lead, 4 Vols., Report No. EPA-600/8-83/028cF, Environmental Criteria and Assessment Office, Research Triangle Park, NC, June 1986.
3. **Mushak, P.**, Biological monitoring of lead exposure in children: overview of selected biokinetic and toxicological issues, in *Lead Exposure and Child Development: An International Assessment*, Smith, M., Grant, L. D., Sors, A., Eds., Kluwer Press, Dordrecht, Netherlands, 1989, 129.
4. U.S. Occupational Safety and Health Administration, OSHA criteria for laboratory proficiency in blood lead analysis, *Arch. Environ. Health*, 37, 58, 1982.
5. **Boutwell, J. H.**, Accuracy and quality control in trace element analysis, in *Accuracy in Trace Analysis: Sampling, Sample Handling, Analysis*, Vol. 1, LaFleur, P. D., Ed., NBS Special Publication No. 422, National Bureau of Standards, U.S. Department of Commerce, Washington, D.C., 1976, 35.
6. **Cali, S. P. and Reed, W. P.**, The role of the National Bureau of Standards: standard reference materials in accurate trace analysis, in *Accuracy in Trace Analysis: Sampling, Sample Handling, Analysis*, Vol. 1, LaFleur, P. D., Ed., NBS Special Publication No. 422, National Bureau of Standards, U.S. Department of Commerce, Washington, D.C., 1976, 41.
7. **Machlan, L. A., Gramlich, J. W., Murphy, T. J., and Barnes, I. L.**, The accurate determination of lead in biological and environmental samples by isotope dilution mass spectrometry, in *Accuracy in Trace Analysis: Sampling, Sample Handling, Analysis*, Vol. 1, LaFleur, P. D., Ed., NBS Special Publication No. 422, National Bureau of Standards, U.S. Department of Commerce, Washington, D.C., 1976, 929.
8. **Hessel, D. W.**, A simple and rapid quantitative determination of lead in blood, *At. Absorp. Newsl.*, 7, 55, 1968.
9. **Delves, H. T.**, A micro-sampling method for the rapid determination of lead in blood by atomic absorption spectrophotometry, *Analyst*, 95, 431, 1970.
10. **Hinderberger, E. J., Kaiser, M. L., and Koirtyohann, S. R.**, Furnace atomic absorption analysis of biological samples using the L'vov platform and matrix modification, *At. Spectrosc.*, 2, 1, 1981.
11. **Matson, W. R. and Roe, D. K.**, Trace metal analysis of natural media by anodic stripping voltammetry, *Anal. Instrum.*, 4, 19, 1966.
12. **Delves, H. T. and Campbell, M. J.**, Measurement of total lead concentrations and isotope ratios in whole blood by use of inductively coupled plasma source mass spectrometry, *J. Anal. At. Spectrom.*, 3, 343, 1988.
13. **Park, C. J., Van Loon, J. C., Arrowsmith, P., and French, J. B.**, Sample analysis using plasma source spectrometry with electrothermal sample introduction, *Anal. Chem.*, 59, 2191, 1987.
14. **Hinners, T. A., Heithmar, E. M., Splitter, T. M., and Henshaw, J. M.**, Inductively coupled plasma mass spectrometric determination of lead isotopes, *Anal. Chem.*, 59, 2658, 1987.
15. **Skogerboe, R. K., Hartley, A. M., Vogel, P. S., and Koirtyohann, S. R.**, Monitoring for lead in the environment, in *Lead in the Environment*, Bogges, W. R., Ed., NSF Report No. NSF/RA-770214, National Science Foundation, Washington, D.C., 1977, 33.

16. **Stevens, R. K., Dzubay, T. E., Russwurm, G., and Rickel, D.,** Sampling and analysis of atmospheric sulfates and related species, *Atmos. Environ.,* 12, 55, 1978.
17. **Jones, J. W. and Boyer, K. W.,** Analysis of foods and related materials using inductively coupled plasma, in *Applications of Inductively Coupled Plasmas to Emission Spectroscopy:* 1977 Eastern Analytical Symposium, Barnes, R. M., Ed., Franklin Institute Press, Philadelphia, PA, 1978, 83.
18. World Health Organization, Principles and procedures for quality assurance, in *Environmental Pollution Exposure Monitoring,* Report No. EFP/HEAL/84.4, World Health Organization, Geneva, Switzerland, 1984.
19. **Duggan, M. J. and Inskip, M. J.,** Childhood exposure to lead in surface dust and soil: a community health problem, *Public Health Rev.,* 13, 1, 1985.
20. **Brunekreef, B. D.,** The relationship between air lead and blood lead in children: a critical review, *Sci. Total Environ.,* 38, 79, 1984.
21. **Annest, J. L., Pirkle, J. L., Makric, D., Neese, J. W., Bayse, D. D., and Kovac, M. G.,** Chronological trend in blood lead levels between 1976 and 1980, *N. Engl. J. Med.,* 308, 1373, 1983.
22. **Skogerboe, R. K.,** *Monitoring Trace Metal Particulates: An Evaluation of the Sampling and Analysis Problems,* Report No. ASTM/STP/555/, American Society for Testing and Materials, Philadelphia, PA, 1974, 125.
23. Ambient Air Quality Surveillance, Code of Federal Regulations, CFR 40: 858, 1982.
24. **Jelinek, C. F.,** Levels of lead in the United States food supply, *J. Assoc. Off. Anal. Chem.,* 65, 942, 1982.
25. **Pennington, J. A. T.,** Revision of the total diet study food lists and diets, *J. Am. Diet. Assoc.,* 82, 166, 1983.
26. **Van Loon, S. C.,** *Selected Methods of Trace Metal Analysis: Biological and Environmental Samples,* John Wiley & Sons, New York, 1985.
27. **Rains, T. C., Rush, T. A., and Butler, T. A.,** Innovations in atomic absorption spectrophotometry with electrothermal atomization for determining lead in foods, *J. Assoc. Off. Anal. Chem.,* 65, 994, 1982.
28. **Fiorino, A., Moffitt, R. A., Woodson, A. L., Gajan, R. J., Huskey, G. E., and Scholz, R. G.,** Determination of lead in evaporated milk by atomic absorption spectrophotometry and anodic stripping voltammetry. Collaborative study II, *J. Assoc. Off. Anal. Chem.,* 56, 1246, 1973.
29. **Capar, S. G., Gajan, R. J., Madzsar, E., Albert, R. H., Sanders, M., and Zyren, J.,** Determination of lead and cadmium in foods by anodic stripping voltammetry. II, *J. Assoc. Off. Anal. Chem.,* 65, 978, 1982.
30. **Kuennen, R. W., Wolnick, K. A., Fricke, F. L., and Caruso, J. A.,** Pressure dissolution and real sample matrix calibration for multi-element analysis of raw agricultural crops by inductively coupled plasma atomic emission spectrometry, *Anal. Chem.,* 54, 2146, 1982.
31. Agency for Toxic Substances and Disease Registry, The Nature and Extent of Lead Poisoning in Children in the United States: A Report to Congress, DHHS Document No. 99-2966, U.S. Department of Health and Human Services, Public Health Service, Atlanta, GA, 1988.
32. **Milar, C. R. and Mushak, P.,** Lead contaminated house dust: hazard, measurement and decontamination, in *Lead Absorption in Children: Management, Clinical and Environmental Aspects,* Chisolm, J. J., Jr. and O'Hara, D. M., Eds., Urban and Schwartzenberg, Baltimore, MD, 1982, 143.
33. **Bornschein, R. L., Succop, P. A., Krafft, K., Clark, C. S., Peace, B., and Hammond, P. B.,** Exterior surface dust lead, interior house dust lead and childhood lead exposure in an urban environment, in *Trace Substances in Environmental Health — XX,* Hemphill, D. D., Ed., University of Missouri, Columbia, MO, 1987, 322.
34. **Chaney, R. L. and Mielke, H. W.,** Standards for soil lead limitations in the United States, in *Trace Substances in Environmental Health — XX,* Hemphill, D. D., Ed., University of Missouri, Columbia, MO, 1987, 357.
35. **Freiberg, C., Molepo, J. M., and Jansoni, B.,** Comparative determinations of lead in soils by X-ray fluorescence, atomic absorption spectrometry and atomic emission spectrometry, *Fresenius Z. Anal. Chem.,* 327, 304, 1987.
36. **Coetzee, P. P., Hoffmann, P., Speer, R., and Lieser, K. H.,** Comparison of trace element determination in powdered soil and grass samples by energy-dispersive XRF and by ICP-AES, *Fresenius Z. Anal. Chem.,* 323, 254, 1986.
37. **Bowman, W. S., Faye, E. H., Sutarno, R., McKeague, S. A., and Kodama, H.,** Soil samples SO-1, SO-2, SO-3, and SO-4 — certified reference materials, CANMET Reports, 79-3, 1979. *(Geostandards Newsl.,* 3, 109, 1979).
38. **Pszonicki, L., Hanna, A. N., and Suschny, O.,** Report on intercomparison IAEA/soil 7 of the determination of trace elements in soil, Report No. IAEA-RL-112, International Atomic Energy Agency, Seibersdorf, Austria Laboratories, 1984.
39. **Skerfving, S.,** Biological monitoring of exposure to lead, in *Biological Monitoring of Toxic Metals,* Rochester Series on Environmental Toxicity, Clarkson, T. W., Friberg, L., Nordberg, G. F., and Sager, P. R., Eds., Plenum Press, New York, 1988, 169.

40. **Tola, S., Hernberg, S., Asp, S., and Nikkanen, J.,** Parameters indicative of absorption and biological effect in new lead exposure: a prospective study, *Brit. J. Ind. Med.,* 30, 134, 1973.
41. **Christoffersson, J. O., Ahlgren, L., Schütz, A., Skerfving, S., and Mattsson, S.,** Decrease of skeletal lead levels in man after end of occupational exposure, *Arch. Environ. Health,* 41, 312, 1986.
42. **Delves, H. T., Sherlock, J. C., and Quinn, M. J.,** Temporal stability of blood lead concentrations in adults exposed only to environmental lead, *Human Toxicol.,* 3, 279, 1984.
43. **Rabinowitz, M., Leviton, A., and Needleman, H.,** Variability of blood lead concentrations during infancy, *Arch. Environ. Health,* 39, 74, 1984.
44. **Chamberlain, A. C.,** Effects of airborne lead on blood lead, *Atmos. Environ.,* 17, 677, 1983.
45. **Piomelli, S., Rosen, J. F., Chisolm, J. J., Jr., and Graef, J. W.,** Management of childhood lead poisoning, *J. Pediatr.,* 105, 523, 1984.
46. **Rosen, J. F., Markowitz, M. E., Bijur, P. E., Jenks, S. T., Wielopolski, L., Kalef-Ezra, J. A., and Slatkin, D. N.,** L X-ray fluorescence of cortical bone lead compared with the $CaNa_2$ EDTA test in lead-toxic children: public health implications, *Proc. Natl. Acad. Sci. U.S.A.,* 86, 685, 1989.
47. **Weinberger, H. L., Post, E. M., Schneider, T., Helu, B., and Friedman, J.,** An analysis of 248 initial mobilization tests performed on an ambulatory basis, *Amer. J. Dis. Child.,* 141, 1266, 1987.
48. **Schütz, A., Skerfving, S., Christoffersson, J. O., and Tell, I.,** Chelatable lead versus lead in human trabecular and compact bone, *Sci. Total Environ.,* 61, 201, 1987.
49. **Grandjean, P.,** Lead poisoning: hair analysis shows the calendar of events, *Human Toxicol.,* 3, 223, 1984.
50. **U.S. Centers for Disease Control,** Preventing Lead Poisoning in Young Children: A Statement by the Centers for Disease Control, DHHS Publication No. 99-2230, U.S. Department of Health and Human Services, Public Health Service, Atlanta, GA, 1985.
51. **Hammond, P. B., Bornschein, R. L., and Succop, P.,** Dose-effect and dose-response relationships of blood lead to erythrocytic protoporphyrin in young children, *Environ. Res.,* 38, 187, 1985.
52. **Boone, J., Hearn, T., and Lewis, S.,** Comparison of interlaboratory results for blood lead with results from a definitive method, *Clin. Chem.,* 25, 389, 1979.
53. **Fachetti, S. and Geiss, F.,** Isotopic Lead Experiment: Status Report, Publ. # EUR-8352EN, Commission of the European Communities, Luxembourg, 1982.
54. **Miller, D. T., Paschal, D. C., Gunter, E. W., Stroud, P. E., and D'Angelo, J.,** Determination of lead in blood using electron thermal atomisation atomic absorption spectrometry with a L'vov platform and matrix modifier, *Analyst,* 112, 1701, 1987.
55. **Berman, E.,** Heavy metals, *Lab. Med.,* 12, 677, 1981.
56. **Roda, S. M., Greenland, R. D., Bornschein, R. L., and Hammond, P. B.,** Anodic stripping voltammetry procedure modified for improved accuracy of blood lead analysis, *Clin. Chem.,* 34, 563, 1988.
57. **Friberg, L.,** Quality assurance, in *Biological Monitoring of Toxic Metals,* Rochester Series on Environmental Toxicity, Clarkson, T. W., Friberg, L., Nordberg, G. F., and Sager, P. R., Eds., Plenum Press, New York, 103, 1988.
58. **Ong, C. N., Phoon, W. O., Lee, B. L., Lim, L. E., and Chua, L. H.,** Lead in plasma and its relationships to other biological indicators, *Ann. Occup. Hyg.,* 30, 219, 1986.
59. **Lauwerys, R., Buchet, J.-P., Roels, H., Berlin, A., and Smeets, J.,** Intercomparison program of lead, mercury and cadmium analysis in blood, urine, and aqueous solutions, *Clin. Chem.,* 21, 551, 1975.
60. **Paschal, D. C. and Kimberly, M. M.,** Determination of urinary lead by electrothermal atomic absorption with the stabilized temperature platform furnace and matrix modification, *At. Spectrosc.,* 6, 134, 1985.
61. **Griffin, R. M.,** Biological monitoring for heavy metals: practical concerns, *J. Occup. Med.,* 28, 615, 1986.
62. **Needleman, H. L., Gunnoe, C., Leviton, A., Reed, R., Peresie, H., Maher, C., and Barrett, P.,** Deficits in psychologic and classroom performance of children with elevated dentine lead levels, *N. Engl. J. Med.,* 300, 689, 1979.
63. **Grandjean, P., Hansen, O. N., and Lyngbye, G.,** Analysis of lead in circumpulpal dentine of deciduous teeth, *Ann. Clin. Lab. Sci.,* 14, 270, 1984.
64. **Fergusson, J. E. and Purchase, N. G.,** The analysis and levels of lead in human teeth: a review, *Environ. Pollut.,* 46, 45, 1987.
65. **Bloch, P., Garavaglia, E., Mitchell, G., and Shapiro, I. M.,** Measurement of lead content of children's teeth *in situ* by X-ray fluorescence, *Phys. Med. Biol.,* 20, 56, 1976.
66. **Keating, A. D., Keating, J. L., Halls, D. J., and Fell, G. S.,** Determination of lead in teeth by atomic absorption spectrometry with electrothermal atomization, *Analyst,* 112, 1381, 1987.
67. **Westerman, M. P., Pfitzer, E., Ellis, L. D., and Jensen, W. N.,** Concentration of lead in bone in plumbism, *N. Engl. J. Med.,* 273, 1246, 1965.
68. **Schütz, A., Skerfving, S., Christoffersson, J. O., Ahlgren, L., and Mattsson, S.,** Lead in vertebral bone biopsies from active and retired lead workers, *Arch. Environ. Health,* 42, 340, 1987.
69. **Drasch, G. A., Böhm, J., and Bauer, C.,** Lead in human bones. Investigations on an occupationally non-exposed population in southern Bavaria (F.R.G.), *Sci. Total Environ.,* 64, 303, 1987.

70. **Somervaille, L. J., Chettle, D. R., and Scott, M. C.,** *In vivo* measurement of lead in bone using X-ray fluorescence, *Phys. Med. Biol.,* 30, 929, 1985.
71. **Wielopolski, L., Ellis, K. J., Vasawani, A. V., Cohn, S. H., Greenberg, A., Puschett, J. B., Parkinson, D. K., Fetterolf, D. E., and Landrigan, P. J.,** *In-vivo* bone lead measurements: a rapid monitoring method for cumulative lead exposure, *Amer. J. Ind. Med.,* 9, 221, 1986.
72. **Somervaille, L. J., Chettle, D. R., Scott, M. C., Tennant, D. R., McKiernan, M. J., Skilbeck, A., and Trethowan, W. N.,** *In-vivo* tibia lead measurements as an index of cumulative exposure in occupationally exposed subjects, *Brit. J. Ind. Med.,* 45, 174, 1988.
73. **Marver, H. S., Tschudy, D. P., Perlroth, M. G., Collins, A., and Hunter, E. J. R.,** The determination of aminoketones in biological fluids, *Anal. Biochem.,* 14, 53, 1966.
74. **MacGee, J., Roda, S. M. B., Elias, S. O., Lington, E. A., Tabor, M. W., and Hammond, P. B.,** Determination of δ-aminolevulinic acid in blood plasma and urine by gas-liquid chromatography, *Biochem. Med.,* 17, 31, 1977.
75. **Piomelli, S., Seaman, C., Zullow, D., Curran, A., and Davidow, B.,** Threshold for lead damage to heme synthesis in urban children, *Proc. Natl. Acad. Sci. U.S.A.,* 79, 3335, 1982.
76. **Piomelli, S. and Davidow, B.,** Free erythrocyte protoporphyrin concentration: a promising screening test for lead poisoning, *Pediatr. Res.,* 16, 366, 1972.
77. **Chisolm, J. J., Jr. and Brown, D. H.,** Microscale photofluorometric determination of "free erythrocyte protoporphyrin" (protoporphyrin IX), *Clin. Chem.,* 21, 1669, 1975.
78. **Lamola, A.-A., Joselow, M., and Yamane, T.,** Zinc protoporphyrin (ZPP): a simple, sensitive fluorometric screening test for lead poisoning, *Clin. Chem.,* 21, 931, 1975.
79. **Balamut, R., Doran, D., Giridhar, G., Mitchell, D., and Soule, S.,** Systematic error between erythrocyte protoporphyrin in proficiency test samples and patients' samples as measured with two hematofluorometers, *Clin. Chem.,* 28, 2421, 1982.
80. **Mitchell, D. G. and Doran, D.,** Effect of bias in hematofluorometer measurements of protoporphyrin in screening programs for lead poisoning, *Clin. Chem.,* 30, 386, 1985.
81. **Piomelli, S. and Graziano, J.,** Laboratory diagnosis of lead poisoning, *Pediatr. Clin. North Amer.,* 27, 843, 1980.
82. **Schwartz, S., Zieve, L., and Watson, C. J.,** An improved method for the determination of urinary coproporphyrin and an evaluation of factors influencing the analysis, *J. Lab. Clin. Med.,* 37, 843, 1951.

Chapter 4

LEAD IN THE ENVIRONMENT: FROM SOURCES TO HUMAN RECEPTORS

Cliff I. Davidson and Michael Rabinowitz

TABLE OF CONTENTS

I.	Introduction	66
II.	The Flow of Lead from Sources Through the Environment	66
	A. Sources of Airborne Lead	66
	1. Natural Sources	66
	2. Anthropogenic Sources	67
	B. Characteristics of Airborne Lead	69
	1. Vapor-Phase Lead	69
	2. Particulate Lead: Chemical Composition and Airborne Concentrations	69
	3. Particulate Lead: Size Distributions	71
	C. Atmospheric Deposition of Lead	73
	1. Wet Deposition	73
	2. Dry Deposition	76
III.	The Flow of Lead from the Environment to Humans	78
	A. Lead in Environmental Media	78
	1. Dust and Soil	79
	2. Water	79
	3. Food	80
	B. Uptake by Humans	81
	1. Inhalation	81
	2. Ingestion	81
IV.	Summary	81
References		82

I. INTRODUCTION

Of all trace metal contaminants in the environment, lead has been the most widely studied. Recent summaries of the literature indicate that several thousand journal articles have been published on this element over the last two decades. Although the amount of lead research has decreased somewhat since 1980, several hundred articles are still being added to the information base each year.

Research on environmental lead has covered a wide variety of topics. One area of interest involves identifying the various sources contributing this metal to the environment, including natural emissions as well as anthropogenic pollutants. Other research has studied the physical and chemical characteristics of lead emitted to the environment. This has involved quantification of the partitioning between particulate and vapor phase lead, as well as identification of the many chemical compounds containing this element. Still other research has considered transport pathways from sources to receptors. Of particular interest are human receptors, who may experience adverse effects from exposure to lead. Although our knowledge is far from complete, we now have an understanding of the most important sources, environmental pathways, and exposure routes to humans.

This chapter summarizes current knowledge of several aspects of environmental lead. The chapter is divided into two parts. The first addresses lead from the perspective of environmental flows; the main topics covered are sources contributing lead to the atmosphere as well as atmospheric transport and transformation. These are the most important parts of the biogeochemical cycle influencing humans. The second part considers lead from the perspective of human exposure, examining several environmental media by which uptake can occur. These media include air, dust, water, and food, with a brief discussion of lead in paint.

II. THE FLOW OF LEAD FROM SOURCES THROUGH THE ENVIRONMENT

There are a number of possible media by which contaminants may enter the environment, e.g., the atmosphere, oceans, fresh water bodies, and the earth's crust. Once emitted, the contaminants may be transported among various media, eventually reaching humans or other receptors. Previous work has shown that the atmosphere is by far the most important compartment by which lead enters the environment and through which lead is dispersed before reaching humans.

In this part of the chapter, we begin by considering natural and anthropogenic sources which contribute lead to the atmosphere. We then examine the characteristics of airborne lead, addressing both vapor-phase and particulate species. Finally, we consider processes which remove lead from the atmosphere, namely wet and dry deposition.

A. SOURCES OF AIRBORNE LEAD
1. Natural Sources

Lead may be released to the atmosphere from erosion of the earth's crust, volcanic eruptions, production of seaspray, emissions from vegetation, and forest fires. Table 1 summarizes two sets of published lead emissions estimates from these sources. Note the significant differences between these estimates. In all cases, the values reported by Settle and Patterson[1] are smaller than those reported by Pacyna.[2] The former study is based on those few measurements of lead which have incorporated the strictest contamination control procedures. The estimates of Pacyna are based on a much larger set of research results from a wide variety of investigators to obtain better representation; however, some of these results are known to be erroneous due to contamination during sampling, sample handling, or

TABLE 1
World-Wide Natural Emissions of Lead[a]

	Pacyna (1986)[2]	Settle and Patterson (1980)[1]
Wind-blown dust	10; (3—35)	2[b]
Volcanoes	6.4; (0.4—96)	1
Seaspray	0.1; (0.001—5)	<1
Vegetation	1.6; (0.2—21)	<0.1
Forest fires	0.5; (0.04—6.8)	—
TOTAL	18.6	~2

[a] In units of 10^6 kg/year.
[b] The windblown dust value from Settle and Patterson includes the influence of volcanic particles; their value for volcanoes is based on sulfur emissions as discussed in the text.

TABLE 2
World-Wide Anthropogenic Emissions of Lead

Mining, nonferrous metals	8.2
Primary nonferrous metal production	76.5
Secondary nonferrous metal production	0.8
Iron and steel production	50
Industrial applications	7.4
Coal combustion	14
Oil combustion (including gasoline)	273
Wood combustion	4.5
Waste incineration	8.9
Manufacture, phosphate fertilizer	0.05
Miscellaneous	5.9
TOTAL	449

[a] In units of 10^6 kg/year.

After Nriagu, J. O., *Nature*, 279, 409, 1979.

analysis. There are also different assumptions involved in the derivations of these values. For example, the estimate of Pacyna for volcanic emissions is based on Nriagu,[3] who uses the product of 10^{10} kg/year of volcanic dust and an average lead concentration in the dust of 640 µg/g. Settle and Patterson[1] use the product of 6×10^9 kg/year of volcanic sulfur emissions and a Pb/S ratio of 2×10^{-7}.

Although there are substantial differences in these estimates, either set of values shows that natural emissions are negligible compared with anthropogenic emissions of lead on a global scale. These anthropogenic emissions, and independent historical record data supporting the dominance of anthropogenic lead, are discussed in the next section.

2. Anthropogenic Sources

The most thorough study of global anthropogenic lead emissions is that of Nriagu,[3] whose results are summarized in Table 2. The estimates show that combustion of oil and its derivatives, especially leaded gasoline, is responsible for the majority of lead emissions world-wide. Production of nonferrous metals such as lead, zinc, and copper, and production

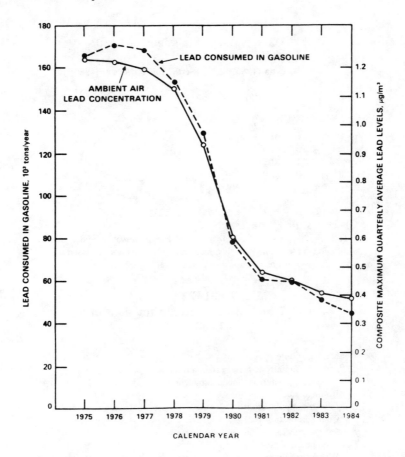

FIGURE 1. Lead consumed in gasoline and ambient lead concentrations, 1975 to 1984. (From U.S. Environmental Protection Agency, EPA/600/8-83/028bF, June, 1986, chap. 5 and 7).

of iron and steel, are also important contributors. Additional discussion of the values in Table 2 is provided by Settle and Patterson[1] and Pacyna.[2]

It is important to note that values in this table are based on 1975 data. Because of the decreasing lead content in gasoline in many countries, the emissions through the 1980's have been reduced. For example, Figure 1 shows that gasoline lead consumption and average airborne lead concentration have both decreased substantially in the U.S. between 1975 and 1984. Nevertheless, global lead emissions from combustion of gasoline remain important because of the continued use of large quantities of leaded fuels in many parts of the world.

Comparing Tables 1 and 2 shows that emissions of lead to the atmosphere from anthropogenic sources greatly exceed those from natural sources. This is substantiated by Figure 2, which shows global lead production over the past 5500 years. The general trend in the figure has been demonstrated using historical environmental records in glaciers, freshwater sediments, marine sediments, tree rings, and other media. As an example, Figure 3 shows historical data from several studies. The available information demonstrates that anthropogenic activities have greatly altered the natural global cycle of lead, emitting orders of magnitude more lead into the atmosphere compared with preindustrial times. Data from ocean corals, however, show a recent decrease in deposited lead probably due to decreasing use of leaded gasoline.[4] In the next section, we examine the chemical and physical forms of this lead which can lend insight into the environmental pathways involved.

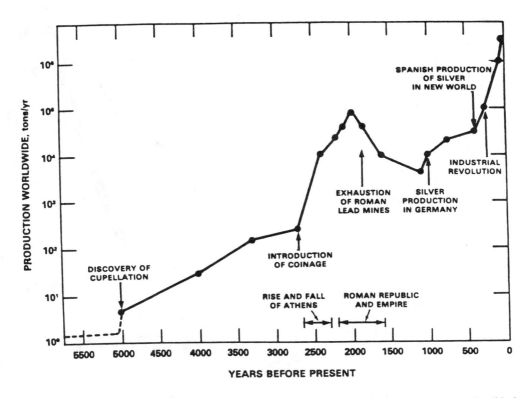

FIGURE 2. The global lead production has changed historically in response to major economic and political events. Increases in lead production (note log scale) correspond approximately to historical increases in lead emissions. (From U.S. Environmental Protection Agency, EPA/600/8-83/028bF, June, 1986, chap. 5 and 7).

B. CHARACTERISTICS OF AIRBORNE LEAD
1. Vapor-Phase Lead

Lead in the atmosphere is found in both vapor and particulate phases. Radojevic and Harrison[5] report that vapor-phase lead makes up only a few percent of the total lead concentration in the ambient atmosphere. However, the vapor-phase compounds can play an important role in the environmental cycling of lead. Much of the vapor-phase lead is derived from tetraalkyllead (R_4Pb) added to gasoline. The alkyl group R in this context may be a methyl group (CH_3) or an ethyl group (CH_3CH_2). Tetraalkyllead and other alkyllead compounds are released to the atmosphere during the manufacture of the gasoline additive, evaporation at filling stations, direct emissions from vehicle tailpipes, and other minor sources.[5] Once airborne, alkyllead species will generally decompose to inorganic lead by successive loss of alkyl groups: $R_4Pb \rightarrow R_3Pb^+ \rightarrow R_2Pb^{2+} \rightarrow RPb^{3+} \rightarrow Pb^{2+}$. This decomposition occurs by direct photolysis in the presence of sunlight, or by reaction with the hydroxyl radical (OH). Typical half-lives have been reported as 5 to 10 h for Me_4Pb and 0.6 to 2 h for Et_4Pb under photochemically active conditions in summer, with values 3 to 4 times as long in winter.[6] Both ethyl and methyl forms of RPb^{3+} have half-lives as long as several days, suggesting the importance of these species as relatively stable intermediates during the decomposition. Eventually, all vapor-phase organic lead will decompose to inorganic Pb aerosol.

2. Particulate Lead: Chemical Composition and Airborne Concentrations

Most of the lead emitted from combustion of leaded gasoline, as well as from other anthropogenic and natural sources, is in the form of an aerosol. Table 3 lists the predominant

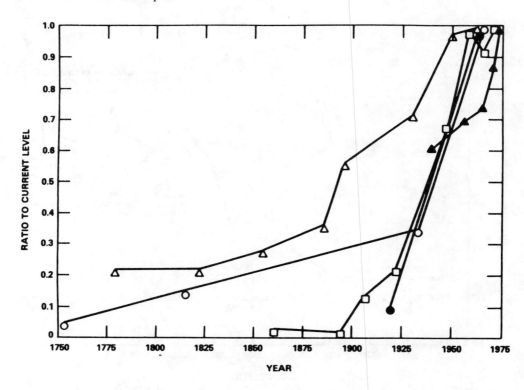

FIGURE 3. Chronological record of the relative increase of lead in snow strata, pond and lake sediments, marine sediments, and tree rings. The data are expressed as a ratio of the latest year of the record and should not be interpreted to extend back in time to natural or uncontaminated levels of lead concentration. ○ = Reference 74; □ = Reference 81; △ = Reference 82; ▲ = Reference 83; ● = Reference 84. (From U.S. Environmental Protection Agency, EPA/600/8-83/ 028bF, June, 1986, chap. 5 and 7.)

TABLE 3
Composition of Lead Aerosols from Pollutant Sources

Source	Composition of particulates
Automotive	$PbCl_2$, $PbBr_2$, $PbBrCl$, $Pb(OH)Br$, $Pb(OH)Cl$, $PbCl_2 \cdot BrCl$, $PbO \cdot PbBr_2$, $PbO \cdot PbBrCl$, $PbO \cdot PbCl_2$, PbO_x, $PbSO_4$, $PbO \cdot PbSO_4$, PbP_2O_7, $Pb_3(PO_4)_2 \cdot PbBrCl$, $Pb_5(PO_4)_3(Cl,Br)$, $Pb_3(PO_4)_2$, $Pb_4O(PO_4)_2$, $2NH_4Cl \cdot PbBrCl$, α-$NH_4Cl \cdot 2PbBrCl$, β-$NH_4Cl \cdot 2PbBrCl$, $PbCO_3$, $PbO \cdot PbCO_3$
Mining activities	PbS, $PbCO_3$, $PbSO_4$, $Pb_5(PO_4)_3Cl$, $PbS \cdot Bi_2S_3$, PbO_x, Pb-silicates
Base metal smelting and refining	Elemental lead, PbO_x, $PbSO_4$, $PbCO_3$, $PbO \cdot PbSO_4$, $(PbO)_2PbCO_3$, Pb in other metal oxides, Pb-silicates
Coal-fired power plants	Surface complex in fly ash, PbO_x, $PbSO_4$, $Pb(NO_3)_2$, $PbO \cdot PbSO_4$
Cement manufacture	$PbCO_3$, $Pb_5(PO_4)_3Cl$
Fertilizer production	$Pb_5(PO_4)_3Cl$, PbO_x, $PbCO_3$
Ferroalloys	Elemental lead, lead alloy chips
Manufacture and utilization of lead products	Lead arsenate, antimonate, chromate, cyanamide, iodide, fluorosilicate, molybdate, nitrate, selenide, silicates, titanate, vanadate, etc.

From Nriagu, J. O., in *The Biochemistry of Lead in the Environment,* Vol. 1, Nriagu, J. O., Ed., Elsevier, Amsterdam, 1978, 137—183. With permission.

chemical forms of lead aerosol emitted from many of the sources given in Table 2, taken from Nriagu.[7] It is of interest that lead halides represent a major constituent of exhaust from the combustion of leaded gasoline. These lead halides generally decompose in the atmosphere after emission, eventually becoming lead oxides, carbonates, or sulfates.

A wealth of data exist on airborne concentrations of lead in urban, rural, and remote areas. Nriagu[7] lists published airborne lead levels from 147 different sites (or groups of sites) in North America, South America, and Europe from urban and rural locations. Most of these data were obtained between 1965 and 1975, and show urban values typically 1,000 to 10,000 ng/m^3 and rural values typically 100 to 1000 ng/m^3. Wiersma and Davidson[8] have summarized published remote area data obtained during 1973 to 1981 from 36 sites and sampling periods world-wide. They report values in the range 0.027 to 97 ng/m^3. A representative collection of data is provided by the U.S. Environmental Protection Agency (EPA)[9] for urban, rural, and remote areas, and is summarized in Table 4.

It is important to note that many of the published lead concentrations in air, as well as lead concentrations in other environmental media, are erroneous due to contamination during sampling, sample handling, and analysis. This is particularly true for many of the earliest measurements, and for data obtained from rural and remote areas where concentrations are low. Many of the values reported in the studies mentioned above are therefore subject to error. The data in Table 4 have been assembled with careful attention to reliability of the studies, although even here some of data may be erroneously high due to contamination. Discussions of contamination problems during collection and analysis of environmental samples for lead have been presented by Patterson[10] and Boutron.[11]

3. Particulate Lead: Size Distributions

The airborne concentration data discussed above have been obtained using particulate filters, which capture particles of a wide range of sizes with high efficiency. The values reported thus represent the total particulate lead concentrations. Some investigators have sampled airborne particles for subsequent chemical analysis using impactors, which fractionate particles by size. Examples of several size distributions for lead are shown in Figure 4, taken from Davidson and Osborn.[12] The ordinate of each graph is the normalized mass distribution function $(\Delta C/C_T)/\Delta \log d_p$, where ΔC is the airborne mass concentration of Pb in a given size range (which extends from $d_{p_{min}}$ to $d_{p_{max}}$), C_T is the total concentration of Pb in all size ranges, and $\Delta \log d_p$ is the difference $\log d_{p_{max}} - \log d_{p_{min}}$. The aerodynamic diameter of a particle d_p represents the size of a unit density sphere whose aerodynamic transport characteristics are identical to those of the original particle. Note that the area under the curve between any two particle diameters is proportional to the fraction of airborne mass in that size interval. Data to construct these graphs have been obtained from differential or cumulative distribution plots, or from tables, in the original references. Personal communication with the authors was necessary in several instances to clarify or update the data.

All distributions have been plotted as though the particles are uniformly distributed in $\log d_p$ in each size range. The resulting histograms are thus only approximations to the true smooth-curve distributions which are unknown, although techniques for estimating the true size spectra have been reported in the literature.[13,14] For consistency, overall minimum and maximum aerodynamic diameters of 0.05 and 25 μm, respectively, have been assumed after Hidy[15] and Davidson,[16] except for a few cases where investigators have taken special measures to characterize distribution endpoints. This style of graphing has been chosen to allow direct observation of those size ranges containing the bulk of the airborne mass, and to emphasize differences in the shapes of the distributions. Such differences would be more difficult to identify in cumulative log probability plots.

Overall, the plots indicate that Pb is predominantly submicron. Except for the few distributions obtained without backup filters (numbers 9, 14, 17, 18, and 37 in Figure 4),

TABLE 4
Atmospheric Lead in Urban, Rural, and Remote Areas of the World[9]

Location	Sampling period	Lead conv. ($\mu g/m^3$)	Ref.
Urban			
New York	1978—79	1.1	62
Boston	1978—79	0.8	62
St. Louis	1973	1.1	62
Houston	1978—79	0.9	62
Chicago	1979	0.8	62
Los Angeles	1978—79	1.4	62
Ottawa	1975	1.3	63
Toronto	1975	1.3	63
Montreal	1975	2.0	63
Brussels	1978	0.5	64
Turin	1974—79	4.5	65
Riyadh, Saudi Arabia	1983	5.5	66
Rural			
New York Bight	1974	0.13	67
United Kingdom	1972	0.13	23
Italy	1976—80	0.33	65
Belgium	1978	0.37	64
Illinois	1973—74	0.23	68
Tennessee	1977	0.11	69
Remote			
White Mtn., CA	1969—70	0.008	70
High Sierra, CA	1976—77	0.021	71
Olympic Natl. Park, WA	1980	0.0022	72
Great Smoky Mtns. Natl. Park, TN	1979	0.015	72
Glacier Natl. Park, MT	1981	0.0046	72
South Pole	1974	0.000076	73
Thule, Greenland	1965	0.0005	74
Thule, Greenland	1978—79	0.008	75
Prins Christiansund, Greenland	1978—79	0.018	75
Dye 3, Greenland	1979	0.00015	76
Eniwetok, Pacific Ocean	1979	0.00017	77
Kumjung, Nepal	1979	0.00086	78
Bermuda	1973—75	0.0041	79
Abastumani Mtns., U.S.S.R.	1979	0.019	80

From Reference 9.

most of the Pb mass is associated with the lower end of the spectrum in each case. This has important implications for the transport of atmospheric Pb: small particles are slow to be scavenged by precipitation or to deposit onto natural surfaces, and hence may remain airborne for long periods of time. Distances of transport are thus likely to be large.

Figure 5 shows averages of several Pb size distributions by sampling location, after Milford and Davidson.[17] These averages were computed using 41 size distributions reported in the literature. The distributions included 36 from Figure 4 (those without backup filters were excluded), plus 5 additional distributions reported too recently for inclusion in Figure 4.

The three categories of the Pb distributions all show similar shapes, with a peak between 0.5 and 1 μm aerodynamic diameter. The similarity in the distributions reflects the fact that

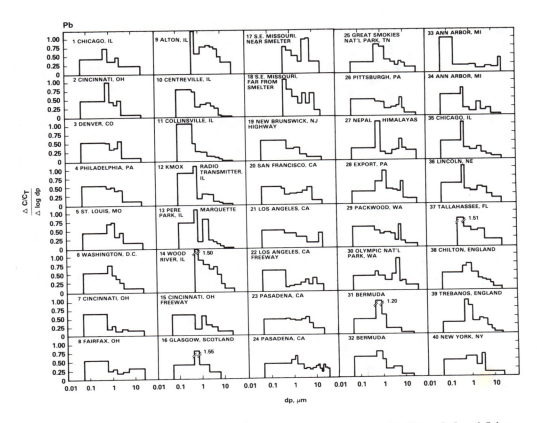

FIGURE 4. Airborne size distributions of lead reported in the literature. (From Davidson, C. I. and Osborn, J. F., in *Toxic Metals in the Atmosphere*, Nriagu, J. O. and Davidson, C. I., Eds., John Wiley & Sons, New York, 1986, 355. With permission.)

anthropogenic combustion processes are the primary sources of airborne Pb at virtually all sites sampled, including those far from emissions. In contrast, average spectra for Na and Al, also shown in Figure 5, show generally larger particles and greater variability. Na is derived from crustal weathering and seaspray, while Al is derived mostly from crustal weathering.

It is important to note that the impactors used to obtain the data of Figures 4 and 5 are highly imperfect devices. Problems include inefficient collection of large particles, particle bounceoff and wall losses inside the impactor, and nonideal collection efficiencies associated with the impactor stages. These and several other problems have been discussed by Milford and Davidson,[17] and require that the data of Figure 4 be considered merely as rough approximations to the true distributions.

C. ATMOSPHERIC DEPOSITION OF LEAD
1. Wet Deposition

Lead can be removed from the atmosphere by wet deposition, i.e., precipitation scavenging, or by dry deposition. In the first category, several studies have examined the lead content of precipitation in urban, rural, and remote areas. Table 5 summarizes the results of two literature reviews on the topic.[18,19] The data show the greatest concentrations in urban areas, with the median value close to the 50 µg/l lead standard in drinking water set by the U.S. EPA. The remote area values pertain to snow and ice core data from Greenland and Antarctica. As discussed earlier for airborne concentration measurements, much of the data

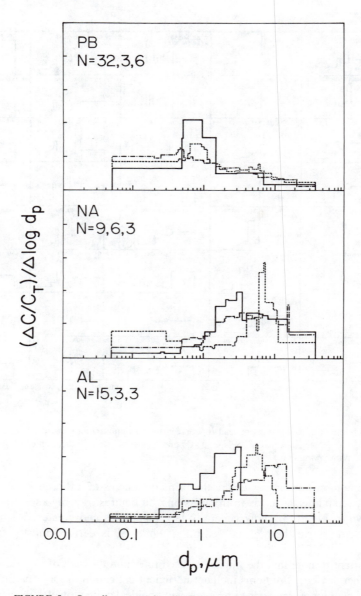

FIGURE 5. Overall average size distributions of Pb, Na, and Al reported in the literature. (—·—) represents urban data; (—) shows data from marine environments; and (– – –) represents nonurban continental data. The three values of N refer to the number of distributions averaged in each of these categories, respectively. (From Milford, J. B. and Davidson, C. I., *J. Air Pollut. Control Assoc.*, 35, 1249, 1985. With permission).

for lead in precipitation has to be considered with caution due to contamination problems. This is especially true for rural and remote areas.

Little information is available on the forms of lead in precipitation. Gatz and Chu[20] have reported that much of the lead in precipitation in the Chicago area is in water-soluble form. More than half of their precipitation samples had >80% soluble lead. For dry deposition and bulk wet plus dry deposition, however, most of the samples had <10% soluble lead. Such solubilities may have a marked effect on ecosystem effects and on geochemical cycling of lead.

Very few studies have examined the wet deposition rate, defined as the product of lead

TABLE 5
Concentration of Lead in Precipitation Reported in the Literature[a]

From Literature Review of Galloway et al. (1982)[18]

Urban (N = 8)	
Range	5.4—147
Median	44
Rural (N = 32)	
Range	0.59—64
Median	12
Remote (N = 6)	
Range	0.02—0.41
Median	0.09

From Literature Review of Barrie et al. (1987)[19]

Rural	
Southern Sweden	8.8
Northern Sweden	3.4
S.E. U.S.	3.2—5.6
S.W. Ontario	7.0
Sudbury, Ontario	4—14
Dorset, Ontario	15
Lewes, Delaware	3.0
Remote	
Enewetak, Pacific	<0.04
N. Pacific	0.15
Bermuda[a]	0.72, 0.77
N.W. Ontario	2.3
Northern Sweden	2.39

[a] All values are in units of $\mu g/l$.
[b] Two separate studies.

concentration in precipitation and the precipitation rate. One of the only studies to consider this issue on a large scale is a network of 32 stations throughout the U.S. which sampled precipitation during 1966 and 1967.[21] Analyses were conducted for lead, zinc, copper, and iron; results for lead are shown in Figure 6. The data indicate much greater wet deposition rates in Eastern U.S. than in the West, consistent with the geographical distribution of lead emissions.

It is also of interest to consider the rate of scavenging of lead by precipitation. Slinn[22] has conducted a detailed review of precipitation scavenging mechanisms for airborne contaminants, considering in-cloud as well as below-cloud processes. This information suggests that submicron particles, such as those predominant in airborne lead size distributions, have smaller scavenging efficiencies than larger particles. The inefficient scavenging of lead is reflected in values of the scavenging ratio reported in the literature (concentration in precipitation/concentration in air, both measured in $\mu g/g$). For example, Cawse[23] reports an annual average value of 240 for lead in the U.K., compared with 1000 to 2000 for several crustal elements. Gatz[24] reports a value of 100 for lead, with values of 500 to 1000 for iron, manganese, and magnesium in St. Louis, MO. Davidson et al.[25] reports an average value of 160 for lead in Greenland snowstorms, compared with average values of 1200 to 2000 for crustal elements. These data suggest that in both populated and remote areas, lead is not as efficiently removed by rain and snow as crustal elements associated with larger particles.

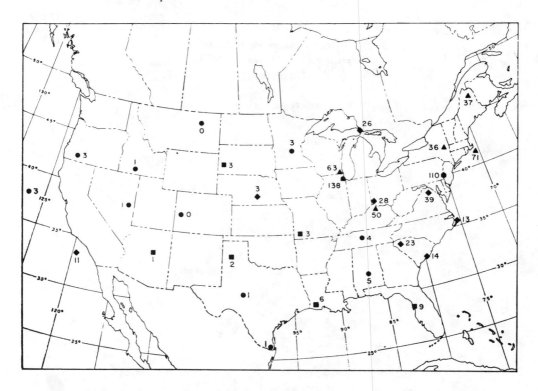

FIGURE 6. Lead values averaged from September 1966 to March 1967. Plotted numerals are g/ha/month. Symbols are g/ha/cm of precipitation. (●) 0—0.4; (■) 0.5—0.9; (♦) 1—4; (▲) 5—9; (●) 10—14; (⬤) > 15. (From Lazrus, A. L. et al., *Environ. Sci. Tech.*, 4, 55, 1970. With permission.)

2. Dry Deposition

Unlike wet deposition for which there are established measurement methods, determining dry deposition of atmospheric contaminants onto natural surfaces is fraught with difficulty. At present, there are no widely accepted methods for directly measuring dry deposition of lead onto natural surfaces of interest. Some investigators have used foliar extraction techniques to wash deposited lead off the surface of leaves or conifer needles. Others have measured lead concentrations in precipitation above and within vegetation canopies, attributing the difference in concentration to dry deposited material washed off the canopy surfaces (throughfall). Measurement of lead concentrations in surface snow has also been used to estimate dry deposition. Methods for inferring dry deposition from airborne concentration data, e.g., by measuring vertical gradients above the surface or by assessing reductions in concentration with distance downwind, have been attempted on some occasions.

Rather than direct measurement of deposition onto the natural surfaces of interest, most dry deposition studies involving lead have used sampling devices which are surrogates to natural surfaces. Examples include filter paper, petri dishes, Teflon® plates, plexiglass plates, and dustfall buckets of various types. In some cases, the surrogate surfaces used have been aerodynamically designed to control the surface boundary layer.

Figure 7 shows a summary of dry deposition data for lead reported in the literature, taken from the review of Davidson and Wu.[26] The values are reported as deposition velocities (deposition flux in $\mu g\ cm^{-2}\ s^{-1}$/airborne concentration in $\mu g\ cm^{-3}$). The figure shows a wide range of values, with a mean of about 0.3 cm/s.

It is of interest to compare these measured values with estimates based on dry deposition models reported in the literature. Figure 8 shows a graph of dry deposition velocity versus particle diameter for several deposition models as summarized by Milford and Davidson.[27]

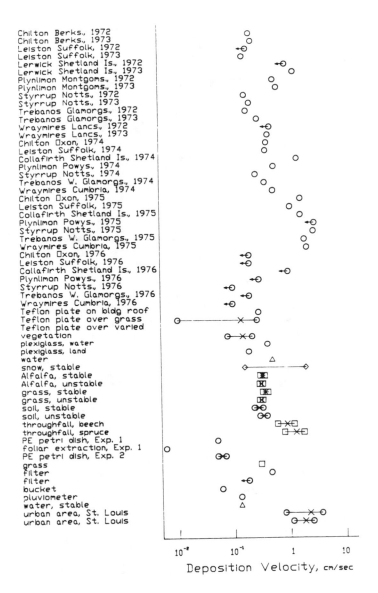

FIGURE 7. Dry deposition velocities of lead reported in the literature. (From Davidson, C. I. and Wu, Y. L., in *Acid Precipitation, Sources, Deposition, and Canopy Interactions*, Vol. 3, Adriano, D. C., Ed., Springer-Verlag, New York, 1990. With permission).

These models have been used with the individual size spectra that were averaged to produce Figure 5 for all three elements (41 for Pb, 18 for Na, and 21 for Al). Details of the calculations are presented by Davidson and Wu.[26] The results show dry deposition velocities for each size spectrum for each curve in Figure 8. Averaged over all size spectra, the deposition velocities for lead ranged from 0.26 cm/s for water to 0.56 cm/s for orchard grass. The deposition velocities for Na ranged from 0.86 to 2.0 cm/s for these surfaces, while the values for Al ranged from 0.98 to 2.4 cm/s. Dry deposition measurement data reported in the literature for Na and Al, comparable to the data in Figure 7 for Pb, showed mean values of 1.9 cm/s and 2.6 cm/s, respectively, for these two elements. It is apparent that dry deposition velocities of lead are smaller than those of crustal elements, analogous to the differences reported for wet deposition. The differences are due to the particle sizes involved:

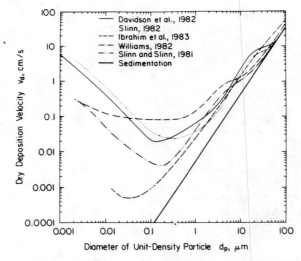

FIGURE 8. Dry deposition velocity as a function of particle diameter for the models of: Davidson et al.[102] for a canopy of Dactylus glomerata (orchard grass) (canopy height = 57 cm, friction velocity = 64 cm/s, roughness height = 5 cm, zero-plane displacement = 14 cm); Slinn[103] for a Eucalyptus forest (canopy height = 17.4 m, friction velocity = 75 cm/s, roughness height = 1.86 m, zero-plane displacement = 21.6 m); Ibrahim et al.[104] for snow (reference height = 10 cm, mean windspeed = 2.5 m/s, relative humidity = 90%); Williams[105] for water (reference height = 10 m, mean windspeed = 10 m/s, relative humidity = 99%); Slinn and Slinn[106] for water (reference height = 10 m, mean windspeed = 5 m/s, relative humidity = 99%). The curve for sedimentation velocity is also shown. (From Milford, J. B. and Davidson, C. I., *J. Air Pollut. Control Assoc.*, 35, 125, 1987. With permission.)

the Pb distributions in Figure 5 have an overall average mass median aerodynamic diameter of 0.55 μm, compared with 3.7 and 4.8 μm for Na and Al, respectively.

Finally, it is of interest to consider measurements of the bulk deposition of lead, and the relative importance of wet and dry deposition. Bulk deposition measurements involve sampling with open buckets or similar collectors which receive dry deposited material as well as precipitation. Nriagu[7] and Galloway et al.[18] have summarized bulk deposition data for lead reported in the literature. In general, values range from <1 mg/m² year in remote areas to >1000 mg/m² year near large stationary sources. Typical values in urban areas are 10 to 100 mg/m² year.

Few estimates are available of the relative importance of wet and dry deposition. Galloway et al.[18] reports that dry deposition is responsible for roughly 60% of the total deposition of lead in marine areas, 30% in rural areas, and 20% in urban areas. Davidson et al.[25] estimate that dry deposition accounts for 10 to 35% of total deposition of lead in the Arctic, with the upper end of the range applying to the highest latitudes where annual precipitation is small. Overall, the available information suggests that there is considerable variability in the relative importance of wet and dry deposition, depending mostly on proximity to sources and on precipitation rate.

III. THE FLOW OF LEAD FROM THE ENVIRONMENT TO HUMANS

A. LEAD IN ENVIRONMENTAL MEDIA

The sources and abundance of lead in those portions of the environment that are pertinent to human exposure are now presented, with the exception of atmospheric lead which was discussed in the previous section.

1. Dust and Soil

Lead is a natural constituent of soil, the upper regolith which overlies bedrock, and of dust, fine wind-borne debris. Typical concentrations are in the range of 10 to 50 µg/g in uncontaminated settings. However, urban values are often 10 to 200 times greater as a result of human activity. These large amounts of lead are available to humans by a variety of routes.

Lead is generally immobile in soil; the organic matter in soil has a large capacity to bind lead firmly in most cases. An exception is sandy soil with low organic content and acidic pH. As a result of the immobility, deposited lead accumulates in soil and is removed only when the soil itself is physically transported, for example, by eolian processes, and soil turnover by worms, landscaping, or tilling. These can be slow processes, and contamination of soil with lead persists many years after deposition has stopped.[28]

Urban soils may contain lead from a variety of sources including automobile exhaust, industrial activity, and deteriorating or intentionally removed lead paint.[30] Roadside soils typically contain 50 to 2000 µg/g lead within 25 m of the curb, with concentrations that decrease with distance. Higher lead levels are associated with increased traffic density, speed, and inclination of the road.[29] Within 3 m of houses with lead-based paint, soil concentrations typically exceed 1000 µg/g. Even soil from urban parks, without any buildings, have been shown to contain from 200 to over 3000 µg/g lead as a result of decades of accumulated deposition.[31] These elevated soil lead levels present the risk of lead exposure through urban vegetable gardening[32] and by direct, if unintentional, ingestion of soil.[33]

The concentration of lead in dust tends to be considerably higher than soil lead levels in the same general setting.[34] This is partially because lead is associated with the smallest particle sizes in soil: half of the lead resides in clay-sized particles which constitute only 5% of the mass.[35] Once resuspended, these particles can remain airborne for longer periods than larger soil particles.

Several of the important sources of lead in dust produce fine particles, such as combustion processes and the weathering of lead-based paints. This combination of high lead levels and small particle size may result in a particular hazard, because of the difficulty in visual detection and cleaning.[36] These fine particles may accumulate on surfaces or be temporarily trapped in carpet and upholstery.

Indoor dust represents a highly variable and mobile source of lead exposure. Blood lead concentrations of urban infants correlate very strongly with indoor dust lead levels even after adjusting for soil lead levels at the home.[37] Also, attempts to control indoor dust lead loading have resulted in lowering of elevated blood lead levels.[38] These illustrate the importance of dust as a vector for human lead exposure.

2. Water

Lead is present in trace amounts in natural surface water and ground water. Atmospheric inputs of lead by deposition or surface runoff usually form insoluble salts or are sorbed onto particles and enter the sediment. Consequently, most water supplies are low in lead, with concentrations less than 5 µg/l. However, drinking water can be markedly higher due to contact with lead bearing pipes, solder, and plumbing fixtures. Waters that are particularly corrosive (plumbosolvent) because of acidity or softness can be troublesome and can result in elevated lead levels in the consumed water, especially in hot, stagnant, or "first draw" water.[39] In these circumstances, drinking water can be the overwhelming source of lead, reaching well over 300 µg/l. Fully flushing typical distribution systems even with corrosive water and fresh lead solder will usually lower water lead levels to below 20 µg/l. Current standards for lead in drinking water are 50 µg/l but are likely to be reduced.

The extent of the problem of lead contamination by the water distribution system is difficult to estimate. It has been estimated that in the U.S., 16% of partially flushed water systems exceed 20 μg/l.[40] The use of lead pipes has long been prohibited and such pipes are being replaced, but they are still used in many rural areas and in some cities. The use of lead solder in residential water supplies has been outlawed by the 1986 Amendments to the Safe Drinking Water Act, but existing systems usually contain some lead, estimated to be about 10 kg per house. Because lead in contact with some waters gradually acquires a partially passivating internal carbonate coating, older houses have less lead in the tap water than newer homes.[41] For example, standing water from homes with copper pipes and lead solder newer than 5 years averaged 31 μg/l, while homes less than 18-months old averaged 74 μg/l. In contrast, homes with galvanized pipes averaged less than 6 μg/l. New, exposed solder can raise standing water lead levels to above 100 μg/l within 40 min.[42] When lead pipes or lead cisterns are used, much higher levels of lead are possible, occasionally resulting in excessive exposures.[43] Faucets and tap fixtures made of brass often have lead present as a contaminant in the alloy, and this can be a source of high lead levels in the first draw.[44] The addition of lye or lime to the water supply before distribution raises the pH to above 7.5 and lowers its corrosiveness.[45] This approach has been shown to be a useful system-wide remedy.[46]

3. Food

Ingestion of lead in food represents the majority of the daily intake of lead for most individuals. A survey of residents of Britain who were not exposed to excessive amounts of lead estimated that the daily lead intake from food is 45 to 90% of the total for adults and 55 to 90% of the total for children.[47] Similar values have been reported in the U.S.[48] Although this represents a major pathway for most people, it is not the route typically responsible for marked elevations of an individual's lead burden. Rather it may account for a population-wide increase above natural levels.

The lead content of food is mostly the result of numerous additive steps. Lead is a natural component in soil, but there is relatively little translocation up to the edible portion of crops. Root vegetables can account for about 10% of daily lead intake.[49] Foliar uptake of atmospheric lead by plants has been estimated to be 16% of daily intake.[50] Because lead is not preferentially stored in the edible portions of animals and does not become enriched along the food chain, animal products represent less than 1% of daily lead intake. The contamination of food during harvesting and processing may be the result of contact with industrial machinery and greases. Perhaps 30% of the total lead in canned goods is the result of prepacking steps.[51] In general, highly processed foods have greater lead content than foods that have less processing. For example, dry milk has more lead than fresh milk. In making butter, the lead in milk tends to go into the buttermilk.[52] Also, foods can absorb lead from water during preparation.[53]

The greatest potential source of lead in food affecting large numbers of people is lead solder in cans used for packing. In a careful study of canned fish, Settle and Patterson[1] have shown that processing increases lead levels in tuna by a factor of 20 over raw levels and canning within lead solder further increases lead levels by a factor of 40,000 above natural levels. This contamination is not only from the leaching of lead from the solder on the wall but also from the splashing of molten solder droplets into the can during manufacturing. The storage of acidic foods in open cans further increases the contamination.[54]

There are several other sources of lead in food. For example, the addition of lead in the kitchen by fallout from urban dusts may be important. The use of lead-containing glazes in pottery for use with food is also a well-recognized although sporadic problem, which can introduce enough lead to cause overt poisoning. Fortunately, there have been recent decreases in the lead content in foods. The average American diet now contains significantly less lead than it did a decade ago,[55] in part because of changes in canning practices, and due to the lowering of the lead content of motor fuels resulting in decreased airborne lead levels. For example, a substantial reduction in the lead content of infant foods was largely the result of changing to lead-free cans and glass containers. Models of daily intake from multiple foods have shown infant dietary lead dropping from 45 to 50 µg/d in 1978 to 13 µg/d in 1985. The decline of lead in adult food, from over 250 µg/d in pre-1975 to less than 50 µg/d at present, have been achieved by the concerted efforts of the U.S. Food and Drug Administration, the National Food Processors Association, and the Can Manufacturers Institute.[56] These trends can be expected to continue.

B. UPTAKE BY HUMANS

The physiological portals by which lead enters the body are discussed in this section. Only inhalation and ingestion are reviewed; percutaneous absorption is negligible, except for organic lead.[57]

1. Inhalation

Of the lead inhaled, about 40 to 50% is deposited and absorbed. Most of the rest is exhaled, and a small fraction is trapped in the upper respiratory tract and later swallowed. The particle size of the aerosol is the primary determinant of its fate. Particles larger than 1 to 2 µm tend to be trapped by the upper respiratory tract, but smaller particles are more efficiently deposited in the lower lung and absorbed. There is extensive information about the relationship between airborne lead exposure and human blood lead response from epidemiological, inhalation, and tracer studies with radioactive and stable isotopes of lead.[58]

2. Ingestion

Approximately 5 to 15% of ingested lead is absorbed by adults. This lead includes not only contaminants from food and beverage, but also any inhaled particles trapped in the airways which are eventually swallowed. Experiments with human volunteers who consumed diets enriched with lead or a tracer have demonstrated that endogenous fecal excretion is a minor route for lead, accounting for about 5% of the lead in feces. Higher gut absorption rates approaching 50% have been observed in infants[59] and also in adults when the lead is administered while fasting.[60] The chemical form of the lead, e.g., chloride, sulfide, alginate, nitrate, carbonate, or acetate, seems to have no effect on absorption rates. Dietary deficiencies of calcium, iron, zinc, copper, and vitamin D have been shown to enhance lead absorption, as does dietary lactate, while dietary fiber and phytate tend to reduce absorption.[61] Most of this information is from the study of animals, but may be applicable to humans.

IV. SUMMARY

This chapter has reviewed various aspects of the occurrence of lead in the environment. The material is present in two categories, from the perspective of environmental flows and from the perspective of human exposure, representing two separate parts of the chapter.

In the first category, attention is focused on airborne lead since the atmosphere is the most important compartment of the biogeochemical cycle of lead which influences humans. Global anthropogenic emissions of lead to the atmosphere are estimated to be 449×10^6 kg/year based on 1975 data; in contrast, two different investigators have estimated emissions from natural sources to be 2×10^6 kg/year and 18.6×10^6 kg/year world-wide.

Most of the lead emitted to the atmosphere is the form of particulate matter, although the few percent of airborne lead emitted in the vapor phase can play a significant role in environmental cycling. Vapor phase emissions are generally alkyllead compounds derived from tetraalkyllead added to gasoline; particulate emissions, which depend on the source type, included lead halides, oxides, carbonates, sulfates, and other compounds.

Airborne lead concentrations are typically 1,000 to 10,000 ng/m³ in urban areas, 100 to 1000 ng/m³ in rural areas, and <1 to 100 ng/m³ in remote regions. Most of this lead is associated with submicron particles; size distributions reported in the literature generally have a mass median aerodynamic diameter in the range 0.5 to 1 μm.

Lead may be removed from the atmosphere by wet or dry deposition. Concentrations of lead in rainwater have a median value of 44 μg/l in urban areas, 12 μg/l in rural areas, and 0.09 μg/l in remote regions, according to one literature review. Dry deposition fluxes are highly variable, although published studies show dry deposition velocities averaging about 0.3 cm/s. The relative importance of wet and dry deposition in removing lead from the atmosphere is also highly variable, with published estimates of dry deposition/total deposition ranging from 10 to 60%.

In the second part of the chapter, contamination of the human environment by lead is discussed. Soil and dust may contain high concentrations of lead, particularly in urban areas and near certain industrial activities. This excessive lead may enter humans by direct ingestion and by contamination of foods. Further additions of lead to the food supply may occur during processing, packaging, and cooking. Lead, unlike mercury, does not increase in concentration from the lower to higher levels of the food chain; there is no bio-amplification.

The lead content of drinking water, although usually quite low at the source and in treatment plants, can be significantly elevated at the tap by corrosion products from lead solder, lead pipes, and other fixtures which contain lead. In some locations, water storage cisterns made of lead metal or painted on the inside with leaded paint may result in excessive concentrations. In cases where there is soft or acidic water exposed to lead, drinking water may be the most important source of daily lead intake.

The efficiency of uptake of environmental lead by humans depends on the route of exposure and a number of host factors. In general, respiration of submicron particles can result in absorption rates of nearly 50%. Absorption of ingested lead from food, water, soil, or dust approaches 10%. However, absorption rates may be several times greater when the lead is ingested by infants or by individuals with nutritional deficiencies.

REFERENCES

1. **Settle, D. M. and Patterson, C. C.,** Lead in albacore: guide to lead pollution in Americans, *Science,* 207, 1167—1176, 1980.
2. **Pacyna, J. M.,** Atmospheric trace elements from natural and anthropogenic sources, in *Toxic Metals in the Atmosphere,* Nriagu, J. O. and Davidson, C. I., Eds., John Wiley & Sons, New York, 1986, 33—52.
3. **Nriagu, J. O.,** Global inventory of natural and anthropogenic emissions of trace metals to the atmosphere, *Nature,* 279, 409—411, 1979.
4. **Shen, G. T. and Boyle, E. A.,** Lead in corals: reconstruction of historical industrial fluxes to the surface ocean, *Earth Planetary Sci. Lett.,* 82, 289—304, 1987.
5. **Radojevic, M. and Harrison, R. M.,** Concentrations and pathways of organolead compounds in the environment: a review, *Sci. Total Environ.,* 59, 157—180, 1987.
6. **Hewitt, C. N. and Harrison, R. M.,** Atmospheric concentrations and chemistry of alkyllead compounds and environmental alkylation of lead, *Environ. Sci. Tech.,* 21, 260—266, 1987.
7. **Nriagu, J. O.,** Lead in the atmosphere, in *The Biogeochemistry of Lead in the Environment,* Vol. 1, Nriagu, J. O., Ed., Elsevier, Amsterdam, 1978, 137—183.
8. **Wiersma, G. B. and Davidson, C. I.,** Airborne concentrations of trace elements in remote areas, in *Toxic Metals in the Atmosphere,* Nriagu, J. O. and Davidson, C. I., Eds., John Wiley & Sons, New York, 1986, 201—266.
9. U.S. Environmental Protection Agency, Air Quality Criteria for Lead, EPA/600/8-83/028bF, June 1986, chapt. 5 and 7.
10. **Patterson, C. C.,** An alternative perspective — lead pollution in the human environment: origin, extent, and significance, in *Lead in the Human Environment,* National Academy Press, Washington, D.C., 1980, 265—249.

11. **Boutron, C. F.,** Atmospheric lead, cadmium, mercury, and arsenic in Antarctic and Greenland: recent snow and ancient ice, SCOPE/UNEP Metals Cycling Workshop, New Delhi, India, February 16—20, 1987.
12. **Davidson, C. I. and Osborn, J. F.,** The sizes of airborne trace metal containing particles, in *Toxic Metals in the Atmosphere,* Nriagu, J. O. and Davidson, C. I., Eds., John Wiley & Sons, New York, 1986, 355.
13. **Esmen, N. A.,** An iterative impactor data analysis method, paper presented at the 51st Colloid and Interface Science Symp., Grand Island, NY, June 19—22, 1977.
14. **Raabe, O. G.,** A general method for fitting size distributions to multi-component aerosol data using weighted least-squares, *Environ. Sci. Technol.,* 12, 1162—1167, 1978.
15. **Hidy, G. M.,** Characterization of Aerosols in California, Final Rep., Air Resources Board Contract No. 358, Science Center, Rockwell International, Tulsa, OK, 1974.
16. **Davidson, C. I.,** The deposition of trace metal-containing particles in the Los Angeles area, *Powder Technol.,* 18, 117—126, 1977.
17. **Milford, J. B. and Davidson, C. I.,** The sizes of particulate trace elements in the atmosphere: a review, *J. Air Pollut. Control Assoc.,* 35, 1249—1260, 1985.
18. **Galloway, N. J., Thornton, J. D., Norton, S. A., Volchok, H. L., and McLean, R. A. N.,** Trace metals in atmospheric deposition: a review and assessment, *Atmos. Environ.,* 16, 1677—1700, 1982.
19. **Barrie, L. A., Lindberg, S. E., Chan, W. H., Ross, H. B., Arimoto, R., and Church, T. M.,** On the concentration of trace metals in precipitation, *Atmos. Environ.,* 21, 1133—1135, 1987.
20. **Gatz, D. F. and Chu, L. C.,** Metal solubility in atmospheric deposition, in *Toxic Metals in the Atmosphere,* Nriagu, J. O. and Davidson, C. I., Eds., John Wiley & Sons, New York, 1986, 391.
21. **Lazrus, A. L., Lorange, E., and Lodge, J. P., Jr.,** Lead and other metal ions in United States precipitation, *Environ. Sci. Tech.,* 4, 55—58, 1970.
22. **Slinn, W. G. N.,** Precipitation scavenging, in Atmospheric Sciences and Power Production, Randerson, D., Ed., Technical Information Center, U.S. Department of Energy, Oak Ridge, TN, 1984.
23. **Cawse, P. A.,** A survey of atmospheric trace elements in the U.K. (1972—73), AERE-R7669, Environ. Med. Sci. Div., AERE Harwell, Oxfordshire, England, 1974.
24. **Gatz, D. F.,** Wet deposition estimates using scavenging ratios, in Proc. First Specialty Symp. on Atmospheric Contributions to the Chemistry of Lake Water, International Association for Great Lakes Research, 28 Sep.—1 Oct., 1975, 21—29.
25. **Davidson, C. I., Santhanam, S., Fortmann, R. C., and Olson, M. P.,** Atmospheric transport and deposition of trace elements onto the Greenland Ice Sheet, *Atmos. Environ.,* 19, 2065—2081, 1985.
26. **Davidson, C. I. and Wu, Y. L.,** Dry deposition of particles and vapors, in *Acid Precipitation, Sources, Deposition, and Canopy Interactions,* Vol. 3, Adriano, D. C., Ed., Springer-Verlag, New York, 1990.
27. **Milford, J. B. and Davidson, C. I.,** The sizes of particulate sulfate and nitrate in the atmosphere — a review, *J. Air. Pollut. Control Assoc.,* 35, 125—134, 1987.
28. **Pripic-Majic, D., Meczner, J., Telisman, S., and Kersanc, A.,** Biological monitoring of lead effects in a smelter community before and after emission controls, *Sci. Total Environ.,* 32, 277—288, 1984.
29. **Page, A. and Ganje, T.,** Accumulation of lead in soils for regions of high and low motor vehicle traffic density, *Environ. Sci. Technol.,* 4, 140—142, 1970.
30. **Meilke, H., Anderson, K., Berry, P., Mielke, R., Chaney, R., and Leech, M.,** Lead concentrations in inner-city soils as a factor in child lead problem, *Am. J. Publ. Health,* 73, 1366—1369, 1983.
31. **Duggan, M. and Inskip, M.,** Childhood exposure to lead in surface dust and soil: a community health problem, *Public Health Rev.,* 13, 1—54, 1985.
32. **Spittler, T. and Feder, W.,** A study of soil contamination and plant lead uptake in Boston urban gardens, *Commun. Soil Sci. Plant Anal.,* 10, 1195—1210, 1979.
33. **Clausing, P., Brunekreef, B., and Van Wijnen, J.,** A method of estimating soil ingestion by children, *Int. Arch. Occup. Environ. Health,* 59, 73—82, 1987.
34. **Brunekreef, B., Veenstra, S., Biersteker, W., and Boley, J.,** The Arnhem lead study, *Environ. Res.,* 25, 441—448, 1981.
35. **Dong, A., Chesters, G., and Smisiman, G.,** Metal composition of soil, sediments, and urban dust and dirt samples from the Menomonee River watershed, Wisconsin, *Water Air Soil Pollut.,* 22, 257—275, 1984.
36. **Duggan, M., Inskip, M., Rundle, S., and Moorcroft, J.,** Lead in playground dust and on the hands of schoolchildren, *Sci. Total Environ.,* 44, 65—79, 1985.
37. **Rabinowitz, M., Leviton, A., Needleman, H., Bellinger, D., and Waternaux, C.,** Environmental correlates of infant blood lead levels, *Environ. Res.,* 38, 96—107, 1985.
38. **Charney, E., Kessler, B., Farfel, M., and Jackson, D.,** Childhood lead poisoning: a controlled trial of the effect of dust-control measures on blood lead levels, *N. Engl. J. Med.,* 309, 1089—1093, 1983.
39. **Moore, M. R.,** Lead in drinking water in soft water areas — health hazards, *Sci. Total Environ.,* 7, 109—115, 1977.
40. **Levin, R.,** Reducing Lead in drinking water, U.S. Environmental Protection Agency, Office of Policy, Planning and Evaluation, Washington, 1986.

41. **Sharrett, A. R., Carter, A. P., Orheim, R. M., and Feinleib, M.,** Daily intake of lead, cadmium, copper and zinc from drinking water, *Environ. Res.,* 28, 456—475, 1982.
42. **Lyon, T. D. B. and Lenihan, J. M. A.,** Corrosion in solder-jointed copper tubes resulting in lead contamination of drinking water, *Brit. Corros. J.,* 12, 41, 1977.
43. **Sherlock, J., Smart, G., Forbes, G. I., Moore, M. R., Patterson, W. J., Richards, W. N., and Wilson, T. S.,** Assessment of lead intakes and dose-response for a population exposed to plumbosolvent water supply, *Human Toxicol.,* 1, 115—122, 1982.
44. **Samuels, E. R. and Meranger, J. C.,** Preliminary studies on the leaching of some metals from kitchen faucets, *Water Res.,* 18, 75—80, 1984.
45. **Karalekas, P. C., Ryan, C. E., and Taylor, F. B.,** Control of lead, copper, and iron pipe corrosion in Boston, *J. Am. Water Works Assoc.,* 75, 92—95, 1983.
46. **Sherlock, J. C., Ashby, D., Delves, H. I., Forbes, G. I., Moore, M. R., Patterson, W. J., Pocock, S. J., Quin, M. J., Richards, W. N., and Wilson, T. S.,** Reduction in exposure to lead from drinking water and its effect on blood lead concentrations, *Human Toxicol.,* 3, 383—392, 1984.
47. **Gloag, D.,** Sources of lead pollution, *Brit. Med. J.,* 282(6257), 41—44, 1981.
48. Nutrition Foundation, Assessment of the safety of lead and lead salts in food, Washington, DC, The Nutrition Foundation, 1982.
49. **Gallacher, J. E., Elwood, P. C., Phillips, K. M., Davies, B. E., Ginnever, R. C., Toothill, C., and Jones, D. T.,** Vegetable consumption and blood lead concentrations, *J. Epidemiol. Communit. Health,* 38, 173—176, 1984.
50. **Chamberlain, A. C.,** Fallout of lead and uptake by crops, *Atmos. Environ.,* 17, 693—706, 1983.
51. **Wolnik, K. A., Fricke, F. L., Capar, S. G., Braude, G. L., Meyer, M. W., Satzger, R. D., and Bonnin, E.,** Elements in raw agricultural crops in the United States, *J. Agric. Food Chem.,* 31, 1240—1244, 1983.
52. **Hayashi, M., Ito, O., Ohira, S., and Akazawa, Y.,** Transfer of lead and cadmium from cow milk to butter, *Bull. Environ. Cont. Toxicol.,* 29, 658—664, 1982.
53. **Moore, M. R., Hughes, M. A., and Goldberg, A. J.,** Lead absorption in man from dietary sources, *Int. Arch. Occup. Environ. Health,* 44, 81—90, 1979.
54. **Capar, S. G.,** Changes in the lead content of foods stored in their open cans, *J. Food Safety,* 1, 241—245, 1978.
55. **Flegel, A. R., Smith, D. R., and Elias, R. W.,** Lead contamination in food, in *Environmental Food Contamination, Advances in Environmental Science and Technology,* Nriagu, J. O. and Simmons, M. S., Eds., J. Wiley & Sons, New York, 1988.
56. **Jelinek, C. F.,** Levels of lead in the United States food supply, *J. Assoc. Off. Anal. Chem.,* 4, 942—6, 1982.
57. **Moore, M., Meredith, P., Watson, W., Sumner, D., Taylor, M., and Goldberg, A.,** The percutaneous absorption of lead-203 in humans from cosmetic preparations containing lead acetate as assessed by whole-body counting and other techniques, *Food Cosmet. Toxicol.,* 18, 399—405, 1980.
58. **Chamberlain, A.,** Effect of airborne lead on blood lead, *Atmos. Environ.,* 17, 677—692, 1983.
59. **Ziegler, E., Edwards, B., Jensen, R., Mahaffey, K., and Fomon, S.,** Absorption and retention of lead by infants, *Pediatr. Res.,* 12, 29—34, 1978.
60. **Rabinowitz, M., Wetherill, G., and Kopple, J.,** Effect of food intake and fasting on gastrointestinal lead absorption in humans, *Am. J. Clin. Nutr.,* 33, 1784—1788, 1980.
61. **Mahaffey, K.,** Nutritional factors in lead poisoning, *Nutr. Rev.,* 39, 3537—7, 1981.
62. **NEDS,** National Emissions Data System [data base], Nationwide [lead] emissions reports [for] 1979 and 1980 — United States, Research Triangle Park, NC: U.S. Environmental Protection Agency, Office of Air Quality Planning and Standards, Printout, available for inspection at: U.S. Environmental Protection Agency, Environmental Criteria and Assessment Office, Research Triangle Park, NC, 1982.
63. NAPS, National Air Pollution Surveillance Reports, 1971—1976. Air Pollution Control Directorate, Ottawa, Canada, report series EPS-5-AP.
64. **Roels, H. A., Buchet, J.-P., Lauwerys, R. R., Bruaux, P., Claeys-Thoreau, F., Lafontaine, A., and Verduyn, G.,** Exposure to lead by the oral and the pulmonary routes of children living in the vicinity of a primary lead smelter, *Environ. Res.,* 22, 81—94, 1980.
65. **Facchetti, S. and Geiss, F.,** Isotopic lead experiment: status report, Luxembourg, Commission of the European Communities, publication no. EUR 8352 EN, 1982.
66. **El-Shobokshy, M. S.,** A preliminary analysis of the inhalable particulate lead in the ambient atmosphere of the city of Riyadh, Saudi Arabia, *Atmos. Environ.,* 18, 2125—2130, 1984.
67. **Duce, R. A., Hoffman, G. L., and Zoller, W. H.,** Atmospheric trace metals at remote northern and southern hemisphere sites: pollution or natural?, *Science,* 187, 59—61, 1975.
68. **Hudson, J. L., Stukel, J. J., and Solomon, R. L.,** Measurement of the ambient lead concentration in the vicinity of Urbana-Champaign, Illinois, *Atmos. Environ.,* 9, 1000—1006, 1975.
69. **Lindberg, S. E. and Harriss, R. C.,** Water and acid soluble trace metals in atmospheric particles, *J. Geophys. Res.,* 88, 5091—5100, 1983.

70. **Chow, T. J., Earl, J. L., and Snyder, C. B.,** Lead aerosol baseline: concentration at White Mountain and Laguna Mountain, California, *Science,* 178, 401—402, 1972.
71. **Elias, R. W. and Davidson, C.,** Mechanisms of trace element deposition from the free atmosphere to surfaces in a remote High Sierra canyon, *Atmos. Environ.,* 14, 1427—1432, 1980.
72. **Davidson, C. I., Goold, W. D., Mathison, T. P., Wiersma, G. B., Brown, K. W., and Reilly, M. T.,** Airborne trace elements in Great Smoky Mountains, Olympic, and Glacier National Parks, *Environ. Sci. Technol.,* 19, 17—35, 1985.
73. **Maenhaut, W., Zoller, W. H., Duce, R. A., and Hoffman, G. L.,** Concentration and size distribution of particulate trace elements in the south polar atmosphere, *J. Geophys. Res.,* 84, 2421—2431, 1979.
74. **Murozumi, M., Chow, T. J., and Patterson, C.,** Chemical concentrations of pollutant lead aerosols, terrestrial dusts and sea salts in Greenland and Antarctic snow strata, *Geochim. Cosmochim. Acta,* 33, 1247—1294, 1969.
75. **Heidam, N. Z.,** Studies of the aerosol in the Greenland atmosphere; SAGA I: results 1979—1980, Roskilde, Denmark: Riso National Laboratory; National Agency of Environmental Protection report no. MST-LUFT-A—73, NTIS, Springfield, VA; DE84-751004, 1983.
76. **Davidson, C. I., Chu, L., Grimm, T. C., Nasta, M. A., and Qamoos, M. P.,** Wet and dry deposition of trace elements onto the Greenland ice sheet, *Atmos. Environ.,* 15, 1429—1437, 1981.
77. **Settle, D. M. and Patterson, C. C.,** Magnitude and sources of precipitation and dry deposition fluxes of industrial and natural leads to the North Pacific at Enewetak, *J. Geophys. Res.,* 87, 8857—8869, 1982.
78. **Davidson, C. I., Grimm, T. C., and Nasta, M. A.,** Airborne lead and other elements derived from local fires in the Himalayas, *Science,* 214, 1344—1346, 1981.
79. **Duce, R. A., Ray, B. J., Hoffman, G. L., and Walsh, P. R.,** Trace metal concentration as a function of particle size in marine aerosols from Bermuda, *Geophys. Res. Lett.,* 3, 339—346, 1976.
80. **Dzubay, T. G., Stevens, R. K., and Haagenson, P. L.,** Composition and origins of aerosol at a forested mountain in Soviet Georgia, *Environ. Sci. Technol.,* 18, 873—883, 1984.
81. **Shirahata, H., Elias, R. W., Patterson, C. C., and Koide, M.,** Chronological variations in concentrations and isotopic compositions of anthropogenic atmospheric lead in sediments of a remote subalpine pond, *Geochim. Cosmochim. Acta,* 44, 149—162, 1980.
82. **Edgington, D. N. and Robbins, J. A.,** Records of lead deposition in Lake Michigan sediments since 1800, *Environ. Sci. Technol.,* 10, 266—274, 1976.
83. **Ng, A. and Patterson, C.,** Changes of lead and barium with time in California off-shore basin sediments, *Geochim. Cosmochim. Acta,* 46, 2307—2321, 1982.
84. **Rolfe, G. L.,** Lead distribution in tree rings, *For. Sci.,* 20, 283—286, 1974.
85. **Cawse, P. A.,** A survey of atmospheric trace elements in the U. K.: results for 1974, AERE-R 8038, Environmental and Medical Sciences Division, AERE, Harwell, Oxfordshire, England, 1975.
86. **Cawse, P. A.,** A survey of atmospheric trace elements in the U.K.: results for 1975, AERE-R 8398, Environmental and Medical Sciences Division, AERE, Harwell, Oxfordshire, England, 1976.
87. **Cawse, P. A.,** A survey of atmospheric trace elements in the U.K.: results for 1976, AERE-R-8869, Environmental and Medical Sciences Division, AERE, Harwell, Oxfordshire, England, 1977.
88. **Davidson, C. I. and Elias, R. W.,** Dry deposition and resuspension of trace elements in the remote High Sierra, *Geophys. Res. Lett.,* 9, 91—93, 1982.
89. **Dedeurwaerder, H. L., Dehairs, F. A., Decadt, G. G., and Baeyens, W. F.,** Estimates of dry and wet deposition and resuspension fluxes of several trace metals in the Southern Bight of the North Sea, in *Precipitation Scavenging, Dry Deposition, and Resuspension,* Pruppacher, H. R., Semonin, R. G., and Slinn, W. G. N., Eds., Elsevier, New York, 1983, 1219—1231.
90. **Dolske, D. A. and Sievering, H.,** Trace element loading of southern Lake Michigan by dry deposition of atmospheric aerosol, *Water, Air, Soil Pollut.,* 12, 485—502, 1979.
91. **Dovland, H. and Eliassen, A.,** Dry deposition on a snow surface, *Atmos. Environ.,* 10, 783—785, 1976.
92. **El-Shobokshy, M. S.,** The dependence of airborne particulate deposition on atmospheric stability and surface conditions, *Atmos. Environ.,* 19, 1191—1197, 1985.
93. **Hofken, K. D., Meixner, R. X., and Ehhalt, D. H.,** Deposition of atmospheric trace constituents onto different natural surfaces, in *Precipitation Scavenging, Dry Deposition, and Resuspension,* Pruppacher, H. R., Semonin, R. G., and Slinn, W. G. N., Eds., Elsevier, New York, 1983, 825—835.
94. **Lindberg, S. E. and Harriss, R. C.,** The role of atmospheric deposition in an eastern U.S. deciduous forest, *Water, Air, Soil Pollut.,* 16, 13—31, 1981.
95. **Little, P. J. and Wiffen, R. D.,** Emission and deposition of petrol engine exhaust Pb. I. Deposition of exhaust Pb to plant and soil surfaces, *Atmos. Environ.,* 11, 437—447, 1977.
96. **Pattenden, N. J., Branson, J. R., and Fisher, E. M. R.,** Trace element measurements in wet and dry deposition and airborne particulate at an urban site, in *Deposition of Atmospheric Pollutants,* Georgii, H. W. and Pankrath, J., Eds., Reidel, Dordrecht, Netherlands, 173—184, 1982.
97. **Peirson, D. H., Cawse, P. A., Salmon, L., and Cambray, R. S.,** Trace elements in the atmospheric environment, *Nature,* 241, 252—256, 1973.

98. **Rohbock, E.,** Atmospheric removal of airborne metals by wet and dry deposition, in *Deposition of Atmospheric Pollutants,* Georgii, H. W. and Pankrath, J., Eds., Reidel, Dordrecht, Netherlands, 1982, 159—171.
99. **Servant, J.,** Deposition of atmospheric lead particles to natural surfaces in field experiments, Symp. Atmosphere-Surface Exchange of Particulate and Gaseous Pollutants, September 4—6, 1974, National Technical Information Services, U.S. Department of Commerce, Springfield, VA, 1976, 87—95.
100. **Sievering, H., Dave, M., McCoy, P., and Sutton, N.,** Deposition of sulfate during stable atmospheric transport over Lake Michigan, *Atmos. Environ.,* 13, 1717—1719, 1979.
101. **Young, J. A.,** The rates of change of pollutant concentrations downwind of St. Louis, Pacific Northwest Laboratory Annual Report for 1977 to the DOE Assistant Secretary for Environment, Atmospheric Sciences, PNL-2500PT3 Battelle, Pacific Northwest Laboratory, Richland, WA, 1978.
102. **Davidson, C. I., Miller, J. M., and Pleskow, M. A.,** The influence of surface structure on predicted particle dry deposition to natural grass canopies, *Water, Air, Soil Pollut.,* 18, 25—43, 1982.
103. **Slinn, W. G. N.,** Predictions for particle deposition to vegetative canopies, *Atmos. Environ.,* 16, 1785—1794, 1982.
104. **Ibrahim, M., Barrie, L. A., and Fanaki, F.,** An experimental and theoretical investigation of the dry deposition of particles to snow, pine trees and artificial collectors, *Atmos. Environ.,* 17, 781—788, 1983.
105. **Williams, R. M.,** A model for the dry deposition of particles to natural water surfaces, *Atmos. Environ.,* 16, 1933—1938, 1982.
106. **Slinn, S. A. and Slinn, W. G. N.,** Modeling of atmospheric particulate deposition to natural waters, in *Atmospheric Pollutants in Natural Waters,* Eisenreich, S. J., Ed., Ann Arbor Science, Ann Arbor, MI, 1981, 23—53.

Section II: Neurobiological Factors

Chapter 5

NEUROLOGICAL PERSPECTIVE ON LEAD TOXICITY

Ellen K. Silbergeld

TABLE OF CONTENTS

I.	Introduction	90
II.	Neurotoxic Effects of Lead — The Whole Organism	91
	A. Experimental Studies	92
III.	Neuropathology of Lead	92
	A. Regional Localization of Lead in Brain	93
IV.	Neurophysiology	94
V.	Neurochemistry	95
VI.	Directions for Research	96
VII.	Hypotheses as to the Neurotoxic Mechanisms of Lead	97
References		98

I. INTRODUCTION

Over the past 15 years, a large literature has been published on the neurotoxic effects of lead in children and adults environmentally and occupationally who are exposed to a broad range of lead levels in the environment. These studies have investigated neurobehavioral, neuropathological, neurophysiological, and neurochemical manifestations of lead toxicity over a range of doses. Persons were studied with exposures over the entire period of neurological development, from *in utero* to adulthood (although not into aging). Over the same period, there have been a number of experimental studies exploring cellular and subcellular mechanisms involved in neurotoxicity. Taken together, these studies have been instrumental in redefining what are "acceptable" levels of lead exposure, with a consensus that the neurotoxic effects observed are potentially irreversible.[3,60]

Despite the relatively large amount of clinical and experimental data, there is as yet no unifying hypothesis of the fundamental neurobiological mechanisms of lead toxicity. This lack of an integrating hypothesis contrasts sharply with the convergence developing from the clinical studies as to the specific nature of lead-induced deficits in neurologic function. It is this contrast that stimulated this chapter.

Several hypotheses have been proposed on the molecular mechanisms of lead neurotoxicity. In general, these fall into the following three categories: (1) lead inhibits cell energetics at the level of cytochrome synthesis or by binding to ATP;[27] (2) lead inhibits calcium binding, release, storage, or the responsiveness of calcium-dependent second messenger systems or calciotropic hormones;[30,48,73,88] and (3) lead inhibits cell heme synthesis at the level of δ-amino-levulinic acid dehydrase, resulting in depletion of heme-dependent proteins (including cytochromes)[73] and stimulates the production of heme precursors such as δ-amino-levulinic acid that may be neuroactive.[56,98]

The limitation of these hypotheses is that none of them is sufficient as an integrated hypothesis for lead neurotoxicity at either the organ or organism level. None of these hypothesized cell-level and organelle-level mechanisms are specific to the nervous system. Almost all cells require energy, regulate calcium, and synthesize heme. Thus, effects of lead on these processes do not explain the specifically neurotoxic effects of lead or the apparently specific expressions of lead neurotoxicity. Moreover, with few exceptions, the dose-response for these effects is unknown, so that comparisons with the human dosimetry cannot be made. The neurotoxicity of lead may be due to the sensitivity of the CNS to cell injury and to the greater significance of small perturbations in CNS cell function compared to other organs. If this is the case, then lead can be considered a nonspecific toxin. The specifically neurotoxic expression of low level exposure could reflect the sensitivity of the nervous system more than any specificity of the biological actions of lead. However, this hypothesis would imply that all toxins have their earliest observable effects as neurotoxins, which is not likely to be the case.[58] This hypothesis would not explain the relative specificity of lead neurotoxicity.

The accumulating evidence from the clinical studies indicated that the effects of lead in children are relatively specific and reproducible from study to study. Meta-analysis of the major epidemiological studies conducted in the U.S. and Europe indicate a remarkable congruence of outcomes and overall dose-response across different socioeconomic and cultural groups.[11a] It is the consistency of the neurotoxic effects which challenges research strategies to determine the biological substrates of these altered behaviors and functions.

This chapter reviews the neurobiological evidence for lead neurotoxicity and proposes areas for research to resolve the disparity between the resolution emerging from the clinical studies and the lack of convergence in basic research. It is a conclusion of this chapter that this lack of correspondence results in large part from the failure to examine integrative mechanisms of lead neurotoxicity in appropriate animal models.

II. NEUROTOXIC EFFECTS OF LEAD—THE WHOLE ORGANISM

The neurotoxic effects of lead in the infant and young child involve deficits in cognitive function and delays in achieving neurodevelopmental milestones.[11a] In school-age children, lead-induced deficits can be more quantitatively defined using the metrics of standardized tests of intelligence. Perino and Ernhart,[71] in one of the first studies of relatively low-level lead neurotoxicity, reported that children with relatively high blood lead levels (in excess of 40 µg/dl), intellectual performance on the McCarthy scales was affected across the range of tests, including cognitive, verbal, perceptual, and motor function. In school-age children, qualitative assessments of behavior and school performance, using validated inventories for evaluation developed by Needleman and Rutter, have consistently found dose-related decrements in behavior, measured as increased frequency of inappropriate or uncontrolled behaviors, many of which are similar to constituents of the Connors' scales for assessing Attention Deficit Disorders.[59,79,117] The Bayley scales, which have been applied to young children, yield less definitive data for the purposes of determining specific areas of deficit; however, effects of lead have been reported in several studies using this measure.[11,11a,83]

There are also studies suggesting consistent effects of lead on neurosensory processing. Several studies[59,115,118] have found decreased performance in tests requiring appropriately timed reactions. In young children, Needleman et al.[59] found evidence for altered auditory processing (deficits in the Seashore rhythm test) which may correspond with decreased auditory sensitivity as reported by Schwartz and Otto.[85] Odenbro et al.[63] found decrements in visuomotor performance as a function of blood levels in children aged 3 to 6 years. Thatcher et al.[110] also reported alterations in auditory evoked response patterns in children whose lead exposure was detected by hair lead analyses. Recent studies in Cincinnati have found changes in the proprioceptive pathways involved in balance to be altered by lead exposure in children.[12,25]

Electrophysiologic studies of children have been relatively limited. Peripheral nerve conduction velocity studies, when reanalyzed, seem to indicate possible threshold for effects of lead or at least an inability to detect effects below 30 µg/dl blood lead.[84] In the CNS, Burchfiel et al.[14a] investigating a sample of the Needleman cohort, reported changes in the EEG power function related to lead dose.[14]

The most extensive studies on lead and neurophysiological outcome were undertaken by the EPA in a cross-sectional and long-term follow-up study. Sensory evoked potentials indicated lead effects, but no effect on EEG power spectra were found.[68] At present, however, there are insufficient data and limitations on the interpretation of available data to determine if these results suggest a specific electrophysiological locus for lead effects.

In adults, fewer CNS-related symptoms are reported, so that it is less easy to assess the consistency of reported neurotoxic effects. In a recent study of neurologic and psychiatric signs in lead-exposed workers, Schottenfeld and Cullen[82] reported a spectrum of nonspecific symptoms including fatigue, depression, irritability, decreased libido, and headache. Neuropsychological testing in these workers, with blood lead levels in excess of 40 µg/dl, revealed abnormalities in visuomotor coordination and fine motor control. In workers with chronic exposure, deficits in short-term memory were also found. Continuous performance tasks and tests of short-term memory involving discriminant perception appear to be sensitive markers of CNS toxicity.[8,112] Valciukas[112] also found that tests requiring visual acuity were affected by lead. Somatosensory function (such as detection of vibration and changes in temperature) have been proposed as sensitive indicators of relatively low level occupational exposures.[8]

Some studies in occupational medicine have concluded that the peripheral nervous system is a more specific target for lead in adults.[86] Conduction velocity of motor and sensory nerves have been studied in lead-exposed cohorts, with generally, but not consistently, reported slowing of median motor nerve conduction velocity in persons with blood lead levels in the range of 30 to 60 μg/dl.[36,86]

A. EXPERIMENTAL STUDIES

The data from studies in animals are more difficult to integrate, in part because there has been less efforts directed towards standardization of exposures or comparability of endpoint measurements between investigations (see Cory-Slechta[22] and Rice[77] for recent critical reviews of this field). Different behavioral tasks have been applied and variable designs of exposure reduce the direct comparability across studies. Even in studies of ethological behaviors, or behaviors expressed without conditioning or training by animals, there has been considerable variation in results. Two so-called spontaneous behaviors are motor activity and aggression. Obviously, neither of these is a simple behavior, and experimental condition shape these behaviors as profoundly as do the manipulations introduced in conditioned behavior paradigms. The ability of lead to affect the level of motor activity (measured by a variety of instruments which generally count body displacement in the horizontal plane) is highly dependent upon nutritional status, the timing of exposure, and the age at which the animal is tested.[52,75] Spontaneous alternation in a radial arm maze, also an unshaped behavior,[65] is also reported to be altered by low level lead exposure.[16] Suggestions of increased activity, or failure to inhibit appropriate activity, are also found in studies where higher response rates and decreased interresponse intervals were observed in lead-exposed animals.[2,23,34,37,69,78,109,116]

Changes in aggressiveness have never been quantitatively described, only mentioned in two early studies as observations.[80,93] These observations have not been confirmed in more standardized studies of aggression in rodents.[26] Social behaviors in rats and primates have also been studied, and alterations in investigative behaviors reported in the absence of changes in overall activity.[24] Effects of lead on mother-infant behaviors are disrupted by lead,[10,15,47,57] although in some of the studies it is not clear as to which organism — mother or infant — was the source of behavioral disruption since both were exposed. This is not so different from many situations in human populations, where all members of the family unit are at risk for lead exposure; in such circumstances it makes little epistemological sense to ascribe effects to one entity alone.

The literature on conditioned responses and learning is more complicated. Behaviors elicited under fixed-interval schedules generally show increased rates of responding and perseveration.[21] Those behaviors which require extinction or inhibition seem to be sensitive to relatively low level lead exposure. However, in operant tasks, the details of schedule and reinforcement are important variables in the sensitivity of the test to detecting lead effects.[21]

One area in which some consistency of findings has been reported is the altered response of lead-exposed rodents to stimulant drugs (see review of Winder and Kitchen[114]). In several studies from different laboratories, rodents express marked alterations in response to D-amphetamine, ranging from so-called paradoxical decreases in motor activity[42,51,75,92,100] to attenuation of drug-elicited hyperactivity. These alterations were observed in rodents with baseline hyper-, hypo-, and normal activity levels.

III. NEUROPATHOLOGY OF LEAD

Lead at a relatively high dose has been associated with neuropathological effects on several cellular elements of the CNS, including hemorrhagic encephalopathy and loss of cell content[44,49,53] and reductions in particular brain regions such as hippocampus.[46] At relatively lower doses, the morphologic effects of lead on the CNS are less clearly detectable. In a comprehensive set of studies, Sundstrom et al.[105-107] found no evidence for brain edema, loss of myelination of CNS axons, or changes in myelin basic protein content in rats with blood lead levels below 30 µg/dl. Decreased myelination in CNS has been found at higher levels of exposure.[76,102] The earliest morphological effect of lead appears to be disruption of blood-brain barrier function. This is consistent with the hypothesis first proposed by Goldstein et al.[32] that the capillary endothelial cell is a specific target for low dose lead intoxication. Other studies reported that lead at a low dose reduces the density of synaptic formation and the extent of dendritic arborization when exposure occurs during neurodevelopment, but as with many of the neurotoxic effects observed in animal model, the concomitant or interactive effects of under nutrition need careful control.[52,105]

Studies of organotypic cell cultures exposed to lead suggest that glia may be selectively sensitive to lead.[30,113] Studies on cerebellar explants exposed to lead through the host organism have found histopathologic and physiological evidence for abnormal development of Purkinje cells;[64,70] unfortunately, comparable studies of brain tissue from other regions have not been done.

Specificity of neuropathologic lesions associated with lead exposure might be regional rather than specific to cell type. There have been attempts to determine a regional localization of lead toxicity; several researchers[17,44,72] have proposed that the hippocampus, which is a zinc-rich structure, is specifically sensitive to lead on the basis of morphologic changes in density of cell layers within that structure. However, these studies did not provide conclusive evidence, based upon a comprehensive examination of other regions of the brain to determine that pathophysiological effects were relatively localized. Averill and Needleman[7] examined synaptogenesis in the cortex in rats exposed to relatively low levels of lead through neurodevelopment; again, their examination was not comprehensive so it cannot be concluded that cortical development is particularly sensitive to lead. As discussed below, some of the neurochemical studies may be of more assistance in guiding further ultrastructural examination of the brain.

A. REGIONAL LOCALIZATION OF LEAD IN BRAIN

A standard approach to defining localization for toxic action is to determine concentrations of the toxin within proposed target areas. Studies on the localization of lead in the CNS have been insufficiently refined to determine if lead exposure results in specific accumulations within defined regions or cell types in the brain. Generally, lead levels are higher in hippocampus and cortex; however, as shown in Table 1, the relative enrichment of these areas is not overwhelming.[28,81] The effects of lead dosage have not been comprehensively determined. In one study, relatively high acute exposure increased lead levels with little difference in all areas examined (pons medulla, cerebellum, midbrain, cortex-striatum).[44] But the hippocampus and generally zinc-rich regions of the CNS appear to accumulate lead preferentially.[28]

At present, the mechanisms which are responsible for lead uptake and distribution in the CNS are not known. Since lead at low dose affects the blood-brain barrier function (see above), it is difficult to sort out whether the distribution of lead within the CNS reflects the relative vascularization of different brain regions (with relatively higher distribution of small capillary/neuronal contact of the pyramidal neurons of the hippocampus[62]) as opposed to a specific uptake and distribution process. Also, in many studies the brain was not perfused to remove blood so that measurements may have reflected blood lead concentrations as well as tissue lead concentrations.

Recent studies by Fowler et al.[55,66] suggest that there is a specific lead-binding protein, a modified or processed residue of α-2 microglobin, which is expressed in mammalian brain. This protein may function as a transport mechanism; it is found in all organs where lead-rich intranuclear inclusion bodies are formed.[55] Its relative distribution within brain regions is not yet known. Another possible set of distribution and compartmentation mechanisms

TABLE 1
Localization of Lead in Brain[a]

Brain region	Normal brain[b]	% increase[c]	ppm[d]
Cortex[e]	227		4.6
Hippocampus	487	590—1090[f]	4.9
Amygdala	302		
Thalamus	102		5.8
Striatum	336	690—840	2.3
Colliculus	232		3.5
Hypothalalmus	338		
Brain stem[g]	178	650—880	5.4
Cerebellum	295	450—1070	4.3

[a] Data are from rats, controls or exposed semichronically as adults.
[b] ng/gm dry wt.; data from Scheuhammer and Cherian.[81]
[c] Data are range of increases relative to control, from La-Fauconnier et al.[44]
[d] X-ray analysis data from fixed thin sections, from Cholewa et al.[19]
[e] Frontal cortex and rest of cortex combined.
[f] Described as "midbrain".
[g] Pons medulla.

for lead may be the calciotropic hormones and calcium-binding proteins, including calmodulin, osteocalcin, and osteonectin. These proteins, which are expressed in the brain particularly during development,[61] bind lead tightly.[29,35,73] (Pounds, 1988).

IV. NEUROPHYSIOLOGY

By its nature, neurophysiology might be expected to provide information on lead neurotoxicity at a higher level of integration. Lead affects the electrical phenomena of single cells, cellular processes taken as aggregates, and subanatomic regions of neurons. Single cells of the CNS and PNS have been studied with electrophysiological probes (see review by Audesirk and Audesirk[5]). Lead *in vitro* blocks presynaptic release of acetylcholine at the neuromuscular junction, sympathetic ganglion, and superior cervical ganglion. Lead also affects membrane properties of neurons.[4] Iontophoretic application of lead to Purkinje cells of the cerebellum blocked noradrenergic responses; superfusion of lead in the micromolar range also blocked norepinephrine induced decreases in cell firing.[108]

In vivo chronic exposure to lead has also been examined neurophysiologically in two preparations: a mollusc exposed *in vivo* and mammalian cerebellar explants grafted into the eye of an exposed host. In the mollusc, exposure to lead added to water at a concentration of 5 μM for 6 to 10 weeks (with a resulting circulating level of lead of about 200 $\mu g/dl$), affected several parameters of neuronal function, varying with cell type but commonly expressed as decreased spontaneous firing rates.[6]

Spinal cord neurons have been studied for responses to lead *in vitro*. In single interneurons, lead in a concentration of 10 to 100 μM reduced neuronal response to afferent stimulation and depressed dorsal root potentials.[101]

In animal models, effects of lead on the electrophysiological function of sets or networks of cells have also been reported. Lead affects sensory evoked responses (auditory and visual) acting at several levels in the processing pathway from peripheral receptors to the CNS (see review by Seppalainen[86]). Perfusion of isolated retinal tissue with a lead depressed rod photoreceptor response to light at doses as low as 1 μM.[99] At a higher level of integration, lead reduces conduction velocity down myelinated axons of the peripheral nervous system.[86]

Integrated electrophysiological measurements of the CNS have also been performed, to a more limited extent, on lead-exposed adults and children (previously discussed). Evoked potential studies of lead-exposed workers reported some changes of amplitude, hypothesized to reflect disinhibition in subcortical centers.[86] In animals, more substantial and in some instances apparently persistent effects on auditory and visual evoked potentials have been found.[114a]

These data indicate that lead clearly affects electrophysiological parameters in many cell types of the central and peripheral nervous systems. However, they do not provide an overall integrative hypothesis of mechanism. Most of the studies have invoked some of the neurochemical and biochemical hypotheses, which are discussed next, to explain their findings.[5,6]

V. NEUROCHEMISTRY

The biochemical analysis of nervous tissue from lead-exposed animals might be expected to provide insights on lead neurotoxicity. However, neurochemical studies of lead-exposed animals, while numerous,[6,87,94,114] do not provide a comprehensive or systemic analysis of brain chemistry or a systematic hypothesis for a transmitter-specific substrate for the effects of lead. The changes in biochemical measures have been reported on a regional, cellular, and transmitter-specific basis.

Studies on lead neurotoxicity in the peripheral nervous system, dating back to the elegant work of Kostial and Vouk in 1957,[40] have focused on cholinergic function in the ganglionic and neuromuscular junctions. These studies have demonstrated that lead inhibits the stimulated release of acetylcholine and increases the unstimulated, or resting, release of transmitter at these sites. Further investigations on the mechanisms of lead effects at cholinergic synapses have demonstrated that these effects on quantal release involve interactions with ionic mechanisms controlling transmitter release.[4,18,20,39,91]

For neuromuscular junction and superior cervical ganglion, the *in vitro* studies are consistent with the *in vivo* findings of decreased stimulus-dependent but not with resting transmitter release.[20,91] In this respect, the CNS may differ from the PNS: lead added *in vitro* did not reduce acetylcholine release from minces prepared from rat cortex.[18] Some neurochemical evidence indicates that lead exposure *in vitro* does inhibit CNS cholinergic function.[2,114] Very recent reports indicate that lead may interfere with *n*-methyl-*d*-aspartate (NMDA)-sensitive glutamatergic synapses in the hippocampus.[1] Given the hypothesized role of this pathway in memory and learning, this new finding should be further explored.

In the CNS, the neurochemical effects of lead appear to be region-specific, rather than transmitter-specific. Trabucchi et al.[33,47,50,51] have examined monoaminergic function in basal ganglia, cortex, and mesolimbic system; of these, the mesolimbic system is most sensitive to the effects of lead. A similar finding was reported by Lasley and Lane,[43] who found that lead exposure impaired receptor-mediated regulation of dopamine synthesis in nucleus accumbens but not in caudate. Dopamine levels were also decreased in ventral tegmentum, but not in nucleus accumbens. GABAergic neurochemistry has been examined in several brain regions.[50,94,95] The effects varied both with region and the parameter being investigated.

The data are insufficient, and the studies too variable in treatment and other parameters, to permit analysis of dose-response which might identify which pathway(s) are most sensitive to lead. In general, responses elicited at micromolar and submicromolar concentrations of lead are more likely to be physiologically relevant, given the range of concern for blood lead (e.g., in the range of 0.25 to 1.0 μM, or 5 to 20 μg/dl whole blood). While some attempts have been made to integrate some of these findings, no overall thesis has been proposed which explains all these findings. In addition, many of these findings are not consistently found. Depending upon dose timing, as well as time of measurement, these changes have been observed in both directions as well as no change at all. It is unclear as to whether these differences are due to inconsistencies in exposure or methods of measurement, or whether the neurochemical effects of lead are highly labile and variable due to interactions between treatment and such factors as nutrition.[52]

The neurochemical investigations have only rarely been undertaken in connection with an hypothesis concerning expression of whole-animal behavioral or physiological effects. The findings of altered drug responses in lead-treated animals stimulated research on monoaminergic pathways[51] and the increased sensitivity of lead treated animals to convulsant stimuli (pharmacological and electrical) has provoked some study of GABAergic function.[54,94,95] The results from these studies can be characterized as, at best, not inconsistent with the behavioral findings; they do not, however, provide a unifying or comprehensive set of data upon which to base an integrative hypothesis of neurochemical mechanisms associated with behavioral alterations of abnormal responses to drug challenge. Only one neurochemical study has been reported on lead-exposed children, in whom urinary excretion of catecholamine metabolites was reduced.[90]

In terms of EC_{50} (concentration required to reach 50% maximal effect), aspects of neurotransmission other than those related to neurotransmitter biochemistry may be more sensitive to lead, at least *in vitro*. Lead activates protein kinase activity at *in vitro* concentrations in the picomolar range.[48] Lead also inhibits NMDA-induced electrophysiological responses and the activity of adenylate cyclase at micromolar concentrations.[1] Some of the functions most sensitive to lead involve calcium metabolism and the calcium-related second messengers of neuronal function; lead is a potent inhibitor of calcium binding to calmodulin and other calciotropic proteins.[29,31,35]

VI. DIRECTIONS FOR RESEARCH

There is a significant gap between the data on the neurobehavioral effects of lead at low dose in children and those reported in relevant animal models. Since animal models are a primary source of information in mechanistic research, this has hindered the development of hypotheses on the mechanisms of lead neurotoxicity. Thus, while in comparison with most other neurotoxins, more is known about the effects of lead,[58] it remains the case that the fundamental mechanisms or sites of action of lead within the CNS are not known. Reviewing this relatively unsatisfying picture of basic research suggests a need to incorporate a reconsideration of the clinical data to focus further research.

Certain aspects of toxicity may be more promising for such integrative investigations. Morphologic studies directed at detecting loss of specific cells are unlikely to be useful indicators of low-level lead effects. Lead does not kill neuronal cells nor is lead a general cytotoxin. Only at relatively high doses is there a general loss of neurons (measured either as decreases in protein or DNA content[53] or as decreased thickness of cell layers in subcortical regions[46,72]). Lead may inhibit certain aspects of neurological development, resulting in changes in neuronal architecture, as reported in studies of cortical synaptogenesis.[7,49] However, it is not known if the cortex is the most sensitive or is the sole target of this effect of lead. More comprehensive studies of regional synaptic development are needed to determine the possible specificity of lead effects.

There is a need for the most fundamental measurements of relative distribution of lead in the brain after exposure which are correlated with low level toxicity. Differential distribution is the simplest explanation for the observation of specificity of organ-level effects and it is the most efficient explanation for the interplay of highly specific expression of neurotoxicity with relatively nonspecific mechanisms of cellular action. However, available data on the distribution of lead in the CNS is inadequate to determine regional or cell-specific accumulation of lead in lead-exposed animals. Overall, lead exposure can result in a 10- to 100-fold lead concentration in the brain. Radioisotopic lead — ^{210}Pb — has been used in only one study for purposes of localization;[103] lead uptake into hypothalamus was qualitatively demonstrated autoradiographically in connection with studies on the neuroendocrine effects of lead. X-ray microanalysis of lead has been done to determine subcellular localization of lead *in vitro*.[89,104] In one study of lead distribution in the brain undertaken with X-ray microanalysis, mice, treated as adults, showed relatively higher concentrations of lead in the thalamus, brain stem, hippocampus, and cortex, and low concentrations in the stratum and colliculus.[19]

VII. HYPOTHESES AS TO THE NEUROTOXIC MECHANISMS OF LEAD

Attempts have been made to integrate some of the findings in lead neurotoxicity. Silbergeld and Goldberg[92] proposed that the apparently opposite effects of lead on cholinergic and monoaminergic function in rats were consistent with the so-called neurotransmitter balance or feedback systems proposed for these two pathways. This interaction has been most clearly demonstrated for the basal ganglia.[25a,38] It is probably mediated by other interneurons and neurotransmitters, rather than a simple bimodal negative feedback system. Lead *in vitro* can affect aspects of dopaminergic neurochemistry, such as transmitter release but not cholinergic or GABAergic function. The effects of exposure, *in vivo* on all these systems, may reflect interactions of dopaminergic pathways with these two systems. However, other studies have reported that lead *in vitro* can affect both resting and stimulated release of GABA from cortical nerve terminals.[54] (Minnema, et al. 1986)

Moreover, the effects of lead on neurochemistry are highly dependent upon the region studied. The most striking effects of lead on dopamine content and synthesis are observed in the mesolimbic system rather than the cortex or basal ganglia, which may be consistent with changes in reactivity rather than motor function. In contrast, GABAergic function in lead-exposed rats was affected in basal ganglia and cerebellum while cortical and hippocampal cholinergic function is affected by lead.

It may be that the locus for lead neurotoxicity has not yet been identified and that all the phenomena reported to date are secondary to an unidentified, primary biochemical lesion (or lesions). Or it may be that the specificity of lead neurotoxicity, as expressed behaviorally, is conferred predominantly by the age of the organism and the developmental stage of different pathways in the brain.[23,97] In both clinical and experimental studies, these are major determinants for the expression of lead neurotoxicity, biochemically, electrophysiologically, morphologically, and behaviorally. The neurobiological mechanisms of lead may follow the presence of transport mechanisms, rather than the distribution of target receptors restricted to specific populations of nerve cells. This hypothesis would preserve the current information on the cellular mechanisms of lead toxicity which appear to occur in all cells, with no particular sensitivity conferred by origin or type. If (for instance) lead is fundamentally acting to interfere with ATPases and protein kinases which either control calcium binding and release, or calcium-dependent cell signaling and signal transduction, then these biological events should not vary from cell to cell.

This suggests that the relative specificity of lead neurotoxicity, affecting cognition, language, reaction time, and attention in the clinical studies, depends upon lead localization. Not much is known about lead uptake and distribution in brain. In addition to the mechanisms of calcium transport and specific lead-binding proteins, another factor in the transport of lead which has been proposed is the ability of certain amino acids to chelate lead.[74] Lead an other metals can be bound by choline, tyrosine, L-dopa, dopamine, and tryptophan. In this form, they may provide a transport substrate which coincidentally carries lead into those cells with specialized membrane transporters for these amino acids. Neurons which require these amino acids as precursors for the synthesis of neurotransmitters have specific, sodium-dependent high affinity membrane transport systems for them.[41] There have been few studies directed at determining the potential role of amino acid transport as a mechanism for selective uptake of lead into neurons. However, Silbergeld and Adler[88] reported that drugs which inhibited synaptosomal uptake of dopamine reduced lead-associated effects on neuronal cell calcium metabolism.

An alternative strategy towards developing neurobiological hypotheses on the neurotoxicity of lead is to consider the evidence from the whole-organism studies of behavioral changes associated with lead exposure. Attempts to critically analyze the experimental literature for such purposes are restricted by variations in experimental design and endpoints which affect the nature and significance of observed effects on complex behavior, as reviewed by Cory-Slechta[22] and others. The clinical literature, as discussed above, appears more convergent and consistent. However, the guidance which an integrative analysis of human lead neurotoxicity can provide is limited — given the limited information available on neurobiological substances of behavior and learning. The areas of cognitive function impaired by lead, while specific in terms of psychometric data, still cover a broad range of neurobiological components, and the evidence for somatosensory deficits must also be accommodated.

Some areas may be examined by coordinated neurochemical and behavioral analyses: the elegant demonstration by Olton[65] of the role of hippocampal cholinergic pathways in short-term spatial memory tasks has been used to explore the potential effects of lead on this behavioral parameter. Petit et al.[2,72] reported that although lead can affect hippocampal neurochemistry, including cholinergic function, lead-exposed animals did not show significant deficits in performance of the radial arm maze test of short-term memory. Another means of exploring the linkages between region-specific neurochemistry and behavior is to measure neurochemical and neurophysiological parameters while the intact animal is expressing behaviors, as has been done by Ungerstedt.[111] Another approach would be to attempt to biochemically "repair" some of the neurochemical alterations in lead-exposed animals and to observe changes in lead-associated deficits or abnormal behaviors. Only the simplest attempts have been made in this area, in several early studies on pharmacologic interventions designed to affect motor activity[92] and administration of anticonvulsants to reduce sensitivity to epileptogenic challenge.[94]

Thus, although much is known of the neurobehavioral effects of lead, and several hypotheses for molecular mechanisms of action have been proposed, there remains an important gap in our understanding of lead neurotoxicity. Given the simple nature of lead, as an elemental toxin, and the universal distribution of subcellular events sensitive to the ionic interference of lead, it remains to be determined how lead acts so specifically as a neurotoxin and why its neurotoxic affects appear to be defined in scope as well as particularly expressed in children. The next stage of basic research on lead should be directed towards answering these critical questions.

REFERENCES

1. **Alkondon, M., Costa, A. C. S., Radhakrishnan, V., Aronstam, R. S., and Albuquerque, E. X.**, Selective blockade of NMDA-activated channel currents may be implicated in learning deficits caused by lead. *FEBS Lett.*, 261, 124—130, 1990.
2. **Alfano, D. P. and Petit, T. L.**, Postnatal lead exposure and the cholinergic system, *Physiol. Behav.*, 34, 449—455, 1985.
3. American Academy of Pediatrics, Statement on childhood lead poisoning, *Pediatrics*, 79, 457—465, 1987.
4. **Atchison, W. D. and Narahashi, T.**, Mechanism of action of lead on the neuromuscular junction, *Neurotoxicology*, 5, 267—282, 1984.

5. **Audesirk, G. and Audesirk, T.**, Effects of chronic low level lead exposure on the physiology of individually identifiable neurons, *Neurotoxicology*, 4, 13—26, 1983.
6. **Audesirk, G.**, Effects of lead exposure on the physiology of neurons, *Prog. Neurobiol.*, 24, 199—231, 1985.
7. **Averill, D. R. and Needleman, H. L.**, Neonatal lead exposure retards cortical synaptogenesis in the rat, in *Low Level Lead Exposure; The Clinical Implications of Current Research*, Needleman, H. L., Ed., Raven Press, New York, 1980, 201—210.
8. **Baker, E. L., Feldman, R. G., White, R. A., Harley, J. P., Niles, C. A., Dinse, G. E., and Berkey, C. S.**, Occupational lead neurotoxicity: a behavioral and electrophysiological evaluation. Study design and one year results, *Brit. J. Indust. Med.*, 41, 352—361, 1984.
9. **Banks, H. A. and Stollery, B. T.**, The longitudinal evaluation of verbal reasoning in lead workers, *Sci. Total Environ.*, 71, 469—476, 1988.
10. **Barrett, J. and Livesey, P. J.**, Lead induced alterations in maternal behavior and offspring development in the rat, *Neurobehav. Toxicol. Teratol.*, 5, 557—563, 1983.
11. **Bellinger, D.**, Effects of low level lead exposure on intelligence, *Environ. Health Persp.*, in press.
11a. **Bellinger, D. and Needleman, H. L.**, Neurodevelopmental effects of low-level lead exposure in children, this volume.
12. **Bhattacharya, A., Shukla, R., Bornschein, R., Dietrich, K., and Kopke, J. E.**, Postural disequilibrium qualification in children with chronic lead exposure: a pilot study, *Neurotoxicology*, 9, 327—340, 1988.
13. **Bull, R. J., McCauley, P. T., Taylor, D. H., and Crofton, K. M.**, The effects of lead on the developing central nervous system of the rat, *Neurotoxicology*, 4, 1—18, 1983.
14. **Burchfiel, J. L., Duffy, F. H., Bartels, P. H., and Needleman, H. L.**, The combined discriminating power of quantitative electroencephalography and neuropsychologic measures in evaluating central nervous effects of lead at low levels, in *Low Level Lead Exposure: The Clinical Implications of Current Research*, Needleman, H. L., Ed., Raven Press, New York, 1980, 75.
14a. **Burchfiel, J. L.**, Low level lead exposure: effect on quantitative electroencephalography and correlation with neuropsychologic measures, this volume.
15. **Bushnell, P. J. and Bowman, R. E.**, Effects of chronic lead ingestion of social development in infant Rhesus monkeys, *Neurobehav. Toxicol.*, 1, 207—219, 1979.
16. **Bushnell, P. J. and Levin, E. D.**, Effects of zinc deficiency of lead toxicity in rats, *Neurobehav. Toxicol. Teratol.*, 5, 283—288, 1983.
17. **Campbell, J. B., Woolley, D., Vijakan, V. K., and Overmann, S. R.**, Morphometric effects of postnatal lead exposure on hippocampal development of the 15-day-old rat, *Dev. Brain Res.*, 3, 595—612, 1982.
18. **Carroll, P. T., Silbergeld, E. K., and Goldberg, A. M.**, Alteration of central cholinergic function by chronic lead acetate exposure, *Biochem. Pharmacol.*, 26, 397—402, 1977.
19. **Cholewa, M., Hanson, A. L., Jones, K. W., McNally, W. P., and Fand, I.**, Regional distribution of lead in the brains of lead-intoxicated adult mice, *Neurotoxicology*, 7, 9—18, 1986.
20. **Cooper, G. P., Suszkiw, J. B., and Manalis, R. S.** Heavy metals; effects on synaptic transmission, *Neurotoxicology*, 5, 247—266, 1984.
21. **Cory-Slechta, D.**, Chronic low-level lead exposure: behavioral consequences, biological exposure indices and reversibility, *Sci. Total Environ.*, 71, 433—440, 1988.
22. **Cory-Slechta, D. A.**, The behavioral toxicity of lead: problems and perspectives, *Adv. Behav. Pharmacol.*, 4, 211—255, 1984.
23. **Cory-Slechta, D. A., Weiss, B., and Cox, C.**, Delayed behavioral toxicity of lead with increasing exposure concentrations, *Toxicol. Appl. Pharmacol.*, 71, 342—352, 1983.
24. **Cutler, M. G.**, Effects of exposure to lead on social behavior in the laboratory mouse, *Psychopharmacology*, 52, 279—282, 1977.
25. **Dietrich, K. N., Kraft, K. M., Pearson, D. T., Bornschein, R. L., Hammond, P. B., and Succop, P. A.**, Postnatal lead exposure and early sensorimotor development, *Environ. Res.*, 38, 130—136, 1989.
25a. **Eadie, M. J., and Tyrer, J. H.**, *Biochemical Neurology*, Alan R. Liss, New York, 1983.
26. **Engellener, W. J., Burright, R. G., and Donovick, P. J.**, Lead, age and aggression in male mice, *Physiol. Behav.*, 36, 823—838, 1986.
27. **Ewers, U., and Erbe, R.**, Effects of lead, cadmium, and mercury on brain adenylate cyclase, *Toxicology*, 16, 227—237, 1980.
28. **Fjerdingstad, E. J., Danscher, G., and Fjerdingstad, E.**, Hippocampus: selective concentration of lead in the normal rat brain, *Brain Res.*, 80, 350—354, 1974.
29. **Fullmer, C. S., Edelstein, S., and Wasserman, R. H.**, Lead-binding properties of intestinal calcium-binding proteins, *J. Biol. Chem.*, 260, 6816—6819, 1985.
30. **Goldstein, G. W.**, Lead poisoning and brain cell function, *Environ. Health Persp.*, 89, 91—94, 1990.
31. **Goldstein, G. W. and Ar, D.**, Lead activates calmodulin sensitive processes, *Life Sci.*, 33, 1001—1006, 1983.

32. **Goldstein, G. W., Asbury, A. K., and Diamond, I.**, Pathogenesis of lead encephalopathy, *Arch. Neurol.*, 31, 382—389, 1974.
33. **Govoni, S., Memo, M., Spano, P. F., and Trabucchi, M.**, Chronic lead treatment differentially affects dopamine synthesis in various rat brain areas. *Toxicology*, 12, 343—349, 1979.
34. **Gross-Selbeck, E. and Gross-Selbeck, M.**, Changes in operant behavior of rats exposed to lead at the accepted no-effect level, *Clin. Toxicol.*, 18, 1247—1256, 1981.
35. **Habermann, E., Crowell, K., and Janicki, P.**, Lead and other metals can substitute for Ca^{++} in calmodulin, *Arch. Toxicol.*, 54, 61—70, 1983.
36. **Hanninen, H., Hernberg, S., Mantere, P., Vesanto, R., and Jalkanen, M.**, Psychological performance of subjects with low exposure to lead, *J. Occupt. Med.*, 20, 683—689, 1978.
37. **Hastings, L., Zenick, H., Succop, P., Sun, T. J., and Sekeres, R.**, Relationship between hematopoietic parameters and behavioral measures in lead-exposed rats, *Toxicol. Appl. Pharmacol.*, 73, 416—422, 1984.
38. **Klawans, H. L.**, *The Pharmacology of Extrapyramidal Movement Disorders*, Karger, Basal, 1973.
39. **Kolton, l. and Yonai, Y.**, Sites of action of lead on spontaneous transmitter release from motor nerve terminals, *Israel J. Med. Sci.*, 18, 165—170, 1982.
40. **Kostial, K. and Vouk, V. B.**, Lead ions and synaptic transmission in the superior cervical ganglion of the cat, *Brit. J. Pharmacol.*, 12, 219—222, 1957.
41. **Kuhar, M. J.**, High affinity choline uptake, *Life Sci.*, 13, 1623—1634, 1974.
42. **Lasley, S. M., Greenland, R. D., Minnema, D. J., and Michaelson, I. A.**, altered central monamine response to D-amphetamine in rats chronically exposed to inorganic lead, *Neurochem. Res.*, 10, 933—944, 1985.
43. **Lasley, S. M. and Lane, J. D.**, Diminished regulation of mesolimbic dopaminergic neurotransmissions by lead exposure, *Toxicol. Appl. Pharmacol.*, 95, 474—483, 1988.
44. **LeFauconnier, J. M., Hauw, J. J., and Bernard, G.**, Regressive or lethal lead encephalopathy in the suckling rat. Correlation of lead levels and morphological findings, *J. Neuropathol. Exp. Neurol.*, 42, 177—190, 1983.
45. **Levin, E. D., Schneider, M. L., Ferguson, S. A., Schantz, S. L., and Bowman, R. E.**, Behavioral effects of developmental lead exposure in Rhesus monkeys, *Develop. Psychobiol.*, 21, 371—382, 1988.
46. **Louis-Ferdinand, R. T., Brown, D. R., Fiddler, S. F., Daughtrey, W. C., and Klein, A. W.**, Morphometric and enzymatic effects of neonatal lead exposure in the rat brain, *Toxicol. Appl. Pharmacol.*, 43, 351—360, 1978.
47. **Lucchi, L., Memo, M., Airaghi, M. L., Spano, P. F., and Trabucchi, M.**, Chronic lead treatment induces in rat a specific and differential effect on dopamine receptors in different brain areas, *Brain Res.*, 213, 397—404, 1981.
48. **Markovac, J. and Goldstein, G. W.**, Picomolar concentrations of lead stimulate brain protein kinase C, *Nature*, 334, 71—73, 1988.
49. **McCauley, P. T., Bull, R. J., Tonti, A. P., Lutkenhoff, S. D., Meister, M. V., Doerger, J. V., and Stober, J. A.**, The effect of prenatal and postnatal lead exposure on neonatal synaptogenesis in rat cerebral cortex, *J. Toxicol. Environ. Health*, 10, 639—651, 1982.
50. **Memo, M., Lucchi, L., Spano, P. F., and Trabucchi, M.**, Dose-dependent and reversible effects of lead on rat dopaminergic system, *Life Sci.*, 28, 795—799, 1981.
51. **Memo, M., Lucchi, L., Spano, P. F., and Trabucchi, M.**, Lack of correlation between the neurochemical and behavioral effects induced by D-amphetamine in chronically lead-treated rats, *Neuropharmacology*, 19, 795—799, 1980.
52. **Michaelson, I. A.**, An appraisal of rodent studies on the behavioral toxicity of lead: the role of nutritional status. In *Lead Toxicity*, Singhal, R. and Thomas, J., Eds., Urban and Schwarzenberg, Baltimore, 1980, 302—365.
53. **Michaelson, I. A.**, Effects of inorganic lead on RNA, DNA and protein content in the developing neonatal rat brain, *Toxicol. Appl. Pharmacol.*, 26, 539—548, 1973.
54. **Minnema, D. J. and Michaelson, I. A.**, Differential effects of inorganic lead and D-aminolevulinic acid vitro on synaptosomal G-aminobutyric acid release, *Toxicol. Appl. Pharmacol.*, 86, 1—11, 1986.
55. **Mistry, P., Lucier, G. W., and Fowler, B. A.**, High-affinity lead binding proteins in rat kidney cytosol mediate cell-free nuclear translocation of lead, *J. Pharmacol. Expt. Ther.*, 232, 462—469, 1985.
56. **Moore, M. R., Goldberg, A., and Yeung-Laiwah, A. A. C.**, Lead effects on the heme biosynthetic pathway. Relationship to toxicity, in *Mechanisms of Chemical-Induced Porphyrinopathies*, Silbergeld, E. K. and Fowler, B. A., Eds., *Ann. N.Y. Acad. Sci.*, 514, 191, 1987.
57. **Mullenix, P.**, Effects of lead on spontaneous behavior, in *Low Level Poisoning; The Clinical Implications of Current Research*, Needleman, H. L., Ed., Raven Press, New York, 1980, 211—220.
58. **National Academy of Sciences — National Research Council**, Neurotoxicology: Basic Mechanisms and Risk Assessment, Washington: NAS Press, in press.
59. **Needleman, H. L., Gatsonis, C., Gunnoe, C., Leviton, A., Reed, R., Peresie, H., Maker, C., and Barrett, P.**, Deficits in psychologic lead levels, *N. Engl. J. Med.*, 300, 689—695, 1979.

60. **Needleman, H. L. and Landrigan, P. J.** The health effects of low level exposure to lead, *Ann. Rev. Public Health,* 2, 277—298, 1981.
61. **Nomura, S., Willis, A. J., Edwards, D. R., Heath, J. K., and Hogan, B. L. M.,** Developmental expression of 2ar (osteopontin) and SPARC (osteonectin) RNA as revealed by in situ hybridization, *J. Cell Biol.,* 106, 441—450, 1988.
62. **Norton, S.,** Toxic responses of the central nervous system, *Casarett and Doulls Toxicology,* Klaassen, C., Doull, J., and Amdur, M., Eds., 1985, 359—386.
63. **Odenbro, A., Greenberg, N., Vroegh, K., Bederka, J., and Kihlstrom, J.-E.,** Functional disturbances in lead-exposed children, *Ambio,* 12, 40—44, 1983.
64. **Olson, L., Bjorklund, H., Henschen, A., Palmer, M., and Hoffer, B.,** Some toxic effects of lead, other metals and antibacterial agents on the nervous system — animal experiment models, *Acta Neurol. Scand.,* 70, (Suppl 100), 1984.
65. **Olton, D. S.,** Spatial memory, *Sci. Amer.,* 236, 82—98, 1977.
66. **Oskarsson, A., Squibb, K. S., and Fowler, B. A.,** Intracellular binding of lead in the kidney: the partial isolation and characterization of postnitochondrial lead binding components, *Biochem. Biophys. Res. Commun.,* 104, 290—298, 1982.
67. **Otto, D. A., Benignus, V. A., Muller, K. E., Barton, C. N., Seiple, K., Prah, J., and Schroeder, S.,** Effects of low to moderate lead exposure on slow cortical potentials in young children: two year follow-up study, *Neurobehav. Toxicol. Teratol.,* 4, 733—742, 1982.
68. **Otto, D. A., Benignus, V. A., Muller, K. E., and Barton, C. N.,** Effects of age and body lead burden on CNS function in young children. I. Slow cortical potentials. *Electroencephal. Clin. Neurophysiol.,* 52, 229—239, 1981.
69. **Overmann, S. R.,** Behavioral effects of asymptomatic lead exposure during neonatal development in rats, *Toxicol. Appl. Pharmacol.,* 41, 459—471, 1977.
70. **Palmer, M. R., Bjorklund, H., Freedman, R., Taylor, D. A., Marwha, J., Olson, L., Seiger, A., and Hoffer, B.,** Permanent impairment of spontaneous Purkinje cell discharge in cerebellar grafts caused by chronic lead exposure, *Toxicol. Appl. Pharmacol.,* 60, 431—440, 1981.
71. **Perino, J. and Ernhart, C. B.,** The relation of subclinical lead level to cognitive and sensorimotor impairment in Black preschoolers, *J. Learn. Disabil.,* 7, 26—30, 1974.
72. **Petit, T. L., Alfano, D. P., and LeBoutillier, J. C.,** Early lead exposure and the hippocampus: a review and recent advances, *Neurotoxicology,* 4, 79—94, 1983.
73. **Pounds, J.,** Effect of lead intoxication on calcium homeostasis and calcium-mediated cell function: a review, *Neurotoxicology,* 5, 295—332, 1984.
74. **Rajan, K. S., Davis, J. M., and Colburn, R. W.,** Metal chelates in the storage and transport of neurotransmitters: interactions of Ca^{++} with ATP and biogenic amines, *J. Neurochem.,* 22, 137—147, 1974.
75. **Reiter, L. W. and MacPhail, R. C.** Factors influencing motor activity measurements in neurotoxicology, in *Nervous System Toxicology,* Mitchell, C. L., Ed., Raven Press, New York, 1982, 45—65.
76. **Reyners, M., Gianfelici de Reyners, E., and Maisin, J. R.,** An ultrastructural study of the effects of lead in the central nervous system of the rat. *Proc. Internat. Confer. Heavy Metals in the Environment,* CEP, Edinburgh, 1979, 58—61.
77. **Rice, D. C.,** Effect of lead on schedule-controlled behavior in monkeys, in *Behavioral Pharmacology, the Current Status,* Weiss, B., Ed., Alan R. Liss, New York, 1985, 473—486.
78. **Rosen, J. B., Berman, R. F., Beuthin, F. C., and Louis-Ferdinand, R. T.,** Age of testing as a factor in the behavioral effects of early lead exposure in rats, *Pharmacol. Biochem. Behav.,* 23, 49—54, 1985.
79. **Rutter, M. R.,** Scientific issues and state of the art in 1980, *Lead Versus Health,* Rutter, M. and Russell Jones, R., Eds., John Wiley & Sons, London, 1983, 1—15.
80. **Sauerhoff, M. W. and Michaelson, I. A.,** Hyperactivity and brain catecholamines in lead-exposed developing rats, *Science,* 182, 1022—1024, 1973.
81. **Scheuhammer, A. M. and Cherian, M. G.,** The regional distribution of lead in normal rat brain, *Neurotoxicology,* 3, 85—92, 1982.
82. **Schottenfeld, R. S. and Cullen, M. R.,** Organic affective illness associated with lead intoxication, *Am. J. Psychiatr.,* 141, 1423—1426, 1984.
83. **Schroeder, S. R., Hawk, B., Otto, D. A., Mushak, P. and Hicks, R. E.,** Separating the effects of lead and social factors on IQ, *Environ. Res.,* 38, 144—154, 1985.
84. **Schwartz, J., Landrigan, P. J., Feldman, R. G., Silbergeld, E. K., Baker, E. L., and von Lindern, I. H.,** Threshold effect in lead-induced peripheral neuropathy, *J. Pediatr.,* 112, 12—17, 1988.
85. **Schwartz, J. and Otto, D.,** Blood lead, hearing thresholds, and neurobehavioral development in children and youth, *Arch. Environ. Health,* 42, 153—160, 1987.
86. **Seppalainen, A. M.,** Neurophysiological approaches to the detection of early neurotoxicity in humans, *CRC Crit. Rev. Toxicol.,* 18, 245—298, 1988.

87. **Shellenberger, K.,** Effects of early lead exposure on neurotransmitter systems in the brain: a review with commentary, *Neurotoxicology,* 5, 177—212, 1984.
88. **Silbergeld, E. K. and Adler, H. S.,** Subcellular mechanisms of lead neurotoxicity, *Brain Res.,* 148, 451—467, 1978.
89. **Silbergeld, E. K., Alder, H. S., and Costa, J. L.,** Subcellular localization of lead in synaptosomes, *Res. Commun. Chem. Pathol. Pharmacol.,* 17, 715—725, 1977.
90. **Silbergeld, E. K. and Chisolm J. J.,** Lead poisoning: altered urinary catecholamine metabolites as indicators of intoxication in mice and children, *Science,* 192, 153—155, 1976.
91. **Silbergeld, E. K., Fales, J. T., and Goldberg, A. M.,** The effects of lead on the neuromuscular junction, *Neuropharmacology,* 13, 795—801, 1975.
92. **Silbergeld, E. K. and Goldberg, A. M.,** Pharmacological and neurochemical investigations of lead induced hyperactivity, *Neuropharmacology,* 14, 431—444, 1975.
93. **Silbergeld, E. K. and Goldberg, A. M.,** A lead induced behavioral disorder, *Life Sci.,* 13, 1275—1283, 1973.
94. **Silbergeld, E. K., Hruska, R. E., Bradley, D., Lamon, J. M., and Frykholm, B. C.,** Neurotoxic aspects of porphyrinopathies: lead and succinylacetone, *Environ. Res.,* 29, 459—471, 1982.
95. **Silbergeld, E. K. and Hruska, R. E.,** Neurochemical investigations of low level lead exposure, in *Low Level Lead Exposure: The Clinical Implications of Current Research,* Needleman, H. L., Ed., Raven Press, New York, 1980, 135—157.
96. **Silbergeld, E. K., Miller, L. P., Kennedy, S., and Eng, N.,** Lead, GABA, and seizures: effects of subencephalopathic lead exposure on seizure sensitivity and GABAergic function, *Environ. Res.,* 19, 371—382, 1979.
97. **Silbergeld, E. K.,** Experimental studies of lead neurotoxicity: implications for mechanisms, dose-response, and reversibility, in *Lead Versus Health,* Rutter, M. and Russell Jones, R., Eds., John Wiley & Sons, London, 1983, 191—218.
98. **Silbergeld, E. K.,** Role of altered heme synthesis in chemical injury to the nervous system, in *Mechanisms of Chemical-Induced Porphyrinopathies,* Silbergeld, E. K. and Fowler, B. A., Eds., *Ann. N.Y. Acad. Sci.,* 514, 297—308, 1987.
99. **Silman, A. J., Bolnick, D. A., Bosetti, J. B., Haynes, L. W., and Walter, A. E.,** The effects of lead and of cadmium on the mass photoreceptor potential: the dose-response relationship, *Neurotoxicology,* 3, 179—194, 1982.
100. **Sobotka, T. J. and Cook, M. P.,** Postnatal lead acetate exposure in rats: possible relationship to minimal brain dysfunction, *Am. J. Ment. Defic.,* 79, 5—9, 1974.
101. **Spence, I., Drew, C., Johnston, G. A. R., and Lodge, D.,** Acute effects of lead at central synapses in vitro, *Brain Res.,* 333, 103—109, 1985.
102. **Stephens, M. C. C., and Gerber, G. B.,** Development of glycolipids and gangliosides in lead-treated neonatal rats, *Toxicol. Lett.,* 7, 373—378, 1981.
103. **Stumpf, W. E., Sar, M., and Grant, L. D.,** Autoradiographic localization of ^{210}Pb and its decay products in rat forebrain, *Neurotoxicology,* 1, 593—606, 1980.
104. **Sulzer, D., Piscopo, I., Ungar, F., and Holtzman, E.,** Lead-dependent deposits in diverse synaptic vesicles: suggestive evidence for the presence of anionic binding sites, *J. Neurobiol.,* 18, 467—483, 1987.
105. **Sundstrom, R., Conradi, N. G., and Sourander, P.,** Vulnerability to lead in protein-deprived suckling rats, *Acta Neuropathol. (Berlin),* 62, 276—283, 1984.
106. **Sundstrom, R. and Karlsson, B.,** Myelin basic protein in brains of rats with low dose lead encephalopathy, *Arch. Toxicol.,* 59, 341—345, 1987.
107. **Sundstrom, R., Muntzing, K., Kalimo, H., and Sourander, P.,** Changes in the integrity of the blood-brain barrier in suckling rats low dose encephalopathy, *Acta Neuropathol. (Berlin),* 68, 1—9, 1985.
108. **Taylor, D., Nathanson, J., Hoffer, B., Olson, L., and Seiger, A.,** Lead blockage of norepinephrine-induced inhibition of cerebellar Purkinje neurons, *J. Pharmacol. Exp. Therap.,* 206, 371—381, 1978.
109. **Taylor, D. H., Noland, E. A., Brubaker, C. M., Crofton, K. M., and Bull, R. J.,** Low level lead exposure produces learning deficits in young rat pups, *Neurobehav. Toxicol. Teratol.,* 4, 311—314, 1982.
110. **Thatcher, R. S., Lester, M. L., McAlester, R. and Horst, R.,** Effects of low levels of cadmium and lead on cognitive functioning in children, *Arch. Environ. Health,* 37, 159—166, 1982.
111. **Ungerstedt, U.,** Measurement of neurotransmitter release in vivo by intracranial dialysis, in *Measurement of Neurotransmitter Release In Vivo,* Marsden, C., Ed., John Wiley & Sons, 1984, 81—105.
112. **Valciukas, J. A., Lilis, R., Eisinger, J., Blumberg, W. E., Fischbein, A., and Selikoff, I. J.,** Behavioral indicators of lead neurotoxicity; results of a clinical field survey, *Int. Arch. Occup. Environ. Health,* 41, 217—236, 1978.
113. **Whetsell, W. D., Sassa, S., and Kappas, A.,** Porphyrin-heme biosynthesis in organotypic cultures of mouse dorsal root ganglia: effects of heme and lead on porphyrin synthesis and peripheral myelin, *J. Clin. Invest.,* 74, 600—607, 1984.

114. **Winder, C. and Kitchen, I.,** Lead neurotoxicity: a review of the biochemical, neurochemical and drug induced behavioral evidence. *Prog. Neurobiol.,* 22: 59—87, 1984.
114a. **Winneke, G.,** Nonrecovery of lead-induced changes of visual evoked potentials in rats, *Toxicol. Lett.,* 1, 77, 1980.
115. **Winneke, G., Kramer, V. B., Brockhaus, A., Ewers, V., Kujanek, G., Lechner, H., Janke, W.,** Neuropsychological studies in children with elevated tooth-lead concentrations, *Int. Arch. Occupt. Environ. Health,* 51, 231—252, 1983.
116. **Winneke, G., Lilienthal, H., and Werner, W.,** Task dependent neurobehavioral effects of lead in rats, *Arch. Toxicol. Suppl.,* 5, 84—93, 1982.
117. **Yule, W., Lansdown, R., Millar, I. B., and Urbanowics, M. A.,** The relationship between blood lead concentrations, intelligence and attainment in a school population: a pilot study, *Develop. Med. Child. Neurol.,* 23, 567—576, 1981.
118. **Yule, W. and Lansdown, R.,** Lead and children's development, in *Heavy Metals in the Environment,* Vol. 2, CEP Consultants, Edinburgh, 1983, 912—916.

Chapter 6

VISUAL AND AUDITORY SYSTEM ALTERATIONS FOLLOWING DEVELOPMENTAL OR ADULT LEAD EXPOSURE: A CRITICAL REVIEW

Donald A. Fox

TABLE OF CONTENTS

I. Introduction 106

II. Visual System 107
    A. Retinal Studies 107
        1. Light Microscopy and Quantitative Histology 107
        2. Electron Microscopy 108
        3. Electroretinography 109
        4. Rhodopsin and Retinal Sensitivity Studies 110
        5. Dark and Light Adaptation Studies 110
        6. Cyclic Nucleotide Metabolism and Calcium Studies 111
        7. Na^+, K^+-ATPase Studies 112
    B. Retinal Ganglion Cell Axons or Optic Nerve Studies 114
    C. Visual Cortex Studies 114
    D. Summary and Critique 115
    E. Future Directions 116

III. Auditory System 116
    A. Developmental Exposure 116
    B. Adult Exposure 117
        1. Human Studies 117
        2. Animal Studies 117
    C. Summary, Critique, and Future Directions 118

IV. Conclusions 118

Acknowledgments 119

References 119

I. INTRODUCTION

Visual[1-23] and auditory[1,23-30] system processing deficits and alterations observed in children, monkeys, and rats following lead exposure during perinatal development[1-30] and in man following chronic lead exposure.[31-42] In addition, acute lead exposure of adult experimental animals has been reported to produce retinal degeneration[43-45] and auditory system dysfunction.[46-48] In spite of the functional and structural impairments produced by lead, precious little information exists on the sites and/or mechanisms of action accounting for these sensory system deficits. Furthermore, the effects of low-level lead exposure on sensory systems, especially during the period of perinatal development, remain a virtually unexplored area of neurotoxicological research. As noted by several of the above cited authors, impairments of vision and hearing could significantly contribute to other reports of neurobehavioral deficits (see Reference 1 for review) such as learning disabilities and poor classroom behavior.

The main objectives of this chapter are: (1) to review the available literature on the visual and auditory processing deficits produced by developmental lead exposure; (2) to describe and critique the experiments aimed at elucidating the sites and mechanisms of action responsible for the lead-induced visual and auditory deficits; and (3) to suggest future directions for research aimed at determining the type and degree of sensory (i.e, visual and auditory) deficit and the site and mechanism of action following low-level perinatal lead exposure. Experiments conducted following acute or chronic lead exposure in adult man and experimental animals also will be reviewed since they are environmentally relevant, clinically important, and offer insight into the sites and mechanisms of action of lead.

For purposes of classifying the various experimental models of lead exposure during development and accidental lead exposure to children, the author will use the mean blood lead concentration (PbB) at the peak of lead exposure or when measured. For purposes of this review, low lead exposure will be designated as a PbB of 10 to 20 µg/dl, moderate lead exposure as a PbB of 21 to 60 µg/dl, high lead exposure as a PbB above 61 µg/dl. This classification is consistent with the latest Centers for Disease Control (CDC) and World Health Organization (WHO) reports described in the Agency for Toxic Substances and Disease Registry (ATSDR) report "The Nature and Extent of Lead Poisoning in Children in the United States: A Report to Congress".[1]

An overall comparison of results from lead exposure during development versus adulthood suggests that the determinants of the anatomical site of action (e.g., receptoral, axonal, and/or central) and degree of sensory dysfunction produced by lead exposure might be mediated by independent variables such as, species, dose, and age at time of lead exposure. Therefore, emphasis will be placed on examining these independent variables as well as dependent variables such as blood lead concentration, site(s) of electrophysiological recording, tissues examined biochemically and location(s) of morphological analysis. Although no single electrophysiological, biochemical, or morphological study to date has examined all of these variables, several studies have examined the effects of lead at various levels of the visual[3,6,12,13,17,18,21a,22,34] and auditory[23,28,47,48] system. In addition, the various functional and morphological methodologies utilized to determine the type and extent of sensory system deficit will be examined. A systematic analysis of stimulus properties (e.g., threshold, intensity, wavelength, and frequency) using electrophysiological or psychophysical methodologies is necessary to adequately and completely document sensory deficits. Fortunately, several of the *in vivo* studies have employed these rigorous techniques.[11,13,16,18a,19,21a,21c,21d,36-40,47,48] Furthermore, quantitative histological and morphological techniques at the light and electron microscopic level of analysis are needed to adequately document and compare changes in lead-exposed animals. Only a few studies, however, have employed these techniques.[8,15,20,21,21c,22] Finally, the use of *in vitro* studies to determine whether the effects of lead are direct, to locate the cellular sites of action, or to determine the underlying mechanisms of action of lead have been limited to only a few studies.[12,19,21b-21d,49-54]

These types of studies are important for three reasons. First, if lead exposure produces visual or auditory deficits there are 3 to 4 million children potentially at risk.[1] Second, if the electroretinogram (ERG), visual evoked potential (VEP), or brainstem auditory evoked potential (BAER) is demonstrated to be a sensitive and reliable indicator of early lead exposure it may be employed as an early noninvasive screening tool. Finally, since lead is a selective rod toxicant (vide infra) it can be utilized as a neurobiological tool, like pharmacological agents and inherited retinal degenerations, to study biochemical mechanisms and structure-function relations in rod photoreceptors and rod-cone interactions.

II. VISUAL SYSTEM

A. RETINAL STUDIES

Long-term scotopic deficits following lead exposure during early postnatal development were reported first in monkeys and hooded rats.[2,10,11,13,17,18,18a,21c,21d] As discussed below, some of these deficits are possibly due to central nervous system (CNS) alterations.[5,8-12,22] Early retinal involvement, however, is clearly suggested by reports in occupationally lead-exposed workers describing a decrease in the amplitude of the ERG b-wave and its oscillatory potentials, a decreased in visual sensitivity under mesopic, but not photopic, conditions and no changes in photopic critical flicker fusion frequency (CFF).[31,33-35] Support for a retinal site of action is further supported by the findings of Fox and Sillman[49] and later Sillman and co-workers[50-52] who demonstrated that micromolar concentrations of lead chloride selectively depress the amplitude and sensitivity of the rod, but not cone, photoreceptor potential in isolated, perfused bullfrog retinas. Intracellular recordings from isolated axolotl rods[53] and from mudpuppy horizontal cells[54,55] exposed to lead chloride validated these original findings. More recently, ERG studies by Fox and co-workers[19,21,21c,21d] have confirmed that low or moderate level lead exposure only during development produces selective rod photoreceptor deficits in adult Long-Evans hooded rats: peak PbB at weaning was 19 or 59 μg/dl, respectively, while PbB values as adults were not different from controls. Quantitative histological and morphological studies, in similarly lead-exposed rats, revealed that animals exposed to low levels of lead during development had no retinal abnormalities whereas those exposed to moderate levels of lead during development had a selective rod degeneration as well as altered morphology in the remaining rods.[20,21,21c]

1. Light Microscopy and Quantitative Histology

Light microscopic examination of adult retinas from rats exposed to moderate levels of lead during development revealed several distinct alterations.[20] The two most notable were the distended and disorganized rod outer segments (ROS) and the thinned outer-nuclear layer (ONL). In the lead-exposed rats random patches of swollen and disrupted ROS primarily occurred in the proximal one half to one third of the ROS. This alteration was never observed in the controls or rats exposed to low levels of lead during development.[20,21c] Interestingly, the retinal pigment epithelium (RPE) and the intimate contact between the RPE cells and outer segments appeared relatively normal in all animals. Additional changes observed were a slightly thinned inner nuclear layer (INL) and inner plexiform layer (IPL), gliosis, and glycogen accumulation only in rod photoreceptors. Total retinal thickness was decreased 15% in the superior retina and 21% in the inferior retina. Photoreceptor and INL necrosis with normal RPE cells also was observed in the retinas of adolescent rats chronically exposed to lead since birth.[14] It is thought that the thinned INL and IPL represent transneuronal degeneration resulting from the rod photoreceptor degeneration.[20] Similarly, the retinas of adult rabbits chronically exposed to lead exhibited photoreceptor necrosis, a thinned ONL, displacement of photoreceptor nuclei into the rod inner segment (RIS) region and additionally, a two- to three-fold increase in the normal height of the RPE and an increase in lysosomal inclusions in the RPE.[43-45]

Quantitative histological analysis revealed that moderate lead exposure during development produced a long-term selective loss of 22% of the rods.[20] The loss of rod nuclei was more extensive in the inferior than in the superior retinal hemisphere and in the posterior than in the peripheral retina. Thus, there was a central-peripheral gradient of rod degeneration. In all treatment groups, the ROS were longer in the superior retina (22 to 30%) than in the inferior retina. In the moderately lead-exposed rats, however, there was a consistently greater decrease in ROS length in the inferior (15%) than in the superior (9%) retina which is consistent with the above rod nuclei data showing a greater loss in the inferior than superior retina. These results, in combination with the selective rod loss, most likely account for the 30 to 34% loss of rhodopsin reported by Fox and Rubinstein.[21] These quantitative histological data are in excellent agreement with the ERG and cyclic nucleotide metabolism studies conducted in similarly lead-exposed rats[19] and described below.

2. Electron Microscopy

The retinas of control and developmentally lead-exposed rats were also examined by electron microscopy.[20,21c] The most obvious lead-induced alteration, observed only in moderately lead-exposed rats, was ROS swelling and disorganization.[20] At high magnification, examination of swollen ROS suggests this alteration is due to a recent shearing force (e.g., osmotic swelling during perfusion) rather than a long-term degeneration. This conclusion is enhanced by the diminution of the "treatment-related perfusion artifact" by perfusion with a hyperosmotic buffer[55a] and by the lack of occurrence of ROS vesiculation in these lead-exposed rats as compared to vitamin A deficient rats[56] or light-damaged rats.[57] A morphological comparison of the external limiting membrane and ONL from control and moderately lead-exposed rats also was made. As previously reported, the intercellular junctional complexes (i.e., zonula adherentes) between Muller cells are characterized by an increased electron density of the membranes and neighboring cytoplasm.[58] Although the Muller cells were slightly swollen in the lead-exposed rats, the structure of the external limiting membrane remained relatively normal. In contrast, the ONL exhibited a selective loss of rod photoreceptor nuclei (vide supra), disrupted cytoarchitecture (i.e., increased internuclei spacing) and gliosis. In addition, most of the remaining rod photoreceptor nuclei were shrunken. Furthermore, the OPL appeared more electron lucent.

The retinas from both lead-exposed groups contained large accumulations of monoparticulate (β-type) glycogen particles, predominantly localized in the distal retina (i.e., ellipsoid region of the RIS, ONL, and OPL). Surprisingly, in the distal retina these large glycogen accumulations were localized exclusively in rod mitochondria: in the RIS, rod axons, and rod synaptic terminals. Two ultrastructural features of the glycogen containing mitochondria are worthy of note. The first is that the enclosing double membrane of mitochondria was always clearly present, although it appeared to be broken down in one or more points. Second, in most cases the volume of glycogen was so large that the cristae were either pushed to the periphery or no longer present.

The lead-induced alterations in rods are similar to those observed in animals with inherited and drug-induced selective rod degenerations.[59,60] In these latter two animal models, elevated levels of cyclic GMP (cGMP) due to an inhibition of cGMP-phosphodiesterase (cGMP-PDE), the enzyme responsible for catabolizing cGMP, are thought to be responsible for the selective rod photoreceptor degenerations.[59,60] The permeability of the ROS plasma membrane cation channels to Na^+, Ca^{++} and Mg^{++} is directly mediated by cGMP such that, in darkness the cation channels are held open by cGMP allowing these ions to enter. The pathological similarities with lead-exposed rats and the suggestion that alterations in cGMP might mediate this selective degeneration led Fox and co-workers[19-21,21c,21d] to conduct ERG and cyclic nucleotide metabolism studies in adult rats exposed to low or moderate levels of lead only during early postnatal development. Selective changes in cGMP metabolism, however, may not totally account for the lead-induced rod selective degeneration since it has been suggested that cGMP is also the internal transmitter in cones.[61] The observed rod selective alterations in the ERG could also result from the loss and/or alteration of rhodopsin. Therefore the rhodopsin content and its relation to the ERG threshold and dark adaptation in the developmentally lead-exposed rats also was examined.[21,21c,21d] An alternative, although not mutually exclusive, hypothesis is that lead produces its selective rod degeneration and functional alterations via an inhibition of the retinal Na^+,K^+-ATPase (Na,K-ATPase) activity. This latter hypothesis is consistent with the observed retinal degeneration that occurs following the intraocular injection of ouabain[61a,61b] and with the rod-selective alterations in amplitude, sensitivity and kinetics following inhibition of retinal Na,K-ATPase by ouabain or strophanthidin.[61c,61d]

3. Electroretinography

Three main dose-response effects of low or moderate level lead exposure during development on the a- and b-wave of single-flash ERGs in fully dark-adapted adult hooded rats were observed.[19,21c] First, the mean amplitudes of the a- and b-wave were both significantly decreased (low Pb: 25 and 15%, respectively; moderate Pb: 40 and 27%, respectively) at all luminance intensities. Second, the mean latencies of the a- and b-wave were significantly increased (low Pb: 23 and 16%, respectively; moderate Pb: 47 and 29%, respectively) at all luminance intensities. Similar decreases in a- and b-wave amplitude[16] and increases in b-wave latency were observed in rats with moderate PbB values following exposure to lead during perinatal development.[16,17] Third, the absolute sensitivity of the a- and b-wave was significantly decreased (low Pb: 0.5 and 0.5 log units, respectively; moderate Pb: 1.0 and 0.5 log units, respectively). No gross differences in the ERG waveforms occurred in the lead-exposed rats, except for a slight increase in the b-wave duration at half-amplitude at the highest intensities. Likewise, the general shape of the voltage-log intensity and latency-log intensity curves for the controls and lead-exposed rats were similar. In addition, the mean log relative b-wave threshold (i.e., retinal sensitivity) in low or moderate level lead-exposed rats was elevated 0.5 log units (3.2×) or 1.2 log units (16×), respectively, at 90 d of age.[21] Thus, following lead exposure during early postnatal development single-flash ERGs reveal significant amplitude, latency, and sensitivity alterations in both the a- and b-wave with larger deficits occurring in the a-wave. These data suggest that low or moderate level lead exposure during early postnatal development produces long-term selective rod deficits. To further delineate and establish the selective rod deficit, scotopically balanced ERGs, cone ERGs, and scotopic and photopic flicker fusion functions were examined in control and lead-exposed rats.

Scotopically balanced b-wave ERGs, elicited with blue (470 nm) and yellow-green (570 nm) stimuli, were examined over a 4-log unit range in fully dark-adapted control and lead-exposed rats. In moderately lead-exposed rats, the mean amplitude of the scotopic b-wave was significantly decreased (41 to 43%) using either the blue or yellow-green stimulus. These results, which further support the presence of a rod deficit, are similar to those observed in the single-flash ERG experiments, although of a slightly larger magnitude (42 versus 27%). Full-field cone ERGs were elicited by using 30-Hz white flashes in the presence of a white background adapting light. The cone b-wave implicit time (or latency) and amplitudes were not significantly different in control and lead-exposed rats.[19,21c] The b-wave CFF curves in control[19,62-64] and lead-exposed[19,21c] rats possess two distinct (rod and cone) plateaus. At low (scotopic) stimulus intensities the CFF was between 10 and 20 flashes/s while at high (photopic) stimulus intensities the CFF increased to between 30 and 40 flashes/s. At the four lowest stimulus intensities the mean CFF was significantly decreased after low and moderate level lead exposure (10 and 18%, respectively). In contrast, at the highest three intensities the mean CFF was either unchanged or increased (10%) in the low and moderate level lead-exposed rats. These results, in combination with those from the cone ERG studies, further support the selectivity of the rod deficit and interestingly suggest a possible enhancement of cone functioning in the moderate level lead group.

4. Rhodopsin and Retinal Sensitivity Studies

In normal man and animals there is a log-linear relation between ERG threshold and rhodopsin content[65-67] and a correlation between rhodopsin content and ROS length.[68,69] Similar relationships between retinal sensitivity, rhodopsin content and ROS length exist in humans with retinitis pigmentosa[70] and in experimental animals with vitamin A deficiency,[71] inherited retinal dystrophy,[72,73] and continuous light exposure.[74]

The rhodopsin content per eye in adult control hooded rats is about 2.0 nmol/eye.[21,21c] Compared to controls, the eyes from adult rats exposed to low and moderate level lead only during development contained 12% and 34% less rhodopsin, respectively. No change in the maximum wavelength of absorption of rhodopsin was seen in the retinas from lead-exposed rats. There is a log-linear relationship between the ERG b-wave threshold and rhodopsin content for all experimental conditions. That is, in both control and lead-exposed rats at 1, 3, and 12 months of age, the log threshold rises linearly as the rhodopsin content decreases. The slopes of the regression lines, determined by the method of least squares, were not significantly different for the controls and lead-exposed rats.

5. Dark and Light Adaptation Studies

One study suggests that a lifetime of moderate-level or high-level lead exposure produces dose-response decreases in the rate of dark adaptation.[17] To determine whether similar changes occurred following low or moderate level lead exposure only during development, the time course and rate of recovery of dark adaptation of the ERG b-wave following a complete (>90%) bleach were examined in lead-exposed rats. There are two mechanisms in the rat retina that dark adapt at different rates following a complete bleach: the first is mediated by cones and the second by rods.[74a,74b] Developmental lead exposure produced a selective dose-response decrease in the sensitivity of the rod phase while the cone phase was unaltered. Following the bleach, the threshold of the rod branch (at 300 min) was significantly elevated 0.5 and 1.2 log units in the low and moderate lead-exposed rats, respectively. These increases in rod threshold were similar in magnitude to those observed with both the b-wave voltage-log intensity and increment threshold studies (vide infra). In addition, they effectively shortened the total range of rod adaptation in both lead groups. Furthermore, in the moderate lead group the sensitivity during the early portion of the rod phase was decreased such that, the rod-cone break occurred between 45 and 50 min instead of between 20 and 25 min: a delay of 25 to 30 min. In addition, the rate of recovery was slowed in the moderate lead group by an average of 18% between 30 and 60 min in the dark. This significantly decreased rate of recovery may be related to the 34% decrease in rhodopsin content per eye observed in this group.[21]

These findings demonstrate that the sensitivity and range of dark adaptation were decreased following low and moderate level developmental lead exposure and that the rate of dark adaptation was decreased only following moderate-level developmental lead exposure. In conjunction with the original results reporting dark adaptation deficits in lead-exposed primates,[17] these results further strengthen the conclusion that lead exposure during early development produces long-term selective rod deficits.

To determine if light adaptation was altered following low or moderate level lead exposure only during development ERG b-wave increment threshold functions were examined.[21d] In fully dark-adapted control eyes, the mean log threshold was -6.8 which was similar to that observed in the voltage-log intensity functions.[19,21c] With adapting luminances 1.0 log unit above threshold, the increment threshold followed the classic Weber-Fechner law (calculated slope = 1.0): a result reported by others.[62-64] In dark-adapted rats from the low and moderate lead groups, the mean log threshold significantly increased by 0.60 and 1.1 log units, respectively. These decreases in retinal sensitivity may be related to the decreases in rhodopsin content per eye determined in these same lead exposure groups.[21,21c] Furthermore, the slopes of the increment threshold function in these rats deviated from the Weber-Fechner law with slopes of 0.89 and 0.81, respectively. This deviation was more pronounced at the lower adapting backgrounds which was a result of the response compression observed in the lead-exposed rats at the lower adapting backgrounds in the log intensity vs. b-wave amplitude functions used to construct these increment threshold functions. The mean threshold changes that occurred in both lead groups at the lowest background intensities (i.e., below -3 log units) were significantly different from controls whereas those at higher background intensities were not different from controls. It has been convincingly demonstrated that ERG responses obtained with log backgrounds equivalent to and brighter than our -3 background are contributed by a photopic receptor mechanism[62,64] whereas those occurring at lower levels of illumination are due to a network adaptive process originating proximal to the photoreceptors.[62,74c] These findings further establish that both low and moderate level developmental lead exposure produce dose-dependent alterations in the sensitivity of the scotopic receptor system without affecting the photopic receptor system.

6. Cyclic Nucleotide Metabolism and Calcium Studies

The concentrations of cAMP and cGMP in the dark-adapted and light-adapted state were determined in retinas from control and similarly lead-exposed rats.[19,21c] No significant differences in cAMP concentrations were found in either state in the lead-exposed rats. In marked contrast, the cGMP concentration exhibited significant dose-response increases in lead-exposed retinas in both the dark-adapted (low Pb: 19%; moderate Pb: 42%) and light-adapted state (low Pb: 12%; moderate Pb: 23%). The increase in the retinal cGMP concentration in the lead-exposed rat retinas could be the result of an activation of guanylate cyclase and/or an inhibition of cGMP-PDE. The guanylate cyclase activity, measured in retinas which were either dark-adapted or light-adapted, did not significantly differ between treatment groups. The cGMP-PDE activity, however, measured in light-adapted retinas was significantly decreased in the low and moderate level lead-exposed retinas compared to controls: 15 and 40%, respectively.

To determine if the effects of lead on the enzymes of cyclic nucleotide metabolism were direct, *in vitro* retinal studies were conducted. The activity of guanylate cyclase and cGMP-PDE was measured in adult control retinas incubated with several concentrations of lead chloride for 5 min. The highest *in vivo* retinal lead concentration was about 10^{-6} M, while at the time of ERG testing and biochemical analysis it was about 10^{-7} M. Incubating with 10^{-7} to 10^{-5} M lead did not affect guanylate cyclase activity. At 1×10^{-4} and 5×10^{-4} M lead, however, guanylate cyclase activity was significantly increased 20 and 158%, respectively. The activity of cGMP-PDE was more sensitive to the effects of added lead, which is similar to what is observed with *in vivo* lead exposure. Below 10^{-7} M, lead did not affect cGMP-PDE activity. However, at 10^{-6}, 10^{-5} and 10^{-4} M lead there were significant decreases (10, 25, and 40%, respectively) in activity.

The *in vitro* data suggest that lead exposure in the mature organism may produce similar retinal deficits. Additionally, if rods and blue-sensitive cones in humans exhibit the same sensitivity to a lead-induced inhibition of cGMP-PDE, as they do to the drug-induced inhibition of cGMP-PDE,[77,78] color vision deficits, in addition to scotopic vision deficits, may result from lead exposure.

The elemental calcium concentration ([Ca]) was determined in the rod outer segments (ROS) of dark-adapted rats exposed to low or moderate levels of lead during development[21d] since lead alters the activity or retinal guanylate cyclase and cGMP-PDE[19,21c] and both enzymes are modulated by calcium.[78a,78b] The [Ca] was significantly elevated in the ROSs of both the low and moderate level lead-exposed rats at 21 d of age (13 and 31%, respectively) and at 90 d of age (9 and 18%, respectively). Interestingly, the ROS [Ca] in the controls and low lead rats rose 10 to 14% between 21 and 90 d of age whereas in the moderate level lead-exposed rats it essentially remained the same. This altered pattern of development in the moderate level group probably reflects the fact that there was a selective loss of 22% of the rods and 20 to 25% of the bipolar cells in this lead exposure group[20] whereas in the low level group there was no retinal degeneration.[21d]

The mechanisms responsible for the decreases in sensitivity of the scotopic receptor system and the increases in ROS [Ca] following low and moderate level developmental lead exposure are unknown. Several possible, though not mutually exclusive, alterations could account for both these alterations. The first involves the lead-induced inhibition of light-activated cGMP-PDE activity and the resulting elevation of retinal cGMP in both the dark-adapted and light-adapted state.[19,21c] An elevated level of cGMP in the dark-adapted rod would open more cGMP-activated channels in the ROS plasma membrane leading to an increased depolarization of the rod.[78c] Following light exposure, the lead-induced inhibition of light-activated cGMP-PDE, coupled with the elevated cGMP level, would prevent closure of all the open cGMP-activated channels and thereby would decrease the sensitivity and increase the rod (and maybe bipolar) [Ca]. Similar rod ERG and photoreceptor sensitivity changes occur in toad, cat, and man following exposure to PDE inhibitors and/or elevation of rod cGMP levels.[53,78d-78g] The second alteration is the inhibition of retinal Na^+,K^+-ATPase (Na,K-ATPase) observed in similarly lead-exposed rats.[21b] An inhibition of rod Na,K-ATPase activity leads to an elevation of the intracellular concentration of ionic sodium ($[Na^+]_i$) and an increased rod depolarization.[61c,61d] Following light exposure, the lead-induced elevation of $[Na^+]_i$ would inhibit (or slow) the Na^+/Ca^{2+} (Na/Ca) exchanger from removing $[Ca^{2+}]_i$ and would thereby lead to an accumulation of $[Ca^{2+}]_i$.[78c] Since the Na/Ca exchanger from rods is approximately $5\times$ slower than in cones,[78c] the rods may preferentially increase their $[Ca^{2+}]_i$. This elevated $[Ca^{2+}]_i$ would prolong cGMP-PDE activation[78b] and thus, through the mechanisms noted above, could decrease the rod sensitivity. Similar rod ERG and photoreceptor sensitivity changes occur in frog and toad following exposure to ouabain or strophanthidin.[61c,61d]

7. Na^+,K^+-ATPase Studies

Lead exposure during early postnatal development produced long-term, dose-dependent and significant decreases in retinal Na,K-ATPase whereas the activity of renal Na,K-ATPase remained unchanged.[21b] Retinal Na,K-ATPase activity in nominally Ca^{2+}-free buffer isolated from the low or moderate level lead-exposed rats was decreased 11 and 26%, respectively. *In vitro* Pb^{2+} experiments were conducted to examine the underlying mechanisms of the Pb^{2+}-induced inhibition of retinal Na,K-ATPase. The half-maximal inhibitory dose (I_{50}) of Pb^{2+} for retinal Na,K-ATPase was 5.21×10^{-7} M. The retinal I_{50} value is equivalent to a blood lead concentration of 11 μg/dl which is in the range observed in the blood of the low lead group and of current clinical concern.[1] The Hill coefficient for the retina was 0.42 which is suggestive of negative cooperativity.[78a] One possibility for this apparent negative cooperativity in the retina is that low concentrations of Pb^{2+} inhibit Na,K-ATPase by binding to different α subunits of the enzyme. This is consistent with the fact that the retina expresses two predominant α subunits of Na,K-ATPase.[78i,78j] Alternatively, and more consistent with the known effects of Pb^{2+} on Na,K-ATPase[78k-78p] and the properties of the enzyme,[78q] Pb^{2+} may subtly but differentially alter the turnover rate of the α3, compared to the α1, isozyme.

Studies on enzyme kinetics, reversibility of inhibition and ion interactions were conducted to investigate further the underlying mechanisms of action of Pb^{2+} on retinal Na,K-ATPase. Lineweaver-Burk plots showed that under the assay conditions employed the control retinal enzyme had an apparent K_m for MgATP of 0.16 mM and a V_{max} of 15.29 μmoles

Pi/mg/h. Kinetic studies conducted in the presence of 0.1 or 1.0 μM Pb^{2+} demonstrated that the Pb^{2+}-induced inhibition was competitive.[78h] In the presence of Pb^{2+}, there was a 2- to 3-fold reduction in the apparent affinity of ATP for Na,K-ATPase which resulted in a 53 to 62% decrease in the V$_{max}$/K$_m$ ratio. The K$_i$ value of Pb^{2+}, as determined by graphical analysis from the Dixon plot was 2.10×10^{-7} M. Ackermann-Potter plots revealed that the Pb^{2+}-induced inhibition of retinal Na,K-ATPase activity was reversible. Thus, submicromolar concentrations of Pb^{2+} competitively and reversibly inhibited isolated retinal Na,K-ATPase.[21b]

As noted by Siegel et al.,[78m] the Pb^{2+}-induced inhibition of Na,K-ATPase suggested that the turnover of the enzyme was slowed but it did not point out the partial reaction(s) on which the rate-limiting effects of Pb^{2+} were exerted. Investigations of the effects of various concentrations Na$^+$, K$^+$, Mg^{2+} and ATP on retinal Na,K-ATPase, inhibited by 5.2×10^{-7} M Pb^{2+} (the I$_{50}$ value), were conducted to help elucidate this problem. In the presence of 5 mM K$^+$, increasing concentrations of Na$^+$ (25 to 150 mM) antagonized the Pb^{2+}-induced inhibition of Na,K-ATPase. In contrast, at a constant concentration of Na$^+$ (130 mM), K$^+$ in the range of 5 to 50 mM did not affect the inhibited enzyme. 10^{-5} M Ca^{2+} (I$_{50}$ value) had no additive effect on the Pb^{2+}-inhibited enzyme. In the presence of 2 mM ATP the Pb^{2+}-induced inhibition of Na,K-ATPase was enhanced 11 to 21% as the concentration of MgCl$_2$ was increased from 2 to 6 mM. In marked contrast, at a constant concentration of MgCl$_2$ (2 mM), increasing the concentration of ATP from 1 to 6 mM reduced the inhibition of Na,K-ATPase by 13 to 27%.

The antagonism of the Pb^{2+}-induced inhibition of Na,K-ATPase by Na$^+$,[21b,78r] suggests that Pb^{2+} was displaced by Na$^+$ from the Na$^+$-dependent phosphorylation site on the enzyme. Initial results from Siegel and co-workers[78k-78n] support this conclusion by showing that Pb^{2+} mimics the action of Na$^+$ by stimulating the phosphorylation of rat brain and electroplax microsomal Na,K-ATPase in the absence of Na$^+$. The Pb^{2+}-stimulated phosphoprotein, in contrast to the Na$^+$-stimulated phosphoprotein, was incapable of undergoing K$^+$-stimulated dephosphorylation. In further studies conducted in the presence of half-saturating concentrations of Pb^{2+} and Na$^+$ however, Siegel et al. found that the levels of phosphate incorporation in the electroplax were cumulative.[78n] Thus, the question of how and at what site on the enzyme Pb^{2+} affected the Na$^+$-dependent phosphorylation remains unsolved.

The potentiation of inhibition of retinal Na,K-ATPase by Mg^{2+} and antagonism of inhibition by high concentrations of ATP[21b,78m,78r,78s] probably reflect physical-chemical alterations in the concentration of Pb^{2+} rather than interactions with the enzyme. For example, the potentiating effect of Mg^{2+} excludes the possibility that Pb^{2+}-induced inhibition was due to competition with Mg^{2+} for an enzyme site or for binding to MgATP and suggests that it resulted from displacement of Pb^{2+} from PbATP. Similarly, increasing the concentration of ATP probably resulted in a lower concentration of Pb^{2+} and thus a decreased inhibition of Na,K-ATPase.

One hypothesis suggests that the decreased retinal Na,K-ATPase activity following developmental lead exposure resulted from the selective loss of rod photoreceptors and bipolar cells in lead-exposed rats since these cells almost exclusively express the α3 (high-affinity ouabain) isozyme of Na,K-ATPase.[78i,78j] Although there is a selective loss of rods and bipolars in the rats exposed to moderate levels of lead during development,[20] there is no retinal degeneration in rats exposed to low levels of lead during development.[21c] Since retinal Na,K-ATPase activity is diminished in both lead-exposed groups, this hypothesis does not appear to be supported.

Alternatively, it appears more likely that a preferential lead-induced inhibition of the α3 isozyme of Na,K-ATPase leads to both a selective loss of photoreceptors and bipolar cells and functional alterations in the remaining photoreceptors and bipolar cells. Our preliminary results, presented in an abstract,[78t] suggest that the α(+) (i.e., α3 and α2) isozyme is more sensitive to the inhibitory effects of Pb^{2+} than the α1 isozyme. Thus, photoreceptors and bipolar cells should be most sensitive to inhibition by Pb^{2+}. Indeed, these are the results we observed following *in vivo* or *in vitro* lead exposure. In addition, this proposed mechanism of lead-mediated cell death and neuronal dysfunction is consistent with the reported effects of ouabain on the retina.[61a-61d]

B. RETINAL GANGLION CELL AXONS OR OPTIC NERVE STUDIES

The effects of high levels of lead on the optic nerve were examined in mice[8] and rats[9] exposed only during the suckling period. In 28-d-old mice, lead exposure produced a loss of 10 to 15% of the large diameter retinal ganglion cell axons (i.e., optic nerve) such that the axon circumferences were skewed to smaller diameters.[8] The authors state no evidence was found of degenerating axons in myelin breakdown. Therefore the question remains as to whether this preferential loss of large diameter axons reflects an actual loss of these axons or the failure of smaller diameter axons to increase in size during the lead exposure period. Interestingly, there was no delay in the onset of myelination, no difference in the total number of glial cells or no difference in the number of oligodendroglia or astrocytes in the lead-exposed mice even though the total amount of myelin produced was decreased 50 to 60%. Biochemically, this hypomyelination was characterized by a 60% decrease in myelin basic protein and a 30 to 60% decrease in the activity of cerebroside sulfotransferase and 2′,3′-cyclic nucleotide 3′-phosphodiesterase (CNPase). Similarly, in the optic nerves of rats exposed to high levels of lead there was a 33% decrease in myelin (determined on the basis of proteolipid protein), a 30% decrease in CNPase and no delay in the onset of myelination levels of lead while there were no changes in myelin (determined on the basis of the concentration of the major protein of PNS myelin, Po) or CNPase in the sciatic nerves of the same animals.[9] In addition, there were no alterations in myelin content or CNPase when rats were exposed to lower levels of lead.[9] The biological significance of a decrease in CNPase or its relation to the lead-induced decrease in retinal cGMP-PDE[19] is unknown since the biologically active cyclic nucleotides are those with a 3′,5′ structure. Thus, exposure to high levels of lead during early postnatal development appears to result in a retarded growth and maturation of the retinal ganglion cell axons with a secondary hypomyelination due to a reduction in axon size. This idea is entirely consistent with the data of Krigman et al.[79] obtained from pyramidal tracts of 30-d-old rats with lead encephalopathy.

The results of the lead studies in rats[9] demonstrate that optic nerve loss and demyelination occurs following high, but not moderate, lead exposure. More recently, it has been demonstrated in 6-year-old monkeys that chronic moderate to high lead exposure from infancy does not result in any change in the mean or median myelinated optic nerve fiber diameter or any change in myelin formation.[22] These results suggest that the latency delays observed in the primary components of the visual evoked potential (VEP) of lead-exposed rats[3] and monkeys[18] are probably not due to segmental demyelination. This conclusion is consistent with the suggestion that the latency delays in the VEP primarily result from alterations in retinal photoreceptors as reflected latency increases in the ERG.[16,19] Furthermore, unpublished results from studies using photic- and electrical-stimulation in rats exposed to low to moderate levels of lead during development[3,12] show that 70 to 75% of the latency delay is due to retinal deficits in temporal processing, 15 to 20% is due to decreased conduction velocity and 15 to 20% is due to increased synaptic delays in the geniculocortical pathway.[21a]

C. VISUAL CORTEX STUDIES

Following low to moderate level lead exposure during development, the highest concentration of lead in the nervous system is found in the visual cortex, retina and hippocampus.[12,80] In young rats and adult monkeys following chronic high level lead exposure during development,[81,82] or infancy,[22] there was a 14 to 27% decrease in the thickness of visual[22,81] and somatosensory[81,82] cortical neuropil, a 12 to 45% decrease in dendritic branching in the visual[22,81] and somatosensory[81,82] cortex, and a 23 to 26% decrease in the number of synapses per neuron in visual[81] and somatosensory cortex.[81,82] The reduction in the subdivision of dendrites probably accounts for the reduced cortical neuropil volume. This is consistent with the lead-induced retardation of growth and maturation of the retinal ganglion cell axons, as previously noted.[79] In rats the decrease in dendritic branching occurred at distances greater than 40 μm from the soma[81] while in the monkeys it occurred closer (i.e., within 20 μm) to the cell body.[22] It is well-known that the further a dendritic branch is from the soma, the greater is the density of spines per unit length of dendrite.[83] Thus, the restricted dendritic field observed following lead exposure in rats probably accounts for the decrease in the mean number of synapses per neuron. The reasons underlying the altered branching pattern observed in the monkey visual cortex remains to be explained.

To date, similar studies as those described above have not been conducted in controls and animals exposed to low levels of lead. In order to determine whether the scotopic deficit produced by lead exposure (vide supra) has a visual cortical and/or retinal site of action such quantitative morphometric studies are needed. As suggested by the histopathological studies of methylmercury poisoning, scotopic deficits can result from selective lesions in the calcarine fissure of the visual cortex.[75,76] The finding that the highest concentrations of lead are found in the visual cortex and retina[12,80] suggest that both may be sites of action following developmental lead exposure.

The effects of lead exposure during development resulting in a moderate PbB on the binding of [^{3}H]quinuclidinylbenzilate (QNB), a cholinergic muscarinic antagonist, and on acetylcholinesterase (AChE) activity in the retina, superior colliculus, lateral geniculate nucleus and visual cortex were examined in the adult rat.[12] A large decrease in [^{3}H]QNB binding (38%) and AChE activity (29%) was found only in the visual cortex of the lead-exposed rats. Scatchard plots of saturation binding data revealed a decrease in the density (B_{max}), but not in the affinity (K_d) of the muscarinic receptor. *In vitro* lead had no effect on [^{3}H]QNB binding or AChE activity in visual cortical membrane preparations from control rats. The decrease in [^{3}H]QNB binding may be due to the inhibition of AChE, however, this latter effect does not appear to be direct as evidenced by the *in vitro* data. These results, in combination with psychophysical and pharmacological studies in similarly lead-exposed rats demonstrating scopolamine supersensitivity,[11,13] suggest that the long-term effects of developmental lead exposure are due to an effect of lead on visual cortex cholinergic neurons.

D. SUMMARY AND CRITIQUE

The original study reporting that developmental lead exposure produces a long term scotopic, but not photopic, deficit was conducted in 2 1/2-year-old rhesus monkeys whose PbB had been maintained at 55 and 85 μg/dl for the first year of life.[2] Only monkeys with high PbB of 85 μg/dl exhibited scoptopic deficits. In a review chapter the authors state, but show no data, that no changes were observed in scotopic or photopic ERG or VEP recordings of the high-lead monkeys while there was a suggestion of reduced neuron density in the calcarine fissure region of the visual cortex.[10] Such a lesion in the visual cortex could possibly account for the scotopic deficit, as observed in methylmercury exposed monkeys,[75,76] although more comprehensive visual testing and quantitative histological and morphological studies will be needed to confirm this hypothesis. In contrast, Lilienthal et al.[18] report that 7-year-old rhesus monkeys, exposed to lead throughout their lifetime, with moderate PbB (40 and 60 μg/dl) have altered ERGs and VEPs under both scotopic and photopic luminance conditions with slightly larger changes occurring under scotopic conditions. Interestingly, results from psychophysical and electrophysiological studies in lead-exposed rats reveal greater decreases in visual acuity under scotopic than photopic luminance conditions.[13] Rummo et al.,[7] who examined 45 controls and 45 children aged 4 to 8 with chronic lead exposure resulting in moderate PbB (55μg/dl), also found a decrease in visual acuity in lead-poisoned children. The site of action of lead accounting for this visual acuity deficit, which was presumably obtained under mesopic to photopic luminance conditions, is unknown. The threshold versus intensity studies by Cavalleri et al.[33] in lead-exposed workers definitely support a retinal (i.e., photoreceptor or bipolar cell) location for the scotopic deficit, however, they only examined two background luminance conditions which were specified only as photopic or mesopic.

These combined results suggest that lead exposure only during early development may result in a selective scotopic deficit. The weight of the evidence suggests that site of action is the rod photoreceptors in the retina, however, alterations in the calcarine area of the visual cortex may be responsible. On the other hand, chronic lead exposure either during development or as an adult may produce scotopic and photopic deficits. The site of action of these more severe deficits are unknown, but both retinal and cortical sites have been implicated.

The work of Fox and co-workers[13,19-21,21a-21d,78t] firmly established that developmental lead exposure in rats results in a selective rod degeneration and long-term rod deficits. At mean PbB of 59 μg/dl selective rod deficits were accompanied by rod degeneration. Not only are these PbB levels above the current level of concern (i.e., 10 to 20 μg/dl[1]), the presence of retinal degeneration obviated a clear interpretation of altered function in the remaining rods. In marked contrast, at mean PbB of 19 μg/dl there were no signs of retinal degeneration while selective ERG rod deficits still occurred, albeit at a reduced level. These results suggest that rod deficits occur independently of rod degeneration although both most likely result from the same underlying mechanism: the expression of which is dependent upon both dose of lead and duration of lead exposure.

E. FUTURE DIRECTIONS

To adequately document whether low-level lead exposure during perinatal development produces a scotopic and/or photopic deficit, a decrease in spatial resolution (i.e., decreased visual acuity), a decrease in temporal resolution (i.e., decreased flicker sensitivity), and/or a dark adaptation deficit, it will be necessary to utilize rigorous and well-validated psychophysical and/or electrophysiological (i.e., ERG and VEP) techniques. The following studies should be performed in children and rhesus monkeys: (1) visual field testing; (2) spatial and temporal contrast sensitivity for gratings; (3) increment-threshold spectral sensitivity with white light backgrounds; (4) threshold versus intensity functions over an adaptation range from complete dark adaptation to high photopic levels of light adaptation; (5) increment-threshold spectral sensitivity with chromatic backgrounds; and (6) flicker sensitivity over an adaptation range from complete dark adaptation to high photopic levels of light adaptation. These are all well-tested experimental procedures that have been performed in humans and monkeys with normal and abnormal visual processing. In some cases the anatomical sites, physiological processes, and biochemical mechanisms accounting for the data in normal and abnormal subjects have been determined. Thus, the results from these studies will provide important data concerning the processing of spatial, temporal, luminance, and chromatic information in the visual pathway. Since subtle impairments in visual processing could have profound effects on learning, this data will also be valuable in further assessing the effects of lead on cognitive function. Future studies will then have to focus on the cellular sites and mechanisms of action. The author believes that no one mechanism will account for the diverse effects of lead poisoning. In the final analysis, most likely it will eventually be found that lead exerts its toxic effects at several cellular and subcellular regulatory sites of action.

III. AUDITORY SYSTEM

A. DEVELOPMENTAL EXPOSURE

There are only a few studies that have examined the effects of perinatal lead exposure on auditory processing.[23-30] These studies reveal that complex auditory processing deficits occur in children with acute lead exposure[24] and asymptomatic lead exposure.[23,25,26,28-30] Otto and his co-workers[23,28,29] are the only investigators to use audiometric testing and the BAER to examine the effects of lead exposure on children. Their results reveal that perinatal lead exposure produces a concentration-dependent increase in hearing thresholds (determined by using pure-tone audiometric screening) and an increase in central conduction time in the auditory pathway (determined using "click" stimuli with BAER) lead-exposed children.[28,29] The increase in wave 1 to 5 interpeak latencies was interpreted by the authors to suggest subclinical pathology of the auditory pathway rostral to cochlear nucleus.[23] Evidence of a lead-induced decrease in hearing acuity (i.e., increased threshold) is supported further by an analysis of audiometric data from the large dataset of NHANES II for children aged 4 to 19 years (about 5,000 children).[30] Hearing thresholds in the right and left ear increased significantly with the PbB level when examined at 500, 1000, 2000, and 4000 Hz although the magnitude of the effect was greater in the right ear at all frequencies. As noted by Schwartz and Otto,[30] these findings provided the first evidence of lead-related effects on hearing in the general population and they occurred in children with PbB below 25 μg%.

To the author's knowledge there is only one report on the effects of perinatal lead exposure on the auditory system in experimental animals.[27] This study, conducted in young rhesus monkeys with moderate PbB, used consonant-vowels syllables to elicit a BAER to investigate the ontogeny of auditory perception.[27] The results show that monkeys exposed to lead exhibited poorer discrimination for the /b/ and /g/ syllables than controls.

B. ADULT EXPOSURE
1. Human Studies

Over the past 40 years, clinical research conducted in Europe has identified frequency-dependent auditory deficits in workers exposed to lead.[36-39] Generally, there were no or slight increases in hearing thresholds below 2000 to 3000 Hz and large increases (30 to 60 dB) in hearing threshold above 2000 to 3000 Hz. For example, Balzano[36] found hearing losses in workers chronically exposed to lead such that, there were no threshold changes below 512 Hz, a moderate increase in threshold between 512 and 2000 Hz, and large increase in thresholds above 2000 Hz. No PbB are available for these studies.[36-39] The same hearing losses were evident whether the tone was produced by air or via bone conduction, thereby demonstrating that the losses were not due to alterations in middle-ear transmission (i.e., conductive hearing loss). Although the above investigators adequately controlled for age-related hearing losses they often used sample sizes that were too small and presented no prelead exposure data.

More recently, Repko et al.[40] performed a series of neurobehavioral tests upon 85 workers in the U.S. exposed to lead (mean PbB of 46 μg/dl) on their jobs within the battery manufacturing industry. Pure-tone thresholds among the lead workers, compared to age-matched controls, were significantly higher at 500, 1000, and 4000 Hz in the right ear and at 3000 to 6000 Hz in the left ear. Loss of hearing in the right ear was positively and significantly correlated to decreases in erythrocyte ALA-D activity, thereby suggesting a concentration-dependent decrement. In addition, a tone-decay test, an auditory adaptation measurement accepted as a useful and powerful technique for separating sensory (cochlear) disorders from those of neural origin,[84] was conducted on the right ear. A normal functioning auditory system exhibits little or no tone decay. In lead workers, tone decay occurred earlier than in any of the control subjects with significant between group differences observed at 500, 2000, and 8000 Hz. There was a significant negative correlation between the audible duration of a tone and PbB. The authors[40] state that the results from this study suggest that chronic lead exposure produces retrocochlear pathology.

2. Animal Studies

Behavioral, morphological, and electrophysiological studies conducted in adult sheep, guinea pigs, and rats have demonstrated that subacute lead exposure produces functional and structural auditory system deficits.[46-48] Using an auditory signal detection task, Van Gelder et al.[47] observed that adult sheep consuming lead acetate (100 mg/kg in feed for 9 weeks) had fewer correct responses and increased response variability compared to age-matched controls. After injecting adolescent guinea pigs once weekly with high doses of lead acetate (300 mg/kg, i.p.) for 7 weeks, resulting in PbB of 310 to 420 μg%, segmental demyelination and axonal degeneration was observed in the cochlear (VIII) nerve while the sensory cells (i.e., spiral and vestibular ganglion cells) of the inner ear appeared normal.[43] The authors thus suggest that adult lead exposure produces a central, but not peripheral, auditory degeneration. The results of a comprehensive series of pure-tone electrophysiological studies, evaluating different levels of the peripheral and central auditory system in adult guinea pigs following dose-response high level lead exposure, are entirely consistent with this interpretation.[45] Yamamura et al.[48] showed that high lead exposure (PbB of 80 to 142 μg/dl) produced no changes in the endocochlear potential or cochlear microphonic (analyzed at 2 to 6 kHz) thus suggesting that lead did not induce a dysfunction of the stria vascularis in the wall of the scala media or organ of Corti, respectively. In contrast, there was a dose-dependent increase (10 to 25 dB) in the threshold of the primary component (N1) of the cochlear nerve as well as a dose-dependent decrease in the maximum amplitude of N1.

C. SUMMARY, CRITIQUE, AND FUTURE DIRECTIONS

Without exception, the investigators conducting electrophysiological, behavioral, and morphological studies on lead-exposed children or adult humans, or animals conclude that lead appears to exert its main effect on the central auditory system (e.g., cochlear nerve) and not on the peripheral auditory system (i.e., cochlea or spiral ganglion).[23,28,29,40-45] If this conclusion is valid, and certainly more detailed electrophysiological and morphological experiments on the cochlea and ascending pathways of the auditory system are needed to prove this, it will be concluded that lead is not an ototoxic substance. Such an ototoxic chemical or drug, like the arsenicals and aminoglycoside antibiotics, is defined as an agent that produces a hearing impairment and corresponding structural damage to the peripheral auditory system.[85] This is in indirect contrast to the effect of lead on the visual system where it appears that lead affects the sensory receptor (i.e., photoreceptors) and visual cortex (vide supra).

Three of the main problems with the above experiments are the lack of pure-tone BAER experiments in the lead-exposed children, the complete lack of experimental data in animals exposed to low levels of lead during perinatal development and the absence of any morphological data on the cochlea. These problems also point the way for future research in this extremely important and relevant area of sensory toxicology. To evaluate the integrity of the auditory system at each level of auditory processing it is necessary to utilize a pure tone burst with carefully controlled onsets and offsets that produce a narrow spectral peak at a pure-tone frequency rather than a "click" stimuli. Although clicks are easier to produce and simultaneously stimulate a large number of hair cells, a click spreads its energy over a broad range of audio frequencies. Therefore, spectral peaks will occur at frequencies determined by the pulse width and resonances in the loudspeaker and in the receiving ear's sound delivery pathway. Clicks are useful for screening tests to determine whether or not the auditory system is responsive, but are useless in audiometric analysis. The latter two problems are self explanatory.

IV. CONCLUSIONS

Sensory system toxicology is a research area of immense importance and opportunity. Subtle alterations in visual or auditory processing during development can have profound immediate, long-term and maybe even delayed effects on the mental, social, and physical health and performance of a developing child. In addition, such alterations can markedly affect the quality of life in adults. The relevance and applicability of the visual system data summarized above to low-level pediatric lead poisoning has yet to be firmly established. On the other hand, the mechanisms of the lead-induced retinal and visual system disorders produced in experimental animals are now beginning to be understood. In contrast, it appears that the developing human auditory system is extremely sensitive to low levels of lead. However, the exact sites and mechanisms of action accounting for the auditory processing deficits have not yet begun to be explored. The data above suggest that low to moderate chronic lead exposure in the mature organism may produce similar visual and auditory deficits — a point not often stressed. Furthermore, these data suggest the need for scotopic visual system evaluation and routine audiometric monitoring of children and adults at risk for lead poisoning. Finally, these data on the visual and auditory toxicity of lead clearly support the idea that 3 to 4 million children in the U.S. exposed to environmental sources of lead are at risk of adverse health effects.[1]

ACKNOWLEDGMENTS

I thank April G. Fox, CCC/Sp. for her invaluable discussions on the auditory system. Original work reported herein from Dr. Donald A. Fox's laboratory was supported in part by an NIH Grant RO1 ES 03183 from the National Institute of Environmental Health Sciences.

REFERENCES

1. U.S. Department of Health and Human Services. The Nature and Extent of Lead Poisoning in Children in the Agency for Toxic Substance and Disease Registry, U.S. Department of Health and Human Services, United States: A Report to Congress, 1988.
2. **Bushnell, P. J., Bowman, R. E., Allen, J. R., and Marlar, R. J.,** Scotopic vision deficits in young monkeys exposed to lead, *Science,* 196, 333, 1977.
3. **Fox, D. A., Lewkowski, J. P., and Cooper, G. P.,** Acute and chronic effects of neonatal lead exposure on the development of the visual evoked response in rats, *Toxicol. Appl. Pharmacol.,* 40, 449, 1977.
4. **Winneke, G., Brockhaus, A., and Baltissen, R.,** Neurobehavioral and systemic effects of longterm blood lead elevation in rats, *Arch. Toxicol.,* 37, 247, 1977.
5. **Fox, D. A., Lewkowski, J. P., and Cooper, G. P.,** Persistent visual cortex excitability alterations produced by neonatal lead exposure, *Neurobehav. Toxicol.,* 1, 101, 1979.
6. **Winneke, G.,** Modification of visual evoked potentials in rats after longterm blood lead elevation, *Act. Nerv. Sup. (Praha),* 4, 282, 1979.
7. **Rummo, J. H., Routh, D. K., Rummo, N. J., and Brown, J. F.,** Behavioral and neurological effects of symptomatic and asymptomatic lead exposure in children, *Arch. Environ. Health,* 34, 120, 1979.
8. **Tennekoon, G., Aitchison, B. S., Frangia, J., Price, D. L., and Goldberg, A. M.,** Chronic lead intoxication: effects on developing optic nerve, *Ann. Neurol.,* 5, 558, 1979.
9. **Toews, A. D., Krigman, M. R., Thomas, D. J., and Morell, P.,** Effect of inorganic lead exposure on myelination in the rat, *Neurochem. Res.,* 5, 605, 1980.
10. **Bowman, R. E. and Bushnell, P. J.,** Scotopic visual deficits in young monkeys given chronic low levels of lead, in *Neurotoxicity of the Visual System,* Merigan, W. H. and Weiss, B., Eds., Academic Press, New York, 1980, 219.
11. **Fox, D. A., Wright, A. A., and Costa, L. G.,** Visual acuity deficits following neonatal lead exposure: cholinergic interactions, *Neurobehav. Toxicol. Teratol.,* 4, 689, 1982.
12. **Costa, L. G. and Fox, D. A.,** A selective decrease of cholinergic muscarinic receptors in the visual cortex of adult rats following developmental lead exposure, *Brain Res.,* 276, 259, 1983.
13. **Fox, D. A.** Psychophysically and electrophysiologically determined spatial vision deficits in developmentally lead-exposed rats have a cholinergic component, in *Cellular and Molecular Neurotoxicology,* Narahashi, T., Ed., Raven Press, New York, 1984, 123.
14. **Santos-Anderson, R. M., Tso, M. O. M., Valdes, J. J., and Annau, Z.,** Chronic lead administration in neonatal rats: electron microscopy of the retina, *J. Neuropathol. Exp. Neurol.,* 43, 175, 1984.

15. **Winneke, G., Beginn, U., Ewert, T., Havestadt, C., Krause, C., Thron, H. L., and Wagner, H. M.**, Study to determine the subclinical effects of lead on the nervous system in children with known prenatal exposure in Nordenham, *Schr.-Reine Verein Wabolu,* 59, 215, 1984.
16. **Azazi, M., Kristensson, K., Malm, G., and Wachmeister, L.**, Studies on developmental alterations in the electroretinogram in rats after post-natal exposure to lead, *Acta Ophthalmol.,* 63, 574, 1985.
17. **Hennekes, R., Janssen, K., Munoz, C. and Winneke, G.**, Lead-induced ERG alterations in rats at high and low levels of exposure, *Concepts Toxicol.,* 4, 193, 1987.
18. **Lilienthal, H., Lenaerts, C., Winneke, G., and Hennekes, R.**, Alteration of the visual evoked potential and the electroretinogram in lead-treated monkeys, *Neurotoxicol. Teratol.,* 10, 417, 1988.
18a. **Gilbert, S. G. and Rice, D. C.**, Life-time exposure to lead produces visual deficits, *Toxicologist,* 11, 164, 1991.
19. **Fox, D. A. and Farber, D. B.**, Rods are selectively altered by lead. I. Electrophysiology and biochemistry, *Exp. Eye Res.,* 46, 597, 1988.
20. **Fox, D. A. and Chu, L. W.-F.**, Rods are selectively altered by lead. II. Ultrastructure and quantitative histology, *Exp. Eye Res.,* 46, 613, 1988.
21. **Fox, D. A. and Rubinstein, S. D.**, Age-related changes in retinal sensitivity, rhodopsin content and rod outer-segment length in hooded rats following low-level lead exposure during development, *Exp. Eye Res.,* 48, 237, 1989.
21a. **Fox, D. A.**, Electrophysiological alterations in humans and animals following low-level developmental lead exposure, *Fund. Appl. Toxicol.,* 9, 599, 1987.
21b. **Fox, D. A., Rubinstein, S. D., and Hsu, P.**, Lead inhibits rat retinal, but not kidney, Na^+,K^+-ATPase, *Toxicol. Appl. Pharmacol.,* 109, 482, 1991.
21c. **Fox, D. A., Katz, L. M., and Farber, D. B.**, Low-level developmental lead exposure decreases sensitivity, amplitude and temporal resolution of rods, *Neurotoxicology,* 12, 641, 1991.
21d. **Fox, D. A. and Katz, L. M.**, Developmental lead exposure selectively alters the scotopic ERG component of dark and light adaptation and increases rod calcium content, *Vision Res.,* 31, in press, 1991.
22. **Reuhl, K. R., Rice, D. C., Gilbert, S. G., and Mallett, J.**, Effects of chronic developmental lead exposure on monkey neuroanatomy: visual system, *Toxicol. Appl. Pharmacol.,* 99, 501, 1989.
23. **Otto, D. A., Robinson, G., Baumann, S., Schroeder, S., Mushak, P., Kleinbaum, D., and Boone, L.**, 5 year follow-up study of children with low-to-moderate lead absorption: electrophysiological evaluation, *Environ. Res.,* 38, 168, 1985.
24. **Jenkins, D. and Mellins, R.**, Lead poisoning in children, *Arch. Neurol. Psychiat.,* 77, 70, 1954.
25. **Perino, J. and Ernhart, C. B.**, The relation of subclinical lead level to cognitive and sensorimotor impairment in Black preschoolers, *J. Learn. Disab.,* 7, 26, 1974.
26. **Needleman, H. L., Gunnoe, C., Leviton, A., Reed, R., Peresie, H., Maher, C., and Barrett, P.**, Deficits in psychologic and classroom performance of children with elevated dentine lead levels, *N. Engl. J. Med.,* 300, 689, 1979.
27. **Laughlin, N. K., Bowman, R. E., Levine, E., and Bushnell, P.**, Neurobehavioral consequences of early exposure to lead in rhesus monkeys: effects on cognitive behaviors, in *Reproductive and Developmental Toxicity of Metals,* Clarkson, T. W., Nordberg, G. F., and Sager, P. R., Eds., Plenum Press, New York, 1983, 497.
28. **Robinson, G., Baumann, S., Kleinbaum, D., Barton, C., Schroeder, S. R., Mushak, P., and Otto, D. A.**, Effects of low to moderate lead exposure on brainstem auditory evoked potentials in children, in Neurobehavioral Methods in Occupational and Environmental Health, World Health Organization, Copenhagen, Denmark, *Environ. Hlth. Doc.,* 3, 1985, 177.
29. **Robinson, G., Keith, R. W., Bornschein, R. L., and Otto, D. A.**, Effects of environmental lead exposure on the developing auditory system as indexed by brainstem auditory evoked potential and pure tone hearing evaluations in young children, in *International Conference: Heavy Metals in the Environment,* Vol. 1, Lindberg, S. E. and Hutchinson, T. C., Eds., CEP Consultants, Edinburg, England, 1987, 223.
30. **Schwartz, J. and Otto, D. A.**, Blood lead, hearing thresholds, and neurobehavioral development in children and youth, *Arch. Environ. Health,* 42, 153, 1987.
31. **Guguchkova, P. T.**, Electroretinographic and electrooculographic examinations of persons occupationally exposed to lead, *Vestnik. Oftalmolog.,* 85, 60, 1972.
32. **Hanninen, H., Mantere, P., Hernberg, S., Seppalainen, A. M., and Kock, B.**, Subjective symptoms in low-level exposure to lead, *Neurotoxicology,* 1, 333, 1979.
33. **Cavelleri, A., Trimarchi, F., Gelmi, C., Baruffini, A., Minoia, C., Biscaldi, G., and Gallo, G.**, Effects of lead on the visual system of occupationally exposed subjects, *Scand. J. Work Environ. Health,* 8 (Suppl. 1), 148, 1982.
34. **Signorino, M., Scarpino, O., Provincialli, L., Marchesi, G. F., Valentino, M., and Governa, M.**, Modification of the electroretinogram and of different components of the visual evoked potentials in workers exposed to lead, *Ital. Electroenceph. J.,* 10, 51, 1983.

35. **Betta, A., De Santa, A., Savonitto, C., and D'Andrea, F.**, Flicker fusion test and occupational toxicology: performance evaluation in workers exposed to lead and solvents, *Human Toxicol.*, 2, 83, 1983.
36. **Balzano, I.**, Chronic lead poisoning and changes in the internal ear, *Rass. Med. Ind.*, 21, 320, 1952.
37. **Koch, C. and Serra, M.**, The effects of tetraethyllead poisoning on hearing and vestibular systems, *Acta Med. Ital. Med. Trop. Subtrop. Gastroenterol.*, 17, 77, 1962.
38. **Gammarrota, M. and Bartoli, E.**, Considerations on relations between lead intoxication and cochlear defect, *Clin. Otorhinolaringoratrica*, 16, 136, 1964.
39. **Valcie, I. and Monojlovic, C.**, Results obtained in examination of a group of workers in the chemical industry in regard to effects of hearing, in XVI Int. Congr. Occupational Health, Tokyo, Japan, 1969.
40. **Repko, J. D., Morgan, B. B., Jr., and Nicholson, J. A.**, Behavioral effects of occupational exposure to lead, NIOSH Research Report, Health, Education, and Welfare Publication No. (NIOSH) 75-164, 1975.
41. **Karai, I., Horiguchi, S.-H., and Nishikawa, N.**, Optic atrophy with visual field defect in workers occupationally exposed to lead for 30 years, *J. Toxicol. Clin. Toxicol.*, 19, 409, 1982.
42. **Sborgia, G., Assennato, G., L'Abbate, N., DeMarinis, L., Paci, C., DeNicolo, M., DeMarinis, G., Montrone, N., Ferrannini, E., Specchio, L., Masi, G., and Olivieri, G.**, Comprehensive neurophysiological evaluation of lead-exposed workers, in *Neurobehavioral Methods in Occupational Health*, Gilioli, R., Cassito, M., and Foa, V., Eds., Pergamon Press, New York, 1983, 283.
43. **Hass, G. M., Brown, D. V. L., Eisenstein, R., and Hemmens, A.**, Relation between lead poisoning in rabbit and man, *Am. J. Pathol.*, 45, 691, 1964.
44. **Brown, D. V. L.**, Reaction of the rabbit retinal pigment epithelium to systemic lead poisoning, *Tr. Am. Ophthalmol. Soc.*, 72, 404, 1974.
45. **Hughes, W. F. and Coogan, P. S.**, Pathology of the pigment epithelium and retina in rabbits poisoned with lead, *Am. J. Pathol.*, 77, 237, 1974.
46. **Gozdzik-Zolnierkiewicz, T. and Moszynski, B.**, Eighth nerve in experimental lead poisoning, *Acta Otolaryngol.*, 68, 85, 1969.
47. **Van Gelder, G. A., Carson, T., Smith, R. M., and Buck, W. B.**, Behavioral toxicologic assessment of the neurologic effect of lead in sheep, *Clin. Toxicol.*, 6, 405, 1973.
48. **Yamamura, K., Terayama, K., Yamamoto, N., Kohyama, A., and Kishi, R.**, Effects of acute lead acetate exposure on adult guinea pigs: electrophysiological study of the inner ear, *Fundam. Appl. Toxicol.*, 13, 509, 1989.
49. **Fox, D. A. and Sillman, A. J.**, Heavy metals affect rod, but not cone, photoreceptors, *Science*, 206, 78, 1979.
50. **Sillman, A. J., Bolnick, D. A., Bosetti, J. B., Haynes, L. W. and Walter, A. E.**, The effects of lead and casmium on the mass photoreceptor potential: the dose-response relationship, *Neurotoxicology*, 3, 179, 1982.
51. **Sillman, A. J., Bolnick, D. A., Bosetti, J. B., Haynes, L. W., and Walter, A. E.**, The effect of lead on photoreceptor response amplitude — influence of the light stimulus, *Exp. Eye Res.*, 39, 183, 1984.
52. **Sillman, A. J., Bolnick, D. A., Bosetti, J. B., Haynes, L. W., and Walter, A. E.**, The effect of lead on photoresponse amplitude — influence of removing external calcium and bleaching rhodopsin, *Neurotoxicology*, 7, 1, 1986.
53. **Tessier-Lavigne, M., Mobbs, P., and Attwell, D.**, Lead and mercury toxicity and the rod light response, *Invest. Ophthalmol. Vis. Sci.*, 26, 1117, 1985.
54. **Frumkes, T. E. and Eysteinsson, T.**, The cellular basis for suppressive rod-cone interaction, *Vis. Neurosci.*, 1, 263, 1988.
55. **Eysteinsson, T. and Frumkes, T. E.**, Physiological and pharmacological analysis of suppressive rod-cone interaction in *Necturus* retina, *J. Physiol.*, 61, 866, 1989.
55a. **Fox, D. A.**, unpublished data.
56. **Carter-Dawson, L., Kuwabara, T., O'Brien, P. J. and Bieri, J. G.**, Structural and biochemical changes in vitamin A-deficient rat retinas, *Invest. Ophthalmol. Vis. Sci.*, 18, 437, 1979.
57. **Kuwabara, T. and Gorn, R. A.**, Retinal damage by visible light, *Arch. Ophthalmol.*, 79, 69, 1968.
58. **Noell, W. K.**, The impairment of visual cell structure by iodoacetate, *J. Cell. Comp. Physiol.*, 40, 25, 1952.
59. **Lolley, R. N., Farber, D. B., Rayborn, M. R., and Hollyfield, J. G.**, Cyclic-GMP accumulation causes degeneration of photoreceptor cells: simulation of an inherited disease, *Science*, 196, 664, 1974.
60. **Farber, D. B. and Shuster, T. A.**, Cyclic nucleotides in retinal function and degeneration, in *The Retina, Part 1*, Adler, R. and Farber, D. B., Eds., Academic Press, New York, 1986, 239.
61. **Haynes, L. and Yau, K.-W.**, Cyclic GMP-sensitive conductance in outer segment membrane of catfish cones, *Nature*, 317, 61, 1985.
61a. **Wolburg, H.**, Time- and dose-dependent influence of ouabain on the ultrastructure of optic neurones, *Cell Tissue Res.*, 164, 503, 1975.
61b. **Wolburg, H.**, Axonal transport, degeneration and regeneration in the visual system of the goldfish, *Adv. Anat. Embryol. Cell. Biol.*, 67, 1, 1991.

61c. **Frank, R. N. and Goldsmith, T. H.**, Effects of cardiac glycosides on electrical activity in isolated retina of the frog, *J. Gen. Physiol.*, 50, 1585, 1967.
61d. **Torre, V.**, The contribution of the electrogenic sodium-potassium pump to the electrical activity of toad rods, *J. Physiol.*, 333, 315, 1982.
62. **Dodt, E. and Echte, K.**, Dark and light adaptation in pigmented and white rat as measured by electroretinogram threshold, *J. Neurophysiol.*, 24, 427, 1961.
63. **Dowling, J.**, Visual adaptation: its mechanism, *Science*, 157, 584, 1967.
64. **Green, D. G.**, Scotopic and photopic components of the rat electroretinogram, *J. Physiol.*, 228, 781, 1973.
65. **Dowling, J. E.**, Chemistry of visual adaptation in the rat, *Nature*, 188, 114, 1960.
66. **Rushton, W. A. H.**, Rhodopsin measurement and dark-adaptation in a subject deficient in cone vision, *J. Physiol.*, 156, 193, 1961.
67. **Fulton, A. B. and Baker, B. N.**, The relation of retinal sensitivity and rhodopsin in developing rat retina, *Invest. Ophthalmol. Vis. Sci.*, 25, 647, 1984.
68. **Bonting, S. L., Caravaggio, L. L., and Gouras, P.**, The rhodopsin cycle in the developing retina. I. Relation of rhodopsin content, electroretinogram and rod structure in the rat, *Exp. Eye Res.*, 1, 14, 1961.
69. **Battelle, B. A. and LaVail, M. M.**, Rhodopsin content and rod outer segment length in albino rat eyes: modification by dark adaptation, *Exp. Eye Res.*, 26, 487, 1978.
70. **Ripps, H., Brin, K. P., and Weale, R. A.**, Rhodopsin and visual threshold in retinitis pigmentosa, *Invest. Ophthalmol. Vis. Sci.*, 17, 735, 1978.
71. **Dowling, J. E.**, Night blindness, dark adaptation, and the electroretinogram, *Am. J. Ophthalmol.*, 50, 875, 1960.
72. **Dowling, J. E. and Sidman, R. L.**, Inherited retinal dystrophy in the rat, *J. Cell Biol.*, 14, 159, 1962.
73. **LaVail, M. M. and Battelle, B. A.**, Influence of eye pigmentation and light deprivation on inherited retinal dystrophy in the rat, *Exp. Eye Res.*, 21, 167, 1975.
74. **Rapp, L. M. and Williams, T. P.**, Rhodopsin content and electroretino-graphic sensitivity in light-damaged rat retina, *Nature*, 267, 835, 1977.
74a. **Cicerone, C. M.**, Cones survive rods in the light-damaged eye of the albino rat, *Science*, 194, 1183, 1976.
74b. **Perlman, I.**, Dark-adaptation in abnormal (RCS) rats studied electrophysiologically, *J. Physiol.*, 278, 161, 1978.
74c. **Green, D.G. and Powers, M. K.**, Mechanisms of light adaptation in rat retina, *Vision Res.*, 22, 209, 1982.
75. **Evans, H. L. and Garman, R. H.**, Scotopic vision as an indicator of neurotoxicity, in *Neurotoxicity of the Visual System*, Merigan, W. H. and Weiss, B., Eds., Raven Press, New York, 1980, 135.
76. **Merigan, W. H.**, Visual fields and flicker thresholds in methylmercury poisoned monkeys, in *Neurotoxicity of the Visual System*, Merigan, W. H. and Weiss, B., Eds., Raven Press, New York, 1980, 149.
77. **Zrenner, E. and Gouras, P.**, Blue-sensitive cones of the cat produce a rodlike electroretinogram, *Invest. Ophthalmol. Vis. Sci.*, 18, 1076, 1979.
78. **Zrenner, E., Kramer, W., Bittner, Ch., Bopp, M., and Schlepper, M.**, Rapid effects on colour vision. Following intravenous injection of a new, non-glycoside positive inotropic substance (AR-L 115 BS), *Doc. Ophthalmol. Proc. Ser.*, 33, 493, 1982.
78a. **Koch, K.-W. and Stryer, L.**, Highly cooperative feedback control of retinal rod guanylate cyclase by calcium ions, *Nature*, 334, 64, 1988.
78b. **Kawamura, S. and Murakami, M.**, Calcium-dependent regulation of cyclic GMP phosphodiesterase by a protein from frog retinal rods, *Nature*, 349, 420, 1991.
78c. **Yau, K.-W. and Baylor, D. A.**, Cyclic GMP-activated conductance of retinal photoreceptor cells, *Ann. Rev. Neurosci.*, 12, 289, 1989.
78d. **Lipton, S. A., Ostroy, S. E., and Dowling, J. E.**, Electrical and adaptive properties of rod photoreceptors in Bufo marinus. I. Effects of altered extracellular Ca^{2+} levels, *J. Gen. Physiol.*, 70, 747, 1977.
78e. **Miller, W. H. and Nicoll, G. D.**, Evidence that cyclic GMP regulates membrane potential in rod photoreceptors, *Nature*, 280, 64, 1979.
78f. **Zrenner, E., Kramer, W., Bittner, C., Bopp, M., and Schlepper, M.**, Rapid effects on colour vision following intravenous injection of a new non-glycoside inotropic substance (AR-L 115), *Doc. Opthalmol.*, 33, 493, 1982.
78g. **Capovilla, M., Cervetto, L., and Torre, V.**, The effect of phosphodiesterase inhibitors on the electrical activity of toad rods, *J. Physiol.*, 343, 277, 1983.
78h. **Dixon, M., Webb, E. C., Thorne, C. J. R., and Tipton, K. F.**, *Enzymes*, 3rd ed., Academic Press, New York, 1979, 333.
78i. **Sweadner, K. J. and McGrail, K. M.**, Na,K-ATPase isozyme localization in retina and optic nerve, *Investig. Ophthalmol. Vis. Sci. Suppl.*, 30, 288, 1989.
78j. **Schneider, B. G. and Kraig, E.**, Na^+ + K^+-ATPase of the photoreceptor: selective expression of $\alpha 3$ and $\beta 2$ isoforms, *Exp. Eye Res.*, 51, 553, 1990.

78k. **Siegel, G. J. and Fogt, S. E.,** Lead ion activates phosphorylation of electroplax Na^+- and K^+-dependent adenosine triphosphatase [(NaK)-ATPase] in the absence of sodium ion, *Arch. Biochem. Biophys.,* 174, 744, 1976.

78l. **Siegel, G. J. and Fogt, S. E.,** Inhibition by lead ion of electrophorus electroplax $(Na^+ + K^+)$-adensoine triphosphatase and K^+-p-nitrophenylphosphatase, *J. Biol. Chem.,* 252, 5201, 1977.

78m. **Siegel, G. J., Fogt, S. E., and Hurley, M. J.,** Lead actions on sodium-plus-potassium-activated adenosinetriphosphatase from electroplax, rat brain and rat kidney, in *Membrane Toxicity,* Miller, M. W. and Shamoo, A. E., Eds., Plenum Press, New York, 1977, 465.

78n. **Siegel, G. J., Fogt, S. K., and Iyengar, S.,** Characteristics of lead ion-stimulated phosphorylation of electrophorus electricus electroplax $(Na^+ + K^+)$-adenosine triphosphatase and inhibition of ATP-ADP exchange, *J. Biol. Chem.,* 253, 7207, 1978.

78o. **Bertoni, J. M. and Sprenkle, P. M.,** Inhibition of brain cation pump enzyme by in vitro lead ion: effects of low level [Pb] and modulation by homogenate, *Toxicol. Appl. Pharmacol.,* 93, 101, 1988.

78p. **Rajanna, B., Chetty, C. S., McBride, V., and Rajanna, S.,** Effects of lead on K^+-para-nitrophyl phosphatase activity and protection by thiol reagents, *Biochem. Int.,* 20, 1011, 1990.

78q. **Sweadner, K. J.,** Enzymatic properties of separated isozymes of the Na,K-ATPase, *J. Biol. Chem.,* 260, 11508, 1985.

78r. **Nechay, B. R. and Saunders, J. P.,** Inhibitory characteristics of lead chloride in sodium- and potassium-dependent adenosinetriphosphatase preparations derived from kidney, brain and heart of several species, *J. Toxicol. Environ. Hlth.,* 4, 147, 1978.

78s. **Hexum, T. D.,** Studies on the reaction catalyzed by transport (Na,K) adenosine triphosphatase. I. Effects of divalent metals, *Biochem. Pharmacol.,* 23, 3441, 1974.

78t. **Fox, D. A., Rubinstein, S. D., and Hsu, P.,** Lead exposure preferentially inhibits the $\alpha(+)$-high affinity ouabain isozyme of rat retinal Na,K-ATPase, *Investig. Ophthalmol. Vis. Sci. Suppl.,* 31, 298, 1990.

79. **Krigman, M. R., Druse, M. J., Traylor, T. D., Wilson, M. H., Newell, L. R., and Hogan, E. L.,** Lead encephalopathy in the developing rat: effect upon myelination, *J. Neuropathol. Exp. Neurol.,* 33, 78, 1974.

80. **Stowe, H. D., Goyer, R. A., Krigman, M. R., Wilson, M., and Cates, M.,** Experimental oral lead toxicity in young dogs, *Arch. Pathol.,* 95, 106, 1973.

81. **Petit, T. L. and LeBoutillier, J. C.,** Effects of lead exposure during development on neocortical dendritic and synaptic structure, *Exp. Neurol.,* 64, 482, 1979.

82. **Krigman, M. R. and Hogan, E. L.,** Effect of lead intoxication on the postnatal growth of the rat nervous system, *Environ. Health Perspect.,* 7, 187, 1974.

83. **Valverde, F.,** Apical dendritic spines of the visual cortex and light deprivation in the mouse, *Exp. Brain Res.,* 3, 337, 1967.

84. **Willeford, J.,** The geriatric patient, in *Audiological Assessment,* Rose, D. E., Ed., Prentice Hall, Englewood Cliffs, NJ, 1978, 403.

85. **Prosen, C. A. and Stebbins, W. C.,** Ototoxicity, in *Experimental and Clinical Neurotoxicology,* Spencer, P. S. and Schaumburg, H. H., Eds., Williams and Wilkins, Baltimore, MD, 1980, 62.

Chapter 7

DEVELOPMENTAL NEUROBIOLOGY OF LEAD TOXICITY

Gary W. Goldstein

TABLE OF CONTENTS

I. Introduction 126

II. Synaptogenesis 126

III. Blood-Brain Barrier 128

IV. Second Messenger Metabolism 131

V. Summary 133

Acknowledgments 133

References 134

I. INTRODUCTION

Central nervous system toxicity from inorganic lead varies from acute encephalopathy with convulsions, coma, and marked residual deficits to subtle dysfunction manifest by learning and behavioral disorders.[1,2] While there is no doubt about the causal relationship between exposures to high levels of lead and the acute encephalopathy, it is more difficult to prove a causal association between low levels of exposure and milder neurologic disorders. However, retrospective studies strongly support the relationship and several large prospective investigations are in progress.[2-4] It is generally agreed that children are more likely than adults to suffer central nervous system disorders after exposure to lead.[5]

In this chapter, I will consider selected cellular and biochemical systems which may disrupt brain function when altered by lead. Three sites are of particular interest: (1) synaptogenesis, (2) the blood-brain barrier, and (3) second messenger metabolism. The goal is to develop a comprehensive hypothesis to explain the susceptibility of immature brain to lead.

II. SYNAPTOGENESIS

By the end of the second trimester of human gestation, most neurons have migrated from their site of origin in the periventricular germinal matrix to take up permanent positions in the cerebral cortex and basal ganglia.[6] Initially, synaptic contacts are sparse, particularly those connecting adjacent cortical areas into neural networks.[7] The number and complexity of synaptic connections, however, increases rapidly during the third trimester and for the first several years after birth.[7] The interconnections become progressively more dense so that by the age of 2 years most cortical regions have almost twice as many synaptic contacts as they will at maturity.[8] An apparent synaptic overshoot and pruning occurs during postnatal development that produces a net reduction in the density of the dendritic arborization between the ages of 2 and 15 years (Figure 1).

It is likely that the density of connections at all stages of brain development represents an equilibrium between formation and elimination of synapses. In the first few years of life, the rate of formation exceeds that of elimination and a net increase in complexity occurs. During the remainder of the first decade, the rate of elimination is greater than that of formation and the density of connections falls. After adolescence, formation and elimination appear in balance so that no change is perceived. The remodeling process appears to be most active during the first 5 years of life (Figure 2).

The phasic change in density of brain connections which occurs during childhood is a major anatomic expression of plasticity in the developing nervous system.[9] Although the general scheme of production and elimination of synapses must be under genetic control, it is numerically impossible for individual genes to determine which of the billions of synapses are to be eliminated. Instead, the pattern of neural activity taking place across a particular synapse appears to determine its fate.[10] In this model, genetic programs guide the developmental increase and subsequent decrease in connections while experience and activity determine which particular synapses persist. The postnatal timing of these events is consistent with integrating early life experiences with brain anatomy. The intensity of this process during the first few years of life is also consistent with the marked changes in neurodevelopmental function that occur in infants and toddlers.

There are good reasons to believe that the biochemical systems and neurotransmitter events that mediate developmental changes in synaptic anatomy are sensitive to lead.[11] It is during the time of maximal reorganization of neural connections (toddler years) that children are most exposed to environmental lead and have the highest blood levels.[12] Experiments in both animal models and *in vitro* systems, show that exposure to inorganic lead disrupts

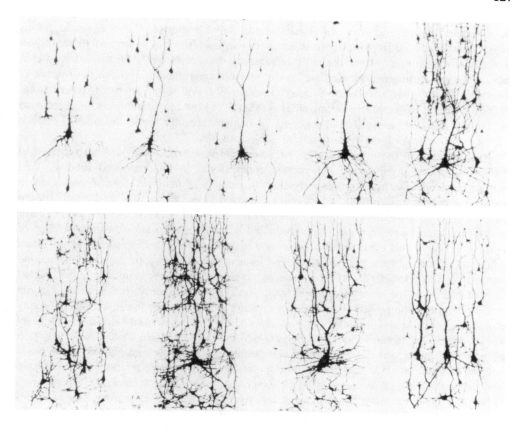

FIGURE 1. Cox-Golgi preparations demonstrating developmental changes in dendritic complexity of the leg area of human motor cortex. Upper row, left to right: 8 months premature; newborn at term; 1 month; 3 months; and 6 months. Lower row, left to right: 15 months, 2 years; 4 years; 6 years. (From Adams, R. and Victor, M., in *Principles of Neurology*, 2nd ed., Laufer, R. S. and Armstrong, T., Eds., McGraw Hill, New York, 1981, chap. 27, 389. With permission.)

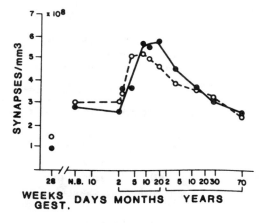

FIGURE 2. Developmental changes in synaptic density in layer IV and layer V/VI of human striate cortex, ○ = combined layers IVa-c; ● = layers V/VI. (From Huttenlocher, P. R. and de Courten, C., *Human Neurobiol.*, 6, 1—9, 1987. With permission.)

the release of neurotransmitters from presynaptic nerve endings.[13-18] For example, when nerve endings isolated from brain are superfused with lead, there is an immediate inhibition of the neurotransmitter release normally induced by depolarization.[17] Furthermore, after a delay of several minutes, an increase in the spontaneous release of neurotransmitters is observed. This biphasic response to lead exposure is found in a variety of experimental settings involving different neurotransmitter systems.[14-18] Low concentrations of lead increase the normal basal release of small amounts of neurotransmitter while the evoked release related to the arrival of an action potential is blocked.

An alteration in the normal pattern of neurotransmitter release would be expected to disrupt use-driven organization of synaptic connections. In this proposed scheme, lead exposure interferes with the process of selective pruning of neuronal connections. Once a neuronal connection is formed, information is passed from one neuron to the next by the release of neurotransmitter molecules from the axon terminal.[19] The neurotransmitters enter the synaptic cleft between the adjacent cells and bind to receptor sites on the dendrites of neuron cell body. This event signals the second neuron and initiates electrical and chemical changes. Depending upon the frequency and intensity of this synaptic activity, the second neuron may be depolarized and release neurotransmitters from its axon terminal to signal the next neuron in the network. Alternatively, and more commonly, the input of a single neuron only alters the possibility of a depolarization which requires the summation of many incoming excitatory and inhibitory signals. In addition to the evoked release of relatively large amounts of neurotransmitter molecules that follow the arrival of a wave of depolarization at the axon terminal, smaller amounts of neurotransmitters are released almost continuously from the axon terminals in the absence of depolarization. This tonic release appears to have important trophic effects on maintaining the integrity of the synaptic contact. The combination of the tonic slow release of neurotransmitters with the phasic release related to axon depolarization, may determine the strength of the synapse and its likelihood of survival during developmental reorganization.[20] The efficiency of this process should be altered by exposure to lead since this metal increases the tonic slow release of neurotransmitter while inhibiting release evoked by depolarization. If, as seems likely, the precision of the pruning process is dependent upon recognition of those synapses transmitting the most meaningful signals, lead may blur the distinctions by increasing background activity and decreasing the transmission of specific signals. The end result may be a nervous system with a normal number and density of contacts among its neurons but with poorly chosen connections. The clinical manifestations would be expected to be subtle since the most primitive networks underlying control of mobility and sensation undergo little change during postnatal development.[21] In contrast, networks important for learning and behavior skills should be more vulnerable. In fact, children exposed to low concentrations of lead during their toddler years seem to have lasting deficits in just these areas.[2]

III. BLOOD-BRAIN BARRIER

The homeostasic mechanisms collectively defined as the blood-brain barrier are also potentially vulnerable sites for the toxic action of inorganic lead in the brain.[22] Injury to the blood-brain barrier plays a major role in the pathogenesis of acute lead encephalopathy but barrier damage may also be involved in the development of the more subtle symptoms associated with low level exposure.

The movement of molecules between the blood stream and interstitial space of the brain is markedly restricted in order to obtain the heightened degree of homeostasis required for synaptic function.[23] Water soluble molecules do not readily enter the brain unless specific transporters are present to mediate their passage across the endothelial barrier. Furthermore, active transport pumps present on the brain surface of the endothelium remove molecules

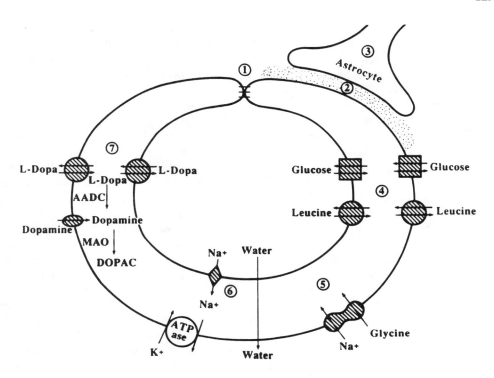

FIGURE 3. Schematic diagram of brain capillary. The *continuous jight junctions* (1) that join endothelial cells in brain capillaries limit the diffusion of large and small solutes across the blood-brain barrier. The *basement membrane* (2) provides structural support for the capillary and, along with the *astrocytic foot processes* (3) that encircle the capillary, may influence endothelial cell function. *Transport carriers* (4) for glucose and essential amino acids facilitate the movement of these solutes into brain, while *active transport systems* (5) appear to cause indirectly the efflux of small, nonessential amino acids from brain to blood. Sodium ion transporters on the luminal membrane and Na^+, K^+-ATPase on the *antiluminal membrane* (6) account for the movement of Na^+ from blood to brain, and this may provide an osmotic driving force for the secretion of interstitial fluid by the brain capillary. The *enzymatic blood-brain barrier* (7) consists of the uptake of neurotransmitter precursors such as L-dopa into the endothelial cells via the large neutral amino acid carrier, and their subsequent metabolism to 3,4-dihydroxyphenylacetic acid (DOPAC) by aromatic amino acid decarboxylase (AADC) and monoamine oxidase (MAO) present within the endothelial cell. Neurotransmitters in the interstitial fluid may also be accumulated and metabolized by the brain capillary. (From Betz, A. L., Goldstein, G. W., and Katzman, R., *Basic Neurochemistry*, 4th ed., Siegel, G. J., Agranoff, B. W., Albers, R. W., and Molinoff, P. B., Eds., Raven Press, New York, 1989, chap. 30, 595. With permission.)

that might disrupt neuronal activity and transport them against a concentration gradient from the brain to blood.

Manifestations of injury to the barrier depend upon severity and may vary from inadequate entry of essential substrates and failure to remove potentially disruptive molecules to a breakdown of barrier exclusion properties with formation of brain edema.[24] Lead may produce a similar graded set of injuries that contribute to synaptic dysfunction in low level exposure and to massive brain edema with high level toxicity.

Endothelial cells in the microvessels of the nervous system express a number of special features which generate the blood-brain barrier (Figure 3).[25] Most important is the continuous belt of complex tight junctions which seal adjacent endothelial cells together.[26] The junctions are formed by a series of ridges and grooves that limit the free diffusion of polar molecules that vary in size from plasma proteins to monovalent cations. Vesicular channels are rare thus limiting the diffusion of substances through the endothelial cells. To compensate for this impermeable barrier and allow for the entry of essential nutrients, the plasma membranes of brain microvascular endothelial cells are enriched in specific transporters that for example mediate the passage of glucose and large neutral amino acids across the barrier.[25] Active transport pumps for ions, neurotransmitters, and organic acids are asymmetrically distributed between the luminal and abluminal surfaces of the endothelial cell to provide for the vectorial transfer of selected molecules across the barrier and against a concentration gradient.[27] These pumps mainly function to remove substances from the brain to the blood.

Astrocytes in the brain play an important role in inducing and then maintaining expression of the blood-brain barrier.[28] The endothelial cells in the brain share a common embryologic origin with the endothelium elsewhere in the body. There is good evidence that astrocytes provide the signals that induce the endothelium to express the blood-brain barrier phenotype. For example, chicken blood vessels growing into fetal quail brain tissue implanted in the abdominal cavity of the chicken are surrounded by quail astrocytes and exhibit barrier features.[29] Also, implantation of astrocytes cultured from rat brain upon the chicken chorioallantoic membrane induce formation of blood vessels with blood-brain barrier properties.[30] These experiments and others suggest that the astrocytes produce a yet to be identified molecular factor that induces the barrier. In mammalian brain, foot processes from astrocytes ensheathe almost the entire circumference of the microvessel. The astrocytes are not sealed together and therefore do not form a physical barrier. Instead they appear to induce the endothelium to form the barrier. This relationship between astrocytes and endothelial cells suggests that injury to the astrocyte by lead could be a mechanism for disrupting the function of the blood-brain barrier.

A number of studies have implicated brain microvessels and the blood-brain barrier in the pathogenesis of lead damage to the nervous system.[31-33] The very nature of the barrier means that the endothelial cells are exposed to lead during the passage of the toxic metal into the brain. Using autoradiography and direct measurements in isolated brain microvessels, the barrier tissue has a marked affinity for lead and accumulates a much higher concentration of the toxin than other brain structures.[32] One consequence this lead accumulation is a disturbance in calcium metabolism which in turn may alter many of the specialized barrier properties.[34]

Our laboratory approaches the biochemical basis for lead toxicity by investigating the effect of lead exposure upon the behavior of blood-brain barrier cells maintained in tissue culture. We found that endothelial cells were actually very resistant to overt toxicity and that they survived and were able to proliferate normally in concentrations of lead that in an intact animal are associated with capillary failure and brain edema.[35] By contrast, astrocytes in culture were quite sensitive to the toxic action of lead and developed cytotoxic changes at relatively low concentrations.[35,36] We interpret this differential susceptibility to suggest that alterations in the function of the blood-brain barrier that follows lead exposure may be caused by a primary injury to astrocytes and a subsequent loss of their ability to maintain expression of the barrier in the endothelial cells.

It is easy to picture how loss of the blood-brain barrier results in brain edema, increased intracranial pressure, and overt symptoms of neurologic damage after exposure to high levels of lead. Could a lead induced lesion in the function of the barrier underlie the more subtle symptoms of low level exposure? The author suggests that this is possible given the role of the barrier in maintaining a precise constancy of interstitial fluid components. Subtle injury to the barrier should first effect the most energy dependent aspect of barrier function — the pumping of ions, organic acids, and neurotransmitters from brain to blood. Normally the concentration of potassium in the interstitial fluid of brain is closely maintained at 2.8 mM despite relatively wide fluctuations in the plasma concentration which normally averages 5.0 mM.[37] The constancy of potassium levels on the brain side of the barrier is maintained in part by an active transport pump that moves potassium against a concentration gradient across the endothelial cells and out of the brain.[38] If this pump mechanism were injured by lead, control of potassium levels would become less constant and this change should interfere with the depolarization of nerve endings that follows an action potential. In a similar fashion, deficient pumping of amino acids and neurotransmitters by the endothelial cells may enhance the concentration of neuroactive molecules and interfere with the efficiency of synapse function. A change in synapse efficiency may translate into less accurate use-dependent organization of synaptic connections during critical phases of brain development.

IV. SECOND MESSENGER METABOLISM

The defects in brain function that occur after low-level lead poisoning are not associated with gross anatomic lesions. Rather, there are functional disturbances that result in learning and behavioral disabilities. This is true both in the human experience and in animal models and suggests that the mechanism of toxicity is not likely to involve those systems that produce cell death and tissue necrosis. Instead a more subtle but lasting disarray of neuronal organization has occurred. The regulatory systems that mediate synaptic function are a likely target whether for the direct effect of lead or an indirect involvement because of altered blood-brain barrier function.

We now know that several intracellular second messenger systems translate the stimulus of neurotransmission into a more lasting change in neuronal activity. These mechanisms appear important in laying down new memories and in the function of neural nets that underlie learning. Modification of intracellular proteins by enzymes that alter their charge and configuration appear to play a central role in this process.[39] Many hormonal systems, including the neurotransmitters, activate protein kinases to produce this effect. The efficiency of this process is important for normal brain function.

When a receptor on the postsynaptic neuron is occupied by a neurotransmitter released from the presynaptic axonal terminal, a sequence of metabolic events is triggered.[19] The most immediate response relates to the opening of ion channels that allow for the movement of sodium, potassium, and chloride ions. The result is a change in membrane potential that makes the postsynaptic neuron either more or less likely to depolarize and generate an action potential to signal through the same mechanism of the next neuron in the circuit. The sensitivity of this mechanism including both the amount of neurotransmitter released from the presynaptic terminal and the degree of response generated in the postsynaptic neuron is dependent upon the state of phosphorylation of many neuronal proteins.[39] The situation is dynamic so that continued use of a given pathway leads to more efficient function in the future and the establishment of preferential circuits. In addition to the immediate change in ion channel permeability, several second messengers may be generated during the course of neurotransmitter occupancy of the postsynaptic receptor. Neurotransmitter receptor sites are also located on the presynaptic axon terminal and this feedback mechanism allows for a second level of control by controlling the amount and ease of neurotransmitter release. Both the transmitting and responding neurons are thus modified by use. This process appears important in the developmental pruning process.

Several classes of protein kinases exist and are activated by different second messengers. The type of second messenger generated in a given cell varies with the neurotransmitter receptor and the properties of the specific cell. The major second messengers include cyclic AMP (cAMP) which is formed from the large intracellular ATP pool upon proper activation of a transduction protein linked to the receptor site and to adenyl cyclase, the enzyme that converts ATP into cAMP.[40] The level of cAMP is a balance between formation by adenyl cyclase and degradation by phosphodiesterase. cAMP, in turn, binds to specific protein kinases that phosphorylate regulatory proteins. In certain cell systems, lead inhibits adenyl cyclase[41] and stimulates phosphodiesterase.[42] The expected result is a lower level of cAMP and a decrement in the phosphorylation of protein substrates of cAMP-protein kinase.

Calcium is another intracellular second messenger.[43] Normally the concentration of calcium inside a cell is 10,000-fold less than that outside a cell. Many cellular functions are regulated by a rise in the intracellular calcium concentration. Calcium may enter a cell after the opening of specific channels in the plasma membrane which allow the rapid entry of calcium down the concentration gradient. Once inside the cell, the calcium can either directly activate certain enzymes or bind to a regulatory protein such as calmodulin.[44] When calcium occupies the binding sites on calmodulin, this protein changes its configuration and becomes capable of activating various enzymes and pump systems. Included among these is a class of protein kinases. This is also the mechanism by which the phosphodiesterase that breaks down cAMP is regulated. Investigations in the author's laboratories and by others have shown that lead can mimic calcium and bind to the calcium sites on calmodulin and cause activation of enzymes normally controlled by the level of intracellular calcium.[42,45] In fact, lead exhibits a greater affinity for the calmodulin binding sites than calcium itself. In this way calcium-calmodulin regulated enzymes may be activated inappropriately in a lead poisoned tissue. One result would be activation of phosphodiesterase and breakdown of cAMP. This would lead to less cAMP-protein kinase activity. On the other hand, calmodulin-protein kinases would be expected to be activated. This imbalance of protein kinase activity may disrupt neuronal function.

Protein kinase C is yet another class of protein kinases.[46] This protein kinase is activated in the absence of calmodulin. It does, however, require calcium and a biochemical product of phospholipid degradation. Recently, it has become clear that the inositol phospholipids, which are a minor class of plasma membrane lipids, represent an important source for second messengers. The polyphosphorylated inositol phospholipids are hydrolyzed by a membrane bound phospholipase which is linked by a transduction protein to receptors for certain hormones and neurotransmitters. The activation process is similar to that described for adenyl cyclase. The substrate in this case is inositol phospholipid. Two products are formed — inositol triphosphate and diacylglycerol. The inositol triphosphate is released into the cytosol where it binds to membrane bound organelles that normally sequester intracellular calcium. The inositol triphosphate causes these organelles to release their calcium into the cytosol of the cell. The calcium can then activate calmodulin, selected enzymes, and also provide a partial signal for protein kinase C. Diacylglycerol, the second product of inositol phospholipid breakdown, is the more direct signal for activating protein kinase C. In the presence of both diacylglycerol and calcium, protein kinase C phosphorylates a specific class of intracellular proteins. These proteins appear particularly important in regulating growth and differentiation. One function of protein kinase C appears to control the sensitivity of the presynaptic terminal to release neurotransmitters and the postsynaptic cell to a lasting change in neuronal conduction.[47] Such changes in long-term potentiation may relate to memory storage. The postnatal development of protein kinase C activity is consistent with its proposed role in use dependent plasticity.[48]

We have found that lead in very low concentrations is capable of substituting for calcium in the activation of protein kinase C (Figure 4).[49] The interaction between lead and diacylglycerol resembled that of calcium and diacylglycerol. This ability of lead to activate protein kinase C was initially demonstrated in a cell free preparation of protein kinase C isolated from brain tissue. In another set of experiments, we showed that intact microvessels prepared from brain also showed activation of protein kinase C after extracellular exposure to lead.[50] Interestingly, the effect of lead upon brain microvessels was only observed when the microvessels were prepared from immature animals.

Taken together, the observations on second messenger metabolism provide a potential biochemical target for the effect of lead upon more subtle neurologic deficits. Disruption of these systems would not be expected to cause cell death. Instead, a less efficient regulatory system might produce a less efficient nervous system. Alterations in neural function during

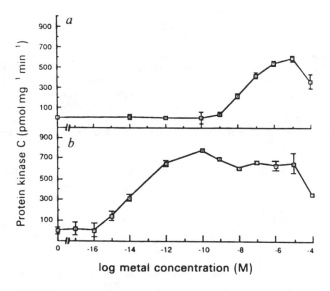

FIGURE 4. Dose-dependent stimulation of protein kinase C by calcium (*a*) and lead (*b*). (From Markovac, J. and Goldstein, G. W., *Nature*, 334, 71-73, 1988. With permission.)

the time of reorganization of synapsis might produce a lasting effect upon the microanatomy of neural circuits and may leave the nervous system inefficient long after lead is cleared from the brain.

V. SUMMARY

At low levels of exposure, lead appears to disrupt several biochemical systems that are important to the developmental remodelling of synaptic connections. These include direct effects upon the release of neurotransmitters and the metabolism of second messengers. It is also possible that alterations in the blood-brain barrier produced by lead cause a secondary disorder in neuron function. Chaos in the regulation of synaptic activity may translate into poor selection of neuronal connections for developmental pruning and a permanently inefficient nervous system. The learning and behavioral disorders found in lead exposed children are consistent with this hypothesis.

Many of the biochemical events implicated in this hypothesis involve the interaction between lead and calcium.[43] In some cases, lead blocks the ability of calcium to reach an active regulatory site. Other times, lead enters the cell and either mobilizes calcium or mimics the action of calcium as regulator of cell function.

Unfortunately, maximal exposure to inorganic lead occurs at the same time that synaptic development is most dynamic — the toddler years. Although neuron proliferation and migration are complete well before birth, connections important for learning, memory, and behavior are formed and extensively remodelled during the first several years of life. This timing allows environmental experiences to influence the anatomy of the brain and in the case of lead exposure possibly leave a lasting adverse effect.

ACKNOWLEDGMENTS

Work described in this chapter from the author's laboratory was supported by Grant ES 02380 from the National Institutes of Health. Also, I thank Dr. Michael V. Johnston for helpful discussion and review of this manuscript.

REFERENCES

1. **Blackman, S. A.,** The lesions of lead encephalitis in children, *Bull. Johns Hopkins Hosp.,* 61, 1, 1937.
2. **Needleman, H. L., Schell, A., Bellinger, D., Leviton, A., and Allred, E. N.,** The long-term effects of exposure to low doses of lead in childhood, *N. Engl. J. Med.,* 322, 83, 1990.
3. **McMichael, A. J., Baghurst, P. A., Wigg, N. R., Vimpani, G. V., Robertson, E. F., and Roberts, R. J.,** Port Pirie cohort study: environmental exposure to lead and children's abilities at the age of four years, *N. Engl. J. Med.,* 319, 468, 1988.
4. **Bellinger, D., Leviton, A., Waternaux, C., Needleman, H., and Rabinowitz, M.,** Longitudinal analyses of prenatal and postnatal lead exposure and early cognitive development, *N. Engl. J. Med.,* 316, 1037, 1987.
5. **Mushak, P., Davis, J. M., Crocetti, A. F., and Grant, L. D.,** Prenatal and postnatal effects of low-level exposure: integrated summary of a report to the US Congress on childhood lead poisoning, *Environ. Res.,* 50, 11, 1989.
6. **Sidman, R. L. and Rakic, P.** Neuronal migration, with special reference to developing human brain, *Brain Res.,* 62, 1, 1973.
7. **Marin-Padilla, M.,** Early ontogenesis of the human cerebral cortex, in *Cerebral Cortex,* Vol. 7, Development and maturation of cerebral cortex, Peters, A. and Jones, E. G., Eds., Plenum Press, New York, 1988.
8. **Huttenlocher, P. R. and deCourten, C.,** The development of synapses in striate cortex of man, *Hum. Neurobiol.,* 6, 1, 1987.
9. **Bear, M. F., Cooper, L. N., and Ebner, F.,** A physiological basis for a theory of synapse modification, *Science,* 237, 42, 1987.
10. **Kleinschmidt, A., Bear, M. F., and Singer, W.,** Blockade of "NMDA" receptors disrupts experience dependent plasticity of kitten striate cortex, *Science,* 238, 355, 1987.
11. **Petit, T. L.,** Developmental effects of lead: its mechanism in intellectual functioning and neural plasticity, *Neurotoxicology,* 7, 483, 496, 1986.
12. **Mushak, P. and Crocetti, A. F.,** Determination of numbers of lead-exposed American children as a function of lead source: integrated summary of a report to the U.S. Congress on childhood lead poisoning, *Environ. Res.,* 50, 210, 1989.
13. **Cooper, G. P., Suszkiw, J. B., and Manalis, R. S.,** Heavy metals: effects on synaptic transmission, *Neurotoxicology,* 5, 247, 1984.
14. **Manalis, R. S. and Cooper, G. P.,** Presynaptic and postsynaptic effects of lead at the frog neuromuscular junction, *Nature,* 243, 354, 1973.
15. **Atchison, W. D. and Narahashi, T.,** Mechanism of action of lead on neuromuscular junctions, *Neurotoxicology,* 5, 267, 1984.
16. **Kostial, K. and Vouk, V. B.,** Lead ions and synaptic transmission in the superior cervical ganglion of the cat, *Br. J. Pharmacol.,* 12, 219, 1957.
17. **Minnema, D. J., Greenland, R. D., and Michaelson, I. A.,** Effect of *in vitro* inorganic lead on dopamine release from superfused rat striatal synaptosomes, *Toxicol. Appl. Pharmacol.,* 84, 400, 1986.
18. **Minnema, D. J., Michelson, I. A., and Cooper, G. P.,** Calcium efflux and neurotransmitter release from rat hippocampal synaptosomes exposed to lead, *Toxicol. Appl. Pharmacol.,* 92, 351, 1988.
19. **Cooper, J. R., Bloom, F. E., and Roth, R. H.,** Cellular foundations of neuropharmacology, in *The Biochemical Basis of Neuropharmacology,* 5th ed., Oxford University Press, New York, 1986, 9.
20. **Matton, M. P.,** Neurotransmitters in the regulation of neuronal cytoarchitecture, *Brain Res. Rev.,* 13, 179, 1988.
21. **Lemire, R. J., Loeser, J. D., Leech, R. W., and Alvord, E. C.,** *Normal and Abnormal Development of the Human Nervous System,* Harper and Row, Hagerstown, 1975.
22. **Goldstein, G. W.,** Brain capillaries: a target for inorganic lead poisoning, *Neurotoxicology,* 5, 167, 1984.
23. **Goldstein, G. W. and Betz, A. L.,** The blood-brain barrier, *Sci. Am.,* 255, 74, 1986.
24. **Goldstein, G. W. and Betz, A. L.,** Blood vessels and the blood-brain barrier, in *Diseases of the Nervous System,* Vol. 1, Ashbury, A. K., McKhann, G. M., and McDonald, W. I., Eds., W. B. Saunders, Philadelphia, 1986, 172.
25. **Betz, A. L., Goldstein, G. W., and Katzman, R.,** Blood-brain CSF barriers, in *Basic Neurochemistry,* 4th ed., Siege, G. J., Albers, R. W., Agranoff, B. W., and Malinoff, P., Eds., Raven Press, New York, 1988, 591.
26. **Brightman, M. W. and Reese, T. S.,** Junctions between intimately apposed cell membranes in the vertebrate brain, *J. Cell Biol.,* 40, 648, 1969.
27. **Betz, A. L. and Goldstein, G. W.,** Polarity of the blood-brain barrier: neutral amino acid transport into isolated brain capillaries, *Science,* 202, 225, 1978.
28. **Goldstein, G. W.,** Endothelial cell-astrocyte interactions: a cellular model of the blood-brain barrier, in *Ann. N.Y. Acad. Sci.,* 1988, 529, 31.

29. **Stewart, P. A. and Wiley, M. J.**, Developing nervous tissue induces formation of blood-brain barrier characteristics in invading endothelial cells: a study using quail-chick transplantation chimeras, *Dev. Biol.,* 84, 183, 1981.
30. **Janzer, R. C. and Raff, M. D.**, Astrocytes induce blood-brain barrier properties in endothelial cells, *Nature,* 325, 253, 1987.
31. **Clasen, R. A., Harmann, J. F., Starr, A. J., Coogan, P. S., Pandolfi, S., Laing, I., Becker, R., and Haas, G. M.**, Electron microscopic and chemical studies of the vascular changes and edema of lead encephalopathy, *Am. J. Path.,* 74, 215, 1973.
32. **Toews, A. D., Kolber, A., Hayward, J., Krigman, M. R., and Morell, P.**, Experimental lead encephalopathy in the suckling rat: concentration of lead in cellular fractions enriched in brain capillaries, *Brain Res.,* 147, 131, 1978.
33. **Goldstein, G. W., Asbury, A. K., and Diamond, I.**, Pathogenesis of lead encephalopathy. Uptake of lead and reaction of brain capillaries, *Arch. Neurol.,* 31, 382, 1974.
34. **Goldstein, G. W., Wolinksy, J. S. and Csejtey, J.**, Isolated brain capillaries: a model for the study of lead encephalopathy, *Ann. Neurol.,* 1, 235, 1977.
35. **Gebhart, A. M. and Goldstein, G. W.**, Use of an *in vitro* system to study the effects of lead on astrocyte-endothelial cell interactions: a model for studying toxic injury to the blood-brain barrier, *Toxicol. Appl. Pharmacol.,* 94, 191, 1988.
36. **Holtzman, D., Olson, J. E., DeVries, C., and Bensch, K.**, Lead toxicity in primary cultured cerebral astrocytes and cerebellar granular neurons, *Toxicol. Appl. Pharmacol.,* 89, 211, 1987.
37. **Bradbury, M. W. B.**, The structure and function of the blood-brain barrier, *Fed. Proc.,* 43, 186, 1984.
38. **Goldstein, G. W.**, Relation of potassium transport to oxidative metabolism in isolated brain capillaries, *J. Physiol.,* 286, 185, 1979.
39. **Hemmings, H. C., Nairn, A. C., McGuinness, T. L., Huganir, R. L., and Greengard, P.**, Role of protein phosphorylation in neuronal signal transduction, *FASEB J.,* 3, 1583, 1989.
40. **Northup, J. K.**, Regulation of cyclic nucleotides in the nervous system, in *Basic Neurochemistry,* 4th ed., Siegel, G. J., Agranoff, B. W., Albers, R. W., and Molinoff, P. B., Eds., Raven Press, New York, 1989, 349.
41. **Nathanson, J. A. and Bloom, F. E.**, Lead-induced inhibition of brain adenylate cyclase, *Nature,* 255, 419, 1975.
42. **Goldstein, G. W. and Ar, D.**, Lead activates calmodulin sensitive processes, *Life Sci.,* 33, 1001, 1983.
43. **Pounds, J. G.**, Effect of lead intoxication on calcium homeostasis and calcium-mediated cell function: a review, *Neurotoxicology,* 5, 295, 1984.
44. **Cheung, W. Y.**, Calmodulin: its potential role in cell proliferation and heavy metal toxicity, *Fed. Proc.,* 43, 2995, 1984.
45. **Richardt, G., Federolf, G., and Habermann, E.**, Affinity of heavy metal ions to intracellular Ca^{2+}-binding proteins, *Biochem. Pharmacol.,* 35, 1331, 1986.
46. **Nishizuka, Y.**, Studies and perspectives of protein kinase C, *Science,* 233, 305, 1986.
47. **Routtenberg, A.**, Synaptic plasticity and protein kinase C, *Prog. Brain Res.,* 69, 211, 1986.
48. **Stichel, C. C. and Singer, W.**, Postnatal development of protein kinase C — like immunoreactivity in the cat visual cortex, *Eur. J. Neurosci.,* 4, 355, 1989.
49. **Markovac, J. and Goldstein, G. W.**, Picomolar concentrations of lead stimulate brain protein kinase C, *Nature,* 334, 71, 1988.
50. **Markovac, J. and Goldstein, G. W.**, Lead activates protein kinase C in immature rat brain microvessels, *Toxicol. Appl. Pharmacol.,* 96, 14, 1988.

Chapter 8

BEHAVIORAL IMPAIRMENT PRODUCED BY DEVELOPMENTAL LEAD EXPOSURE: EVIDENCE FROM PRIMATE RESEARCH

Deborah C. Rice

TABLE OF CONTENTS

I. Introduction 138

II. General Design Consideration 138
    A. Dosing and Blood Levels 138
    B. Behavioral Testing Procedures 140

III. Behavioral Impairment in Monkeys 141
    A. Tests Measuring Learning, Attention, and Memory 141
        1. Discrimination Reversal 141
        2. Spatial Delayed Alternation 144
        3. Hamilton Search Task 144
        4. Delayed Matching to Sample 145
        5. Evidence from Other Primate Laboratories 146
        6. Selected Collaborative Data from the Rat 147
    B. Intermittent Schedules of Reinforcement — Lead and Activity 148
    C. Social Behavior and Environmental Reactivity 150

IV. Summary and Conclusions 150

References 151

I. INTRODUCTION

This chapter is not intended to be a review of the animal literature on the behavioral effects of lead. There are a number of such reviews,[1,2] including an excellent critical review.[3] The present effort focuses on relatively subtle impairment produced by reasonably low-level developmental exposure to lead in monkeys, although selected relevant studies in the rodent are included for comparison. This body of data is remarkably consistent between laboratories and species in the types of deficits detected as a result of developmental lead exposure. Moreover, the deficits observed in monkeys are similar to those observed in children in terms of the underlying behavioral mechanisms presumed to be responsible for the impairment.

Most studies on the effects of lead in children have used some measure of IQ as the main dependent variable. This measure has the advantage of being standardized for the population, as well as providing a global measure of cognitive performance. It has the disadvantage of being nonspecific; a poorer performance on a particular subtest usually provides little information on the behavioral mechanisms responsible for that performance. Moreover, global measures of performance are undoubtedly less sensitive than tests designed to assess specific lead-induced deficits, if indeed these were known. Investigators have attempted to address this issue by including behavioral measures other than IQ in their assessments. Early research using teachers' rating scales implicated various forms of attentional deficits as possible contributors to poor performance on intelligence tests as a consequence of lead exposure.[4,5] Several researchers subsequently designed tests aimed at evaluating attentional processes in lead-exposed children[6,7] including simple reaction time and performance on vigilance tasks. The body of literature presently suggests that developmental exposure to lead in children results in attentional problems, with concomitant increased distractibility, short attention span, and inability to adapt to complex behavioral requirements. The impairment on intelligence tests as a result of lead exposure may be entirely or partly a consequence of these attentional problems; this issue has yet to be addressed.

Much of the research in monkeys has focused on utilizing behavioral procedures sensitive to attentional and/or learning deficits to characterize the effects of developmental lead exposure. Similar to children, monkeys exhibit impaired attention and short-term memory, increased distractibility, and inflexibility in adapting to changes in behavioral requirements. Such effects have been observed by different laboratories, in different species of monkeys, and utilizing different behavioral tasks and equipment. Deficits have been observed as a result of *in utero* and/or postnatal exposure, and findings have been replicated repeatedly within individual laboratories. Thus the effects have proved extremely robust. Moreover, research in other species also produces consistent behavioral impairment as a result of lead exposure, if appropriately sensitive behavioral tasks are used.

II. GENERAL DESIGN CONSIDERATION

A. DOSING AND BLOOD LEVELS

In children, the largest proportion of lead body burden results from ingestion in food and water, although direct inhalation also contributes. In studies described in the present chapter, lead exposure in the monkey was via daily oral dosing. The preponderance of literature on the behavioral effects of developmental lead exposure in monkeys has emerged from two laboratories — the University of Wisconsin Primate Center and our laboratory at the Health Protection Branch in Canada. In our laboratories, monkeys were dosed in milk substitute formula as infants, and in a voluntarily-consumed gelatin capsule after infancy. This regimen resulted in blood lead levels peaking at about 100 d of age, and decreasing after withdrawal from infant formula (but not lead) at 200 d of age (Table 1). A lower steady

TABLE 1
Summary of Results from the Health Protection Branch[a]

Cohort Group

Peak blood lead at 100 d of age (μg/dl)	Steady state blood lead after 300 d of age (μg/dl)	Behavioral impairment[b]
100	40	FI performance (juveniles)
50	30	FI performance (juveniles) Nonspatial discrimination reversal (juveniles) Matching to sample — spatial and nonspatial (young adults)
25 or 15	13 or 11	FI performance (juveniles) DRL performance (juveniles) Nonspatial discrimination reversal with irrelevant cues (juveniles) Delayed alternation (adults) Spatial discrimination reversal with irrelevant cues (adults)

[a] Postnatal exposure only.
[b] See discussion in chapter.

state blood level was then maintained into adulthood. The Wisconsin laboratory typically dosed infants for only a year after birth, in milk substitute formula. Females used for *in utero* experiments at Wisconsin were dosed in drinking water (Table 2). Much of the actual behavioral testing by the Wisconsin group was therefore performed when blood lead levels were similar or identical to those of control monkeys.

When research into the behavioral effects of developmental lead exposure began in the mid-1970s, public health concern centered around exposure of children to lead-based paint or industrial fallout from lead smelters. In 1979, the landmark study by Needlemen et al.[4] implicating universal environmental exposure as a cause of behavioral impairment in children, was published; that same year the first four monkey studies were published. The two papers from the University of Wisconsin reported impairment during infancy and as juveniles in the same groups of rhesus monkeys, with blood lead levels during the first year of life of approximately 50 or 90 μg/dl. Two papers from our laboratory, in cynomolgus monkeys with peak levels early in life of 50 μg/dl and ongoing steady state levels of 30 μg/dl, reported deficits on two very different behavioral tasks. The Needleman study used tooth lead as a marker of body burden in children, but subsequent blood lead data from some of the children revealed the "high lead" group to have blood lead levels of about 35 μg/dl, while those of the comparison "low lead" control group were approximately 25 μg/dl. At the time these investigations were published, 30 μg/dl was considered the low risk cut-off value for children by both the U.S. Environmental Protection Agency (EPA) and the Centers for Disease Control (CDC).

The last decade has witnessed intense research into the behavioral consequences of developmental lead exposure. Epidemiological studies have revealed impaired functioning at lower and lower blood levels, with no apparent threshold for effect. In our laboratory, we were fortunate to have two groups of monkeys, born in the mid 1970s, carrying what turned out to be environmentally relevant blood lead levels (11 to 25 μg/dl and their controls). We have had the opportunity to study this group of monkeys extensively, observing consistent behavioral impairment from the juvenile period through adulthood. Groups with higher blood lead levels have provided complete dose-effect information.

TABLE 2
Summary of Results from the University of Wisconsin
Cohort Group

Average blood lead during dosing (µg/dl)	Behavioral deficits
Postnatal Exposure	
50 or 90 for first year	Size, color, spatial discrimination reversal (infants)
45 or 90 for first year	Discrimination reversal — multidimensional, color, shape, size (infants)
	Spatial discrimination reversal with irrelevant cues (juveniles)
	Hamilton search task (young adults)
30—40 or 60	Social behavior (infants)
	Adaptation to new environment (infants)
110 by 12 weeks of age, decrease thereafter, or	Social behavior (infants)
70—80 for first year, or	Adaptation to new environment (infants)
130 by 12 weeks of age,	
70 for remainder of first year	
300 by 10 weeks of age,	Discrimination reversal—color, size, form (infants)
90 for remainder of first year	Hamilton search task (young adult)
	Spatial delayed alternation (young adult)
55 by 5 weeks of age,	Visual exploration (infants)
35 for remainder of first year	Neonatal behavioral assessment battery (infants)
Prenatal Exposure	
50 in mothers and infants at birth	Social development — no effect Hamilton search task (young adults)
Cross-Fostering	
35—45 during lead exposure	Increased clinging by infant-mother pairs

B. BEHAVIORAL TESTING PROCEDURES

The general testing strategy used by the Wisconsin laboratory and our laboratory at the Health Protection Branch differed in at least one important respect. The Wisconsin laboratory utilized the Wisconsin General Testing Apparatus (WGTA), developed at that institution in the 1940s. In the WGTA testing procedure, the investigator sits across a table from the monkey, initiates the trials, and records the data by hand. The stimuli are three-dimensional objects that are actually handled by the monkey, with the food reward usually directly underneath the correct stimulus. Our laboratory utilized a computer for stimulus display, schedule programming, and data recording in all but one study, which used a WGTA. The stimuli were two-dimensional, back-lit onto the clear response buttons. Reward consisted of apple juice delivered through a spout not associated physically with the stimuli or response manipulanda.

Each system has its advantages. The main advantage of the WGTA, other than low cost and lack of requirement for sophisticated technology, is that monkeys learn readily. The opportunity to manipulate the stimuli and the direct association of stimulus and reward allows the (normal) monkey to learn quickly and easily. The advantages of an automated system include the opportunity to collect and analyze data in greater detail, and removal of the interaction between the investigator and the monkey as a relatively uncontrolled variable. In the case of behavioral toxicological experiments, the fact that the monkey may find learning more difficult in an automated system may render the situation more sensitive to detection of behavioral impairment in toxicant-treated monkeys.

In summary, the research conducted at our laboratory and at the University of Wisconsin differed with respect to species, dosage regimen, and behavioral apparatus. Many of the details of schedule parameters of specific tests also differed between the laboratories, or between groups of monkeys tested in the same laboratory. Moreover, groups of monkeys were born over a number of years in both laboratories, with the inevitable concomitant changes in general rearing procedures. Despite these procedural differences, however, the types of behavioral deficits observed are gratifyingly similar between laboratories, between experiments within a laboratory, and across time. Moreover, the data from different experiments provide orderly dose-effect data when combined. The only other two laboratories known to this reviewer to publish studies on developmental lead exposure in monkeys also reported behavioral deficits.

III. BEHAVIORAL IMPAIRMENT IN MONKEYS

Research concerning the intellectual capabilities of monkeys in a controlled experimental setting has been ongoing for more than 50 years, providing a large body of data from which to choose tests of intellectual ability. Results of epidemiological studies in children, as well as data from rodents, also provided initial clues concerning tests possibly sensitive to lead-induced deficits. As research progressed, the development of a clearer hypothesis concerning at least one of the behavioral mechanisms of lead-induced cognitive deficits, namely attentional impairment, provided even more direction for the choice of promising tests for use with monkeys. It seemed clear almost from the outset, for example, that acquisition of a simple visual discrimination task would not access the behavioral mechanisms damaged by low- to moderate-level lead exposure. For the methodology to be sensitive to lead impairment, it would have to introduce contingencies that challenged the monkeys' processing and reasoning capabilities in a number of ways. These have included introduction of irrelevant stimuli, imposition of delays between the relevant cue and the opportunity to respond, changing the "rules" of the test as it progressed, and requiring the monkey to abstract general principles from the specific presented task. These strategies have proved remarkably successful in detecting impairment produced by developmental lead exposure in monkeys.

A. TESTS MEASURING LEARNING, ATTENTION, AND MEMORY
1. Discrimination Reversal

A behavioral test that has proven sensitive to the effects of developmental lead exposure is the discrimination reversal task. In this paradigm, the monkey is presented with two or more stimuli which vary in one or more ways (form, shape, color, position, etc.). The monkey must respond to a specified stimulus (i.e., always choose the red rather than the green, irrespective of position or shape) in order to be rewarded, usually with a preferred food or juice. When the monkey learns the task to some pre-determined criterion, the "rule" is changed (reversed) so that the previously incorrect stimulus (green rather than red) becomes the correct one. Typically a number of such reversals is instituted. A normal monkey will learn each successive reversal more quickly, displaying a "learning curve". In addition, the task may change in terms of the relevant stimulus dimension; for example "attend to the color and ignore the shape" may change to "attend to the shape and ignore the color". Both of these manipulations tax intellectual capabilities different from those required by the simple acquisition of a discrimination, and hence proved more sensitive to the effects of lead.

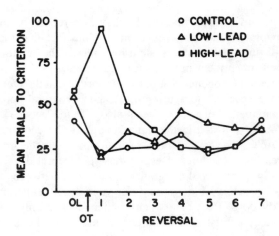

FIGURE 1. Mean number of trials to learn a series of nonspatial form discrimination reversals. Monkeys were tested at approximately 1 year of age, after lead exposure had terminated but while blood lead levels were still elevated. (From Bushnell, P. J. and Bowman, R. E., *Pharmacol. Biochem. Behav.*, 10, 733, 1979. With permission.)

In 1979, researchers from the Primate Center at the University of Wisconsin published two papers reporting impaired reversal learning performance in rhesus monkeys (*Macaca mulatta*) exposed postnatally to lead.[8,9] Blood lead levels of approximately 50 or 90 μg/dl were associated with impairment on a series of tasks including spatial, color, and size discrimination reversal tasks early in life (Figure 1). A subset of these monkeys was found to be impaired on a series of spatial reversal tasks with irrelevant color cues at the age of 4 years, despite the fact that lead exposure ceased at 1 year of age, and blood lead levels at the time of testing were at control levels. In these monkeys acquisition of performance, as well as performance across reversals, was impaired as a result of early lead exposure.

In the same year, we reported deficits on a simple nonspatial form discrimination reversal task in 2- to 3-year-old cynomolgus monkeys (*Macaca fascicularis*) exposed to lead from birth.[10] Blood levels of these monkeys peaked at approximately 50 μg/dl, and decreased after infancy to stable levels of about 30 μg/dl. Lead exposed monkeys were not impaired on the acquisition of the discrimination task, but did not improve over successive reversals as quickly as control monkeys.

This apparently robust effect of lead on discrimination reversal performance was pursued in our laboratory with a group of monkeys (controls and two dosed groups) exposed to lower levels of lead from birth onward. Blood levels peaked during infancy at 25 or 15 μg/dl for the higher and lower dose groups, respectively, then decreased to steady state levels of 13 or 11 μg/dl. The present CDC guidelines consider a child to be at low risk for adverse health effects of lead at blood levels below 25 μg/dl (although this may be revised downward shortly).

These monkeys were tested as juveniles (3 year olds) on a series of nonspatial discrimination reversal problems, including form discrimination, color discrimination with irrelevant form cues, and form discrimination with irrelevant color cues.[11] This afforded the opportunity to change two sets of "rules": the positive and negative stimulus within a pair of relevant stimuli (such as cross versus square); and the relevant stimulus dimension (such as form versus color). Lead-treated monkeys were not impaired on the acquisition of any of the three tasks, but were impaired over the set of reversals on the form discrimination, which was their first introduction to a discrimination reversal task, and the color discrimination with irrelevant cues. On the form discrimination with irrelevant cues, the treated groups as a

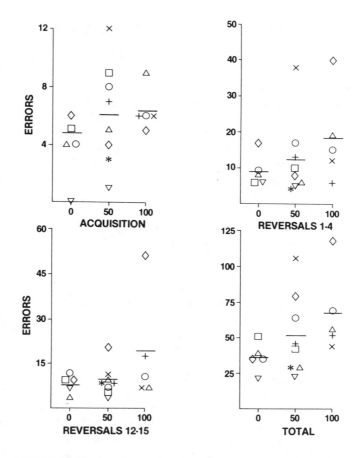

FIGURE 2. Number of errors for acquisition, reversals 1 to 4, reversals 12 to 15, and over the entire series of reversals for a spatial discrimination reversal task with irrelevant form cues. Each *symbol* represents an individual monkey; *horizontal lines* represent group means: 0 = control; 50 = group with blood levels of 11 μg/dl at time of testing; 100 = group with blood levels of 13 μg/dl at time of testing. (From Gilbert, S. G. and Rice, D. C., *Toxicol. Appl. Pharmacol.*, 91, 484, 1987. With permission.)

whole did not differ statistically from the control group, although certain individual treated monkeys were impaired.

When these monkeys were mature adults (9 to 10 years old), they were tested on a series of spatial discrimination reversal tasks: no irrelevant cues, irrelevant form cues, and irrelevant form and color cues.[12] In this task, the monkey was required to respond to a particular position irrespective of what stimuli appeared on the response buttons. Treated monkeys were impaired relative to controls in the presence but not in the absence of irrelevant stimuli (Figure 2). Moreover, the lower dose group was impaired only during the first task after the introduction of irrelevant cues, but not after irrelevant stimuli were familiar. For both the spatial and nonspatial discriminating reversal tasks, there was evidence that lead-exposed monkeys were attending to the irrelevant stimuli in systematic ways, suggesting that this behavior was responsible for, or at least contributing to, the impairment in performance.

There are several generalizations that may be drawn from these studies. It may be stated with some degree of confidence that lead impairs performance on both spatial and nonspatial discrimination reversal performance in monkeys exposed developmentally (postnatally) to lead. It is clear that the discrimination reversal paradigm is more sensitive to lead-induced behavioral impairment than is simply the acquisition of a discrimination task. In the studies just described, impairment in task acquisition was observed only at high blood lead levels, or when the "rules" were changed (i.e., when changing between tasks requiring attention to different stimulus dimensions). It also appears that lead-exposed monkeys exhibit a greater degree of impairment across a set of reversals in the presence of distracting irrelevant stimuli, although performance was sometimes impaired in the absence of irrelevant stimuli, especially at higher doses. It is also often the case that the greatest degree of impairment was observed at the beginning of new reversal tasks, again at higher doses. This constellation of effects points to decreased adaptability and/or increased distractibility in lead-treated monkeys, although other types of impairment may also be present.

2. Spatial Delayed Alternation

Another relatively simple task that has proved sensitive to disruption by lead, both in our laboratory and the Wisconsin laboratory, is spatial delayed alternation. This task requires the monkey to alternate responding between two stimuli, with each correct alternation being reinforced. The stimuli are identical, with no cue to indicate which is the correct stimulus. Delays of various lengths can be instituted between opportunities to respond, presumably rendering the task more difficult. This task is considered to be a test of spatial memory. A study at the University of Wisconsin revealed that rhesus monkeys exposed to lead from birth to 1 year of age, with peak blood levels as high as 300 µg/dl and levels of 90 µg/dl for the remainder of the first year of life, were markedly impaired on this task as adults.[13] The deficit was more marked at shorter than at longer delay values.

In a study in our laboratory, monkeys with steady-state blood lead levels of 11 or 13 µg/dl, previously described, were tested on a spatial delayed alternation task at 7 to 8 years of age, before the spatial discrimination reversal tasks.[14] Our paradigm was different from that employed by the Wisconsin group in what may have been two important respects. First, delay values of different lengths were not mixed within a session as in the Wisconsin study, but rather were increased over the course of the experiment, with the longest delay during the terminal sessions. Second, errors were corrected; that is, if the monkey made an error, it had to respond on the alternate button before continuing with the alternations. (In the Wisconsin study, if the monkey responded incorrectly on the left side, for example, the next correct response would also be on the left side.) Both treated groups were impaired relative to controls, during the initial acquisition of the task as well as at the longer but not shorter delay values (Figure 3). In addition, monkeys in both treated groups displayed marked perseverative behavior, responding in some cases on the same incorrect response button for hours at a time. The opportunity for this perseverative behavior was provided by the inclusion of the correction procedure. This effect may also not have been observed if long delay values had been interspersed throughout the session, rather than comprising the entire session. The severe deficit observed in our monkeys was comparable to that observed following extensive brain lesions in certain parts of the cerebral cortex, and were certainly unexpected in monkeys with a history of such moderate lead exposure.

3. Hamilton Search Task

The Wisconsin group has utilized the Hamilton Search Task to study attention and spatial memory. In this task, a row of boxes is baited with food, and then closed. The monkey lifts the lids to obtain the food. The most efficient performance requires that each box be opened only once, necessitating that the monkey remember which boxes have already been opened. Monkeys exposed to doses of lead sufficient to produce blood lead levels of approximately 45 or 90 µg/dl during the first year of life or 50 µg/dl *in utero* were impaired

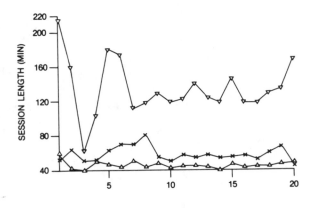

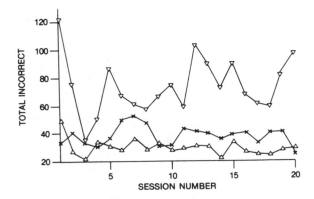

FIGURE 3. Session length and total number of incorrect responses over the course of the sessions on the longest (15 s) delay value of a spatial delayed alternation task. Session length was increased by each error, since a session consisted of 100 correct trials (responses). △ = control; x = 11 μg/dl blood level group, ▽ = 13 μg/dl blood lead group. (From Rice, D. C. and Karpinski, K. F., *Neurotoxicol Teratol.*, 8, 219, 1988. With permission.)

in their ability to perform this task at 4 to 5 years of age.[15] This was replicated in another group exposed postnatally to higher lead levels, and tested at 5 to 6 years of age.[13] The effects on Hamilton Search Task were, in general, less robust than effects on delayed spatial alternation tested in the same monkeys, despite the fact that both tests presumably assess attention and spatial memory. The greater deficit observed on the delayed alternation task may have been due to the requirement for alternation or adaptation of response pattern, an ability which seems to be globally impaired in lead-exposed monkeys.

4. Delayed Matching to Sample

We used a delayed matching to sample paradigm to assess attention and short term memory in the group of monkeys previously discussed, with preweaning blood lead levels of 50 μg/dl and postweaning steady-state blood lead levels of 30 μg/dl.[16] These monkeys began testing at 3 to 4 years of age. They were tested on two different tasks, spatial and nonspatial matching to sample. In the nonspatial task, the monkey was shown a sample stimulus, for example an orange disk. Following a delay period which varied from a few seconds to several minutes, the monkey was shown disks of various colors, and was required to respond on the orange disk to obtain a food reward. The color of the sample disk varied

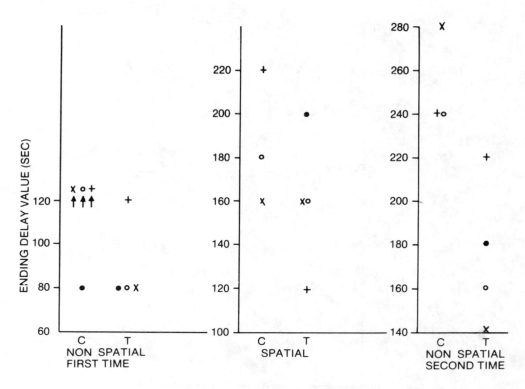

FIGURE 4. Delay value at which monkeys reached chance performance on nonspatial and spatial delayed matching to sample tasks. The nonspatial task was tested twice under different experimental conditions. Each *symbol* represents an individual monkey: C = control; T = monkeys with steady-state blood lead values of 30 μg/dl. For the nonspatial task first time, three control monkeys were performing above chance at 120 s, but were not tested further since all treated monkeys were performing at chance levels. (From Rice, D. C., *Toxicol. Appl. Pharmacol.*, 75, 337, 1984. With permission.)

from trial to trial, and the positions of the colored test disks also varied between trials. For the spatial task, the monkey was required to remember in which position a sample disk appeared, and respond at that same position after the interposition of a delay. Lead-exposed monkeys were impaired on both the spatial and nonspatial versions of this task. They were not impaired in their ability to learn the matching task per se, but were increasingly impaired as the delay between exposure to the sample stimulus and the set of stimuli to be matched was increased (Figure 4). Moreover, investigation of the types of errors revealed that for the nonspatial matching task, lead-exposed monkeys were responding incorrectly on the position that had been correct on the previous trial. This type of perseverative behavior is reminiscent of the perseverative errors observed in other groups on delayed alternation, or attention to irrelevant cues on discrimination reversal tasks.

5. Evidence from Other Primate Laboratories

In a study on the effects of *in utero* exposure to lead performed at the University of Iowa, female cynomolgus monkeys were exposed prior to and during pregnancy at doses sufficient to produce blood lead levels of 60 or 80 μg/dl in the mothers, and slightly lower blood lead levels in the offspring at birth.[17] Offspring tested at 6 to 18 months of age were found to be impaired on a simple visual discrimination task as well as a task that requires the infant to time a response between 10 and 15 s after the onset of a cue light. In both tasks, failure to respond resulted in the termination of the trial. The impairment of lead-treated monkeys on both tasks was the result of letting some trials terminate — simply not

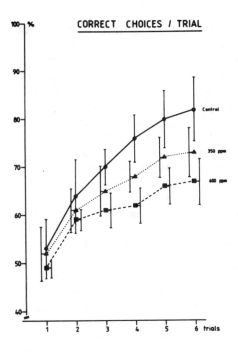

FIGURE 5. Percent correct choices for the first six trials of each visual discrimination test of a learning set task, averaged over the total of 75 d of testing. Monkeys wee exposed *in utero* plus postnatally and tested at approximately 1 year of age. (From Lilienthal, H., Winneke, G., Brockhaus, A., and Molik, B., *Neurobehav. Toxicol. Teratol.*, 8, 265, 1986. With permission.)

performing the task. Treated monkeys were observed looking away from the response panel during the trial, or moving away from the panel altogether.

A final example of the types of intellectual deficits produced in monkeys as a result of developmental lead exposure was performed in Winneke's laboratory in Germany.[18] Rhesus monkeys were exposed to lead *in utero* and continuing during infancy at doses sufficient to produce blood lead values up to 50 µg/dl in the lower dose group and 110 µg/dl in the high dose group. These monkeys were tested on a learning set formation ("learning to learn") task as juveniles. The task is essentially a sequential series of visual discrimination problems; when the monkey learns one task to a preset criterion, another is presented. Different stimulus sets are used for each discrimination. Normal monkeys will learn successive discriminations more quickly as a result of exposure to the learning situation. Lead-exposed monkeys were impaired both in terms of improvement in performance across trials on any given problem, as well as ability to learn successive problems more quickly as the experiment progressed (Figure 5). Such a deficit represents impairment in the ability to take advantage of previous exposure to the same set of task "rules". This deficit is reminiscent of failure of lead-treated monkeys to improve as quickly as controls over a series of discrimination reversals.

6. Selected Collaborative Data from the Rat

A large number of studies have been performed in other species, mostly the rat, over the last 15 or so years. Much of the early research utilized high dose exposure producing overt toxicity or failure to thrive in developmental studies. Insensitive or poorly controlled behavioral endpoints often resulted in inconsistent or contradictory results. A recent review of studies that used learning tasks, largely in the rodent, concluded that deficits as a result of developmental lead exposure at reasonable doses were consistently observed.[3] A couple of pertinent examples will serve to illustrate the species generality of the intellectual deficits produced by developmental lead exposure. As early as 1977, Winneke's laboratory in Germany reported deficits in visual discrimination learning in rats exposed to lead *in utero* at what were at that time considered to be very moderate blood lead levels in the dams — 25 to 30 µg/dl.[19] In that study, lead-exposed rats learned a line orientation discrimination task at the same rate as controls, but were markedly impaired on the acquisition of a more difficult size discrimination task. In a later study from the same laboratory,[20] developmental exposure at similar blood levels resulted in deficits in visual discrimination and spatial memory tasks, but not on a simple avoidance task. These studies underscore the importance of the choice of behavioral endpoint in studying lead toxicity. In these examples, difficult but not easier tasks were sensitive to disruption by developmental lead exposure.

B. INTERMITTENT SCHEDULES OF REINFORCEMENT — LEAD AND ACTIVITY

Intermittent schedules of reinforcement have been studied extensively by experimental psychologists, and performance thus generated is orderly, predictable, and reproducible. (See Reference 21 for review of the use of intermittent schedules in behavioral toxicology.) The performance on intermittent schedules is the same across species, thus ensuring that the performance under study has species generality. An extensive body of literature describes the effects of a wide variety of psychoactive agents, and again is extremely consistent across species and in the hands of different investigators. Thus intermittent schedules provide a powerful tool to the behavioral toxicologist interested in characterizing the effects of a neurotoxic agent such as lead.

The intermittent schedule most studied with lead is the fixed interval (FI) schedule of reinforcement. In this schedule, the subject is rewarded for the first response (typically a lever press) after a specified time has elapsed. Responding before the specified time has no scheduled consequences. Even though there is no requirement for more than a single response at the end of the interval, animals (and humans under certain conditions) typically make many responses during the interval in a very characteristic response pattern. There is a pause at the beginning of the interval, followed by a gradually accelerating rate of response terminating in reinforcement. This pattern lends itself to analysis of a number of important performance parameters. The rate of response, for example, is in some sense a measure of the activity of the animal. The change in pattern of responding across the interval, on the other hand, may be considered a measure of the temporal discrimination, or timing ability, of the animal.

In 1979, we reported that monkeys exposed to lead exhibited an increased rate of response (increased activity) on an FI schedule of reinforcement (Figure 6).[22] This effect was observed in the group already described which had steady state blood lead concentrations of approximately 30 µg/dl. We studied performance on the FI schedule precisely because it is a measure of "activity" — an activity specified by the investigator. When our study was initiated there was evidence from the rodent literature that lead affected activity, although results were conflicting due to a variety of factors. In addition, investigations in children reported indications of what, at that time, was termed "hyperactivity". We reasoned that specifying the activity (lever press) might prove more sensitive to changes produced by lead than some measure of locomotion. In fact, we monitored locomotion in these same monkeys, and observed no differences between lead-treated and control monkeys. On the FI schedule, however, lead-exposed monkeys not only exhibited increases in rate, but responded in short "bursts" interposed with short pauses, an atypical pattern of FI responding. The rate-increasing effect of lead on FI performance was later replicated in monkeys exposed to lower

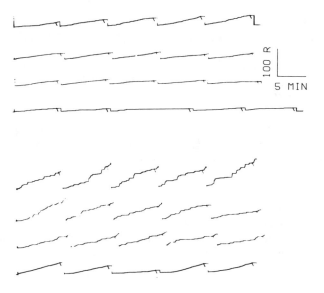

FIGURE 6. Cumulative records for control (top) and lead-exposed (bottom) monkeys performing on an FI schedule of reinforcement. Each lever press stepped the pen vertically, while time is represented horizontally. The reinforced response in each FI is signaled by the downward deflection of the pen. The pen reset to baseline at the end of each FI cycle. Lead-treated monkeys responded at a higher rate than control monkeys, and responded in bursts, making the records appear steplike. (From Rice, D. C., Gilbert, S. G., and Willes, R. F., *Toxicol. Appl. Pharmacol.*, 51, 503, 1979. With permission.)

doses of lead (steady-state blood lead concentrations of 11 or 13 µg/dl).[23] The unusual pattern of responding was not observed at the lower doses. The three dose groups exhibited an orderly dose-related increase in rate of responding as a consequence of lead exposure. This same dose-dependent increase in rate on FI performance has been replicated at the University of Rochester in rats with blood levels comparable to those in monkeys.[24-26]

Exposure to relatively high levels of lead during development, resulting in steady state blood lead levels of 40 µg/dl with a peak early in life of approximately 100 µg/dl, resulted in a change in the pattern of responding over the course of the FI.[27] Both the initial pause time and the pattern of acceleration in rate were changed in lead-exposed monkeys, with responding shifting toward the later part of the interval. The overall rate of response was also higher in the treated monkeys, as had been observed at lower doses.

Since lead-exposed monkeys exhibited higher rates of responding on the FI schedule, which does not specify any particular response rate, we were interested in determining whether these monkeys would inhibit responding if required to do so. We examined this question using the groups exposed to the lowest levels of lead, with long-term steady-state blood lead levels of 11 or 13 µg/dl.[28] Immediately following testing on the FI, the ability of these monkeys to learn a DRL (differential reinforcement of low rate) schedule of reinforcement was assessed. The DRL schedule required that the monkey inhibit responding for at least 30 seconds after the preceding response in order to receive a reward. Responding before the 30 s elapsed reset the clock, and another 30-s wait was in effect. Thus the DRL schedule "punishes" all but very low rates of responding by postponing the opportunity for reinforcement. There is no external cue to signify the end of the wait period; the monkey must resort to internal timing mechanisms. Lead-treated monkeys were able to perform the DRL task in a way that was indistinguishable from controls. However, they learned the task at a slower rate, as measured by the increment in reinforced responses and decrement of nonreinforced responses over the course of the early sessions. This suggests that it may be more difficult for lead-treated monkeys to utilize internal cues, or that they have more trouble learning to inhibit inappropriate responding.

C. SOCIAL BEHAVIOR AND ENVIRONMENTAL REACTIVITY

The effects of developmental lead exposure have been assessed on other behaviors, including mother-infant and peer social behavior and reaction to environmental manipulation, by the group at the University of Wisconsin. Infants exposed to lead either prenatally or postnatally engaged in more clinging to the mother than did control monkeys;[29] this was observed in mothers not exposed to lead but taking care of lead-exposed infants as well as in mothers exposed to lead. Monkeys treated with lead postnatally and raised with peers exhibited suppressed play behavior and increased social clinging.[30] Moreover, these abnormal patterns were exacerbated when these infant monkeys wee introduced to a novel environment. Infant monkeys exposed to lead also exhibited decreased muscle tonus and increased arousal or agitation on a neonatal developmental battery analogous to the Brazelton Neonatal Behavioral Assessment Battery developed for human infants.[31] In a test of visual exploration, these monkeys also displayed decreased looking at novel stimuli. A test similar to this has proved to be highly predictive of future intelligence in human infants.[32]

It appears, then, that lead has important effects on social behavior and reactivity to environmental circumstances, which may be manifested early in life. The latter effect in particular may be an important contributing factor to the deficits on intellectual tasks observed as a result of lead exposure.

IV. SUMMARY AND CONCLUSIONS

It is well established that monkeys exhibit behavioral impairment as a result of developmental exposure to lead. This has been demonstrated in two different species of monkey, in different laboratories, in a number of different groups of monkeys, and on a variety of behavioral tasks. Moreover, behavioral impairment has been observed in every group tested, with no evidence of a threshold for effect. Specifically, research in our laboratory has consistently demonstrated impairment in a group of monkeys with steady-state blood lead levels of 11 µg/dl and a peak early in life of 15 µg/dl, compared to a "control" group with blood lead levels of 3 to 5 µg/dl. Moreover, deficits persisted into adulthood.

The types of deficits observed in monkeys are indicative of impairment in attention, learning, and/or short-term memory. While it seems reasonably certain that these monkeys exhibit specific attentional deficits, it is not clear whether the impairments observed in short-term memory and learning ability are the result of this attentional impairment, or represent additional specific deficits. This issue has not been adequately addressed. It is also the case that lead-exposed monkeys exhibit more "activity" when the type of activity (lever press) is specified by the investigator. Moreover, these effects have also been observed at low blood levels in rodents by researchers willing to investigate the relevant behavioral and toxicological variables.

The types of deficits observed in monkeys as a result of developmental lead exposure are very similar to those observed in lead-exposed children. Moreover, at least one behavioral mechanism, namely attentional deficits, seems to underlie deficits in both species. Deficits have been observed in monkeys at the lowest blood levels examined-comparable to those at which deficits are observed in children. These findings provide evidence that the behavioral deficits correlated with lead exposure in children are in fact the result of lead exposure.

REFERENCES

1. **Bornschein, R., Pearson, D., and Reiter, L.,** Behavioral effects of moderate lead exposure in children and animal models. Part II. Animal studies, *CRC Crit. Rev. Toxicol.,* 8, 101, 1980.
2. **Needleman, H. L., Ed.,** *Low Level Lead Exposure: The Clinical Implications of Current Research,* Raven Press, New York, 1980.
3. **Cory-Slechta, D. A.,** The behavioral toxicity of lead: problems and perspectives, in *Advances in Behavioral Pharmacology,* Vol. 4, Thompson, T. and Dews, P., Eds. Academic Press, New York, 1984, 211.
4. **Needleman, H. L., Gunnoe, C., Leviton, A., Reed, R., Peresie, H., Maher, C., and Barrett, P.,** Deficits in psychologic and classroom performance of children with elevated dentine lead levels, *N. Engl. J. Med.,* 300, 689, 1979.
5. **Yule, W., Lansdown, R., Millar, I., and Urbanowicz, M.,** The relationship between blood lead concentration, intelligence, and attainment in a school population: a pilot study, *Dev. Med. Child. Neurol.,* 23, 567, 1981.
6. **Needleman, H. L.,** Lead at low doses and behavior of children, *Neurotoxicology,* 4, 121, 1983.
7. **Winneke, G., Beginn, V., Ewert, T., Havestadt, C., Kraemer, V., Krause, C., Thon, H. L., and Wagner, H. M.,** Comparing the effects of perinatal and late childhood lead exposure on neuropsychologic outcome, *Environ. Res.,* 38, 155, 1985.
8. **Bushnell, P. J. and Bowman, R. E.,** Reversal learning deficits in young monkeys exposed to lead, *Pharmacol. Biochem. Behav.,* 10, 733, 1979.
9. **Bushnell, P. J. and Bowman, R. E.,** Persistance of impaired reversal learning in young monkeys exposed to low levels of dietary lead, *J. Toxicol. Environ. Health,* 5, 1015, 1979.
10. **Rice, D. C. and Willes, R. F.,** Neonatal low-level lead exposure in monkeys (*Macaca fascicularis*): effect on two-choice non-spatial form discrimination, *J. Environ. Pathol. Toxicol.,* 2, 1195, 1979.
11. **Rice, D. C.,** Chronic low-lead exposure from birth produces deficits in discrimination reversal in monkeys, *Toxicol. Appl. Pharmacol.,* 77, 201, 1985.
12. **Gilbert, S. G. and Rice, D. C.,** Low-level lifetime lead exposure produces behavioral toxicity (spatial discrimination reversal) in adult monkeys, *Toxicol. Appl. Pharmacol.,* 91, 484, 1987.
13. **Levin, E. D. and Bowman, R. E.,** Long-term lead effects on the Hamilton Search Task and delayed alternation in monkeys, *Neurobehav. Toxicol. Teratol.,* 8, 219, 1986.
14. **Rice, D. C. and Karpinski, K. F.,** Lifetime low-level lead exposure produces deficits in delayed alternation in adult monkeys, *Neurotoxicol. Teratol.,* 10, 207, 1988.
15. **Levin, E. D. and Bowman, R. E.,** The effect of pre- or postnatal lead exposure on Hamilton Search Task in monkeys, *Neurobehav. Toxicol. Teratol.,* 5, 391, 1983.
16. **Rice, D. C.,** Behavioral deficit (delayed matching to sample) in monkeys exposed from birth to low levels of lead, *Toxicol. Appl. Pharmacol.,* 75, 337, 1984.
17. **Hopper, D. L., Kernan, W. J., and Lloyd, W. E.,** The behavioral effects of prenatal and early postnatal lead exposure in the primate *Macaca fascicularis, Toxicol. Indust. Health,* 2, 1, 1986.
18. **Lilienthal, H., Winneke, G., Brockhaus, A., and Molik, B.,** Pre- and postnatal lead-exposure in monkeys: effects on activity and learning set formation, *Neurobehav. Toxicol. Teratol.,* 8, 265, 1986.
19. **Winneke, G., Brockhaus, A., and Baltissen, R.,** Neurobehavioral and systemic effects of longterm blood lead-elevation in rats. I. Discrimination learning and open field-behavior, *Arch. Toxicol.,* 37, 247, 1977.
20. **Winneke, G., Collet, W., and Lilienthal, H.,** The effects of lead in laboratory animals and environmentally-exposed children, *Toxicology,* 49, 291, 1988.
21. **Rice, D. C.,** Quantification of operant behavior, *Toxicol. Lett.,* 43, 361, 1988.
22. **Rice, D. C., Gilbert, S. G., and Willes, R. F.,** Neonatal low-level lead exposure in monkeys: Locomotor activity, schedule-controlled behavior, and the effects of amphetamine, *Toxicol. Appl. Pharmacol.,* 51, 503, 1979.
23. **Rice, D. C.,** Effects of lead on schedule-controlled behavior in monkeys, in *Behavioral Pharmacology: The Current Status,* Seiden, L. S. and Balster, R. L., Eds., Alan R. Liss, New York, 1985, 473.
24. **Cory-Slechta, D. A. and Thompson, T.,** Behavioral toxicity of chronic postweaning lead exposure in the rat, *Toxicol. Appl. Pharmacol.,* 47, 151, 1979.
25. **Cory-Slechta, D. A., Weiss, B., and Cox, C.,** Delayed behavioral toxicity of lead with increasing exposure concentration, *Toxicol. Appl. Pharmacol.,* 71, 342, 1983.
26. **Cory-Slechta, D. A., Weiss, B., and Cox, C.,** Performance and exposure indices of rats exposed to low concentrations of lead, *Toxicol. Appl. Pharmacol.,* 78, 291, 1985.
27. **Rice, D. C.,** Schedule-controlled behavior in infant and juvenile monkeys exposed to lead from birth, *Neurotoxicology,* 9, 75, 1988.
28. **Rice, D. C. and Gilbert, S. G.,** Low lead exposure from birth produces behavioral toxicity (DRL) in monkeys, *Toxicol. Appl. Pharmacol.,* 80, 421, 1985.

29. **Schantz, S. L., Laughlin, N. K., Van Valkenberg, H. C., and Bowman, R. E.,** Maternal care by rhesus monkeys of infant monkeys exposed to either lead or 2,3,7,8-tetrachlorodibenzo-p-dioxin, *Neurotoxicology,* 7, 637, 1986.
30. **Bushnell, P. J. and Bowman, R. E.,** Effects of chronic lead ingestion on social development in infant rhesus monkeys, *Neurobehav. Toxicol.,* 1, 207, 1979.
31. **Levin, E. D., Schneider, M. L., Ferguson, S. A., Schantz, S. L., and Bowman, R. E.,** Behavioral effects of developmental lead exposure in rhesus monkeys, *Develop. Psychobiol.,* 21, 371, 1988.
32. **Rose, S. A. and Wallace, I. F.,** Visual recognition memory: a predictor of later cognitive functioning in preterms, *Child. Develop.,* 56, 843, 1985.

Section III: Human Lead Exposure

Chapter 9

OCCUPATIONAL LEAD EXPOSURE

Thomas D. Matte, Philip J. Landrigan, and Edward L. Baker

TABLE OF CONTENTS

I. History of Occupational Lead Exposure from Ancient to Recent Times 156
   A. The Uses of Lead 156
   B. A History of Lead Poisoning 156

II. Current Sources of Occupational Lead Exposure 158
   A. Overview and Current Trends 158
   B. Lead Smelters 160
   C. Battery Plants 161
   D. Other End-Use Industries 161
   E. Small Shops 161
   F. The Unregulated Hazards of the Construction Trades 162
   G. Fouling One's Own Nest: Lead Exposure Among Children of Lead Workers 163
   H. Occupational Lead Exposure in Developing Countries 163

III. Health Effects in Workers 164

IV. Case Management 165

V. Prevention 165

References 166

I. HISTORY OF OCCUPATIONAL LEAD EXPOSURE FROM ANCIENT TO RECENT TIMES

A. THE USES OF LEAD

Lead is an ancient metal. Techniques for the shaping and smelting of lead were known by the third millenium B.C. in Mesopotamia,[1] Troy, and predynastic Egypt, and the smelting of lead was especially highly developed by tribes living on the Black Sea coast near Pontus in Asia. Figurines, statuettes, and other artifacts of lead have been found in the remains of a number of ancient civilizations, including the Chinese, Greco-Roman, and Mayan Indian.[2-4]

Consumption of lead has increased markedly in the past 250 years. A most interesting natural record of trends in lead use is provided by study of the lead content in core samples from the perpetually frozen icecap of northern Greenland.[5] Each year a layer of ice and snow is added to this sediment by precipitation, and incorporated into each year's stratum are particles of lead and other materials that had been air-borne. In the earliest samples examined (deposited in 800 B.C.) the lead content was 0.0005 mg/kg ice; that level probably represents an irreducible natural minimum and reflects the contributions of such background sources as volcanic smoke and meterorites.[6] In the ice sample from 1750, corresponding approximately to the beginning of the Industrial Revolution in Europe, the lead concentration had increased to 0.01 mg/kg; that rise was apparently due to a gradual increase in lead smelting activity in the later middle ages.[7] Since 1750, lead levels in the icecap strata have risen sharply, first as the result of industrialization, and then even more abruptly over the past 40 years as the result of widespread combustion of leaded fuels. Lead levels in the ice of today are 0.21 mg/kg, a 420-fold increase over natural background levels.

World production of refined lead increased from 5.2×10^6 t in 1982 to more than 5.6×10^6 t in 1986, of which slightly less than 1×10^6 t was produced in the U.S.[8] Of lead produced in the U.S., 63% was secondary lead, recycled from previously consumed products, while primary lead, refined from domestic ore and small amounts of imported ore, accounted for the balance. Nearly all lead mined in the U.S. that year came from lead mines in Missouri and lead-producing precious metal or zinc and copper mines in Colorado, Idaho, and Montana. More than 90% of known U.S. lead reserves are found in Missouri, Alaska, Idaho, and Colorado.[9]

U.S. consumption of lead in 1986 was 1.1×10^6 t (the difference between consumption and production being made up by imports of refined lead).[8] Storage battery production accounted for 76% of lead consumed in the U.S. in 1986. Two significant lead-consuming industries, motor fuels and paints, have declined in importance because of environmental regulations. Lead use in gasoline antiknock additives declined from 218,000 t in 1976[9] to <29,000 t in 1986. The use of lead-containing paint pigments has been reduced for many applications and has been banned for interior paints in many countries; 14,000 t were used in paint pigments in 1986. Other major industrial uses of lead in 1986 were: ammunition manufacture (44,000 t); glass and ceramic production (41,000 t); solders (21,000 t); sheet lead for building construction, storage tanks and radiation shielding (17,000 t); and cable covering (17,000 t).[8]

B. A HISTORY OF LEAD POISONING

It is the opinion of Hunter[3] that lead poisoning has been recognized as a specific occupational disease since the third or fourth century B.C. Hunter states that "Hippocrates (370 B.C.) described a severe attack of colic in a man who extracted metals, and was probably the first of the ancients to recognize lead as the cause of the symptoms"; Waldron[4] notes that it has been difficult to locate in the original the precise text cited by Hunter. Lead poisoning was, however, clearly described by Nikander, a Greek poet and physician of the second century B.C. The following account from Nikander's Theriaca and Alexipharmaca (Major) details the adverse consequences of exposure to cerussa (lead carbonate) and specifically notes the occurrence of colic, paralysis, visual disturbance, and encephalopathy:

> The harmful cerussa, that most noxious thing
> Which foams like the milk in the earliest spring
> With rough force it falls and the pail beneath fills
> This fluid astringes and causes grave ills.
> The mouth it inflames and makes cold from within
> The gums dry and wrinkled, and parch'd like the skin
> The rough tongue feels harsher, the neck muscles grip
> He soon cannot swallow, foam runs from his lip
> A feeble cough tries, it in vain to expel
> He belches so much, and his belly does swell
> His sluggish eyes sway, then he totters to bed
> Complains that so dizzy and heavy his head
> Phantastic forms flit now in front of his eyes
> While deep from his breast there soon issue sad cries
> Meanwhile there comes a stuporous chill
> His feeble limbs droop and all motion is still
> His strength is now spent and unless one soon aids
> The sick man descends to the Stygian shades.

Ramazzini (1713)[10] in his de Morbis Artificum Diatriba described lead poisoning in potters and in portrait painters:

> In almost all cities there are other workers who habitually incur serious maladies from the deadly fumes of metals. Among these are the potters...When they need roasted or calcined lead for glazing their pots, they grind the lead in marble vessels...During this process, their mouths, nostrils, and the whole body take in the lead poison...First their hands become palsied, then they become paralytic, splenetic, lethargic, and toothless.
>
> In have observed that nearly all the painters whom I know, both in this and other cities, are sickly...For their liability to disease, there is an immediate cause. I mean the materials of the colors that they handle and smell constantly, such as red lead (Minimum), cinnabar, and white lead (Cerussa).

In Great Britain, Charles Turner Thackrah[11] described chronic occupational lead poisoning in plumbers, white lead manufacturers, housepainters, paperstainers, and potters. Thackrah noted that:

> Plumbers are exposed to the volatilized oxide of lead which rises during the process of casting. The fumes frequently induce at the time nausea and tightness at the chest; and men who are much in this department are soon affected with colic and palsy.
>
> The manufacturers of white lead are subjected to its poison both by the lungs and by the skin. Many soon complain of head-ache, drowsiness, sickness, vomiting, griping, obstinate constipation, and to these succeed colic or inflammation of the bowels, disorders of the urinary organs, and finally, the most marked of the diseases from lead, palsy. We observed the muscles of the fore-arm, more frequently and sooner to suffer than other parts. The eyes are also affected with chronic inflammation, or reduced nervous power.

A major figure in the more recent history of occupational lead poisoning in Great Britain was Sir Thomas Legge, who in 1897 was appointed first Medical Inspector of Factories.[12] Legge[13] stressed that industrial lead poisoning is "due almost entirely to lead dust or fume," and he proposed four axioms pertaining to the control of occupational lead poisoning:

1. Unless and until the employer has done everything — and everything means a good deal — the workman can do next to nothing to protect himself, although he is naturally willing enough to do his share.
2. If you can bring an influence to bear external to the workman (i.e., one over which he can exercise no control) you will be successful; and if you cannot or do not, you will never be wholly successful.
3. Practically all industrial lead poisoning is due to the inhalation of dust and fumes; and if you stop their inhalation you will stop the poisoning.
4. All workmen should be told something of the danger of the material with which they come into contact and not be left to find it out for themselves — sometimes at the cost of their lives.

Under Legge's influence, lead poisoning was made a notifiable disease in Britain in 1899. With the continuing surveillance and control that followed on that action, the number of reported cases fell from 1,058 with 38 deaths in 1900, to 505 in 1910,[14] and to 59 in 1973,[15] despite the considerable increases in lead consumption that occurred during that time.

In the United States, cases of lead poisoning were noted during the 17th and 18th centuries in the pottery, pewter-making, shot-dropping, and lead-smelting industries.[16] A major increase in incidence occurred in the latter half of the 19th century following the discovery of lead deposits in the Rocky Mountain States. McCord[16] estimates that as many as 30,000 cases of lead poisoning may have occurred from 1870 to 1900 in Utah, the state most seriously affected; an important factor in the Utah epidemic was the abundance there of cerussite (lead carbonate) ore, which is more soluble and thus more readily absorbed than the more common ore, galena (lead sulfide).

Little systematic attention was given to the prevention and control of lead poisoning in the U.S. until Dr. Alice Hamilton, late Assistant Professor Emerita of Industrial Medicine at Harvard, began her surveys of lead exposure in 1910.[17] She observed that.[18]

> Only a few years ago, we were most of us under the impression that our country was practically free from occupational poisoning...that our lead works were so much better built and managed, our lead workers so much better paid, and therefore better fed, than the European, that lead poisoning was not a problem here as it is in all other countries...As a matter of fact, the supposed advantages...obtain only in a few of the lead trades...[and] that far from being superior to Europe in the matter of industrial plumbism, we have a higher rate in many of our lead industries than have England or Germany.

Hamilton and her colleagues investigated lead exposure in various industries, among them storage battery manufacture (1915)[19] — in which she found 164 cases of lead poisoning in five plants employing 915 men — (prevalence rate, 17.9%), typefounding (1917),[20] smelting (1919),[21] enameling of sanitary ware (1919),[21] and pigment production (1934).[22] Hamilton had to face considerable opposition "from both employers and members of her own profession, one of whom described her report on lead poisoning as false, malicious, and slandering."[12] Nevertheless, her work had considerable influence on improving conditions of exposure in the North American lead industries.

Although progress in the control of occupational lead exposure has undoubtedly been made in the U.S. over the past 40 years,[23] the condition is not yet nationally notifiable, and in recent years lead poisoning continues to be discovered in a variety of occupations where lead exposure occurs (discussion follows). It is clear that more vigorous enforcement of existing legislation will be required if a degree of control similar to that reported for Great Britain[15] is to be attained.

II. CURRENT SOURCES OF OCCUPATIONAL LEAD EXPOSURE

A. OVERVIEW AND CURRENT TRENDS

Blood lead levels in adults, as in children, correlate with the degree of environmental exposure.[24] However, in contrast to children, reports of clinical lead toxicity in adults from contamination of the general environment are rare. Occupational lead poisoning in adults continues to be described with disturbing frequency. The proportion of elevated blood lead levels in adults attributable to occupational exposure is not precisely known, but an analysis of data from the Second U.S. National Health and Nutrition Evaluation Survey[25] and the National Occupational Hazard Survey, suggests it is high. Of adult males in potentially lead-exposed occupations, 5.8% had blood lead levels of 30 μg/dl and above, compared to 1.2% of males in nonlead exposed occupations. Of adult males with elevated blood leads 98% had potential occupational exposure. Lead exposure from environmental sources may, how-

TABLE 1
Occupations with Potential for Exposure to Inorganic Lead — A Partial Listing

Babbitters	Gold refiners	Patent leather makers
Battery makers	Gun barrel browners	Pearl makers, imitation
Bookbinders	Incandescent lamp makers	Pipe fitters
Bottle cap makers	Insecticide makers	Plastic workers
Brass founders	Insecticide users	Plumbers
Brass polishers	Japan makers	Pottery glaze mixers
Braziers	Japanners	Pottery workers
Brick burners	Jewelers	Putty makers
Brick makers	Junk metal refiners	Radiator repair workers
Bronzers	Lacquer markers	Riveters
Brush makers	Lead burners	Roofers
Cable makers	Lead counterweight makers	Rubber buffers
Cable splicers	Lead flooring makers	Rubber makers
Canners	Lead foil makers	Scrap metal workers
Cartridge makers	Lead mill workers	Sheet metal workers
Ceramic makers	Lead miners	Shellac makers
Chemical equipment makers	Lead pipe makers	Ship dismantlers
Chippers	Lead salt makers	Shoe stainers
Cutlery makers	Lead shield makers	Shot makers
Demolition workers	Lead smelters	Solderers
Dental technicians	Lead stearate makers	Solder makers
Diamond polishers	Lead workers	Steel engravers
Dye makers	Linoleum makers	Stereotypers
Electronic device makers	Linotypers	Tannery workers
Electroplaters	Lithographers	Temperers
Electrotypers	Match makers	Tetraethyl lead makers
Emery wheel makers	Metal burners	Tetramethyl lead makers
Enamel burners	Metal cutters	Textile makers
Enamelers	Metal grinders	Tile makers
Enamel makers	Metal miners	Tin foil makers
Farmers	Metal polishers	Tinners
File cutters	Metal refiners	Type founders
Filers	Mirror silverers	Typesetters
Flower makers, artificial	Motor fuel blenders	Varnish makers
Foundry moulders	Musical instrument makers	Wallpaper printers
Galvanizer	Painters	Welders
Glass makers	Paint makers	Zinc mill workers
Glass polishers	Paint pigment makers	Zinc smelter chargers

Modified from Reference 72.

ever, lower the margin of safety for occupationally exposed adults, placing them at greater risk of lead toxicity.[26]

Because of its physical properties, including a low melting point, corrosion resistance, malleability, and high density, lead finds use in a wide variety of industries and trades. As a result, there is potential for lead exposure in many occupations (Table 1). However, the level of lead exposure varies widely across such occupations and is modified by several factors. Inhalation of air-borne lead particulate is the principal route of lead absorption in occupationally exposed adults. Although blood lead levels tend to increase with increasing air lead concentrations, there is a great deal of variability in blood lead levels at a given air lead level.[27] Because lead accumulates in slowly exchanging body tissues, largely in bone, duration of exposure also influences the blood lead level. Additional variability is related to differences in the size distribution of air-borne lead particulate generated by different processes.[28] In general, smaller particulates remain air-borne longer and are inhaled, retained, and absorbed more efficiently. High temperature processes generate lead fume, a fine particulate formed by condensation of air-borne lead vapor (appreciable amounts of lead fume are generated above about 500°C), and therefore tend to be especially hazardous. Some mechanical processes (e.g., sanding) generate fine lead dust particles and are more hazardous than others (e.g., scraping) that generate larger particles. Blood lead levels are also influenced by lead contamination of the hands and face[29] and of food and cigarettes brought into the workplace. The quantitative contribution of these other factors, while not thoroughly investigated, is certainly smaller than that of inhalation in most workplaces, and is modified by worker hygiene.

In developed countries, regulation, engineering advances, and medical surveillance have reduced the incidence of clinical lead poisoning in the traditional high risk industries, such as smelting and battery manufacture.[30] In the U.S., blood lead monitoring mandated by the 1978 Occupational Safety and Health Administration (OSHA) lead standard[31] means that workers may be referred for clinical evaluation while still asymptomatic.[32] At the same time, advances in the understanding of lead toxicity have shifted concern from clinical lead toxicity to subclinical effects on the central and peripheral nervous systems, reproduction, and hematopoiesis, which result at lower blood lead levels.

Unfortunately, the benefits of reduced exposure and medical monitoring have not reached many workers in less traditional lead-related industries. In particular, small workplaces, generally overlooked by regulatory activities and lacking resources for adequate health and safety measures, may be emerging as a major source of more serious cases of occupational lead toxicity in developed countries. The construction trades are also largely exempted from the provisions of the OSHA lead standard. A lack of regulations and resources for occupational health and safety measures also prevails in many developing countries.[33] Therefore, one might expect occupational lead exposure in such countries to resemble that which prevailed in developed countries before recent preventive measures were instituted. Though data on lead exposure (and other occupational health hazards) in developing countries are sparse, workers in lead-related industries in developing countries have been found to have a high prevalence of unsafe blood lead levels.

The remainder of this section will describe studies of occupational lead exposure in traditional large-scale industries, small shops, the construction industry, among children of lead workers, and in developing countries. While recent studies will be emphasized, some of the most recently published studies of traditional lead producing/using industries antedate the U.S. OSHA lead standard and may not reflect current exposures in those industries. Because we have attempted to complement, rather than duplicate, other recent reviews of occupational lead exposure,[30,34] readers are encouraged to consult those works for additional perspectives.

B. LEAD SMELTERS

Lead smelters are classified as primary (those which refine lead from ore) and secondary (those which reclaim lead from scrap). Primary lead smelters tend to be large-scale operations, but, because of the limited sources of lead ore, they are relatively few in number (five in the U.S.[35]). On the other hand, lead scrap, primarily spent car batteries, is ubiquitous. Secondary smelters therefore tend to be smaller in scale, but are more numerous (60 in the U.S.[35]).

Lead smelting involves three main stages: handling and preparation of ore (sintering) or scrap (battery breaking), furnace smelting, and refining/casting.[36] Lead exposure can occur at each step, but re-entrainment of lead dust that accumulates in the smelter yard can also be a major source of exposure.[37] Primary smelters generate some of the highest and most difficult-to-control ambient lead levels of any industrial process. While the highest air lead levels may occur in ore preparation areas, the high temperature smelting process may produce smaller, more highly respirable lead particulate.[24] Exposures at secondary smelters tend to be lower than those at primary smelters, but are still quite high. In a survey of workers at two secondary smelters conducted during the mid-1970s, Lilis and colleagues[38] found a 76% prevalence of blood lead levels above 60 μg/dl.

C. BATTERY PLANTS

Lead-acid storage battery manufacture constitutes the largest end-user industry. The processes generating lead fume or dust include casting grids of metallic lead or lead alloy, pasting the interstices of the grid with lead oxide paste, and assembling batteries using gas welding.[36] Although some jobs, such as mixing oxide paste, may be at particularly high risk for lead exposure, exposure patterns may differ considerably across plants, and workers may be exposed to multiple processes in smaller plants. At a large U.S. battery manufacturer, environmental and blood lead monitoring between 1974 and 1976 showed that only 6% of blood lead levels were 60 μg/dl or more, but 55% were moderately elevated (40 to 59 μg/dl). The current U.S. permissible exposure limit (PEL) of 50 μg/m^3 was exceeded by 82% of the air samples.

D. OTHER END-USE INDUSTRIES

A few studies of industries that were more recently described as having hazardous lead exposure are worth reviewing. During auto body manufacture and repair, body seams or defects may be filled with a lead alloy solder, which is then machined to achieve a smooth finish. At one autobody assembly process, 20% of employees were found to have blood lead levels of 60 μg/dl and above.[39] More recently, lead exposure was studied at an aluminum forging facility where a product containing lead naphthenate was sprayed on molds to promote separation of the mold and forging. Lead fume was generated by the heat of the forging process, resulting in air lead concentrations as high as 430 μg/m^3 and a mean PbB in workers exposed full-time of 63 μg/dl.[40]

At a plant manufacturing electrical resistors and capacitors, leaded glass was used in a vitreous enameling process. Apparently, plant management was unaware of the potential for lead exposure until an index case of lead poisoning resulted in an investigation of the facility. Air lead levels ranged from 61 to 1700 μg/m^3 and 42% of workers had PbB of 50 μg/dl and above.[41] This report illustrates the importance of physician recognition and reporting in preventing occupational lead poisoning.

E. SMALL SHOPS

Physician awareness plays an especially important role in identifying lead hazards in small workplaces, where the absence of monitoring for air and blood lead levels may result in lead exposure being detected only because of severe toxicity. Some recent investigations of lead exposure in small establishments are described.

Using workers' compensation claims in Ohio and a laboratory-based, heavy metals registry in New York, National Institutes of Occupational Safety and Health (NIOSH) investigators[42] have identified workplaces where lead poisoning cases occurred. Workplaces with lead poisoning cases tended to be in industries where small establishments are prevalent and where lead is not a primary part of the industrial process (e.g., automotive repair and construction). Lead-related OSHA inspections in the same states tended to be concentrated on larger-scale, traditional lead-using industries (e.g., primary and fabricated metal industries).

Auto radiator repair involves using lead solder to seal joints, and exposure occurs during both disassembly and reassembly of radiators. In Finland, where substantial progress has been made in reducing occupational lead poisoning, car radiator repair workers tested during the mid-1970s had a median blood lead level of 38 μg/dl, and about 40% of such workers had blood lead levels above 40 µg/dl.[30] Following the identification of an index case of lead poisoning in a radiator repair shop worker in the Boston area, a survey of shops revealed a 39% prevalence of blood lead levels 40 µg/dl or above.[43] The investigators noted an absence of appropriate ventilation, eating and smoking by workers in the shops, and use of ineffective or poorly maintained respirators.

Occupational lead poisoning in North Carolina was reported in two workers (blood lead levels of 72 and 79 µg/dl) involved in recycling spent car batteries at small-scale operations. The authors stated that 10 such operations were located in one rural North Carolina county.[44] In Japan, a cottage industry involving quench-hardening of steel cutlery in a molten lead bath was implicated in elevated blood lead levels in workers (mean 35 µg/dl, range 13 to 92 µg/dl).[45]

Stained glass work involves heating, drawing, soldering, and polishing lead strips to form the network that supports the colored glass. Increased lead absorption was noted in a group of professional stained glass workers, but the highest blood lead found was 35 µg/dl.[46]

Indoor firing ranges can be sources of lead exposure, especially for employees. Airborne lead fume and dust are generated by hot propellant gasses, friction of bullets in the gun barrel, and fragmentation of bullets striking the target. Blood lead levels in employees at one range varied from 41 to 77 µg/dl, and air lead levels from 14 to 91 µg/m^3.[47] Lead exposure can be reduced by exhaust ventilation and use of jacketed bullets.

F. THE UNREGULATED HAZARDS OF THE CONSTRUCTION TRADES

Workers in a number of the construction trades are occupationally exposed to lead, yet the OSHA lead standard specifically excludes both construction and agricultural work.[31] The potential for lead exposure among plumbers, pipefitters, and painters has long been known, and in recent years, demolition and renovation of old lead-painted structures have emerged as high risk activities.

Efforts to remove lead-based paint from older housing where children are exposed have resulted in lead poisoning among deleading workers.[48] Dry scraping is the only recommended procedure for removal of leaded interior paint. For expediency, contractors have, however, been reported to use sanding, abrasive blasting, wire scraping, or burning for removal of leaded paint, all of which generate high levels of lead fume or fine lead dust. These methods require rigorous containment and assiduous cleanup of the area, as well as worker protection. Residents must be removed from the site and not permitted to return until the property is certified as safe. It is estimated that 52% of all housing units in the U.S. (as of 1980) have lead levels in paint that are a potential hazard to children; pre-1940 housing is almost certain to contain leaded paint.[35] Many states have programs and/or regulations for screening children and abating lead paint hazards, and recent Housing and Urban Development (HUD) regulations require deleading of public housing where children are found to have blood lead levels above 25 µg/dl.[49] While the potential for continuing lead exposure among deleaders is therefore large, monitoring and education of workers in this activity may be hampered by the small size of deleading contractors and employment of temporary and immigrant workers.

In demolition or restoration of lead-painted metal structures, such as bridges, gas torches may be used to remove paint or cut through structural iron. The high temperatures generate readily inhaled lead fume. A survey of New York City iron workers involved in demolition activities found 58% with blood lead levels above 40 µg/dl. Not only those involved in the burning process themselves, but nearby workers as well were affected.[50] Investigators in the Netherlands reported a mean PbB of 93 µg/dl among workers demolishing a lead-painted steel bridge. Air lead levels downwind from the burning operation (range 1,700 to 14,000 µg/m^3) were dramatically higher than upwind levels (range 60 to 80 µg/m^3).[51] Lead paint was removed from a bridge in Massachusetts using abrasive blasting and scraping, instead of burning. Still, high air lead levels (range 10 to 1090 $\mu g/m^3$) were observed. Of the workers who were scraping and repriming a section of the bridge, 58% had blood lead levels above 60 $\mu g/dl$.[52] Subsequent attention to safety practices, such as sealing canvas shrouds around the blasting process, reduced air lead levels, and the authors concluded that safe removal might be possible with proper supervision. Lead poisoning was also reported among workers using abrasive blasting to remove leaded paint from a bridge in Australia.[53]

Because of the need for corrosion resistance, leaded paints have been heavily used in shipbuilding. Air lead levels monitored during a ship overhaul found that much higher levels were associated with sanding to remove paint (range 3 to 1570 $\mu g/m^3$) than with use of a pneumatic chipper (range 2 to 17 $\mu g/m^3$).[54] Also of note was the fact that most of the paints originally used to paint the ship were believed "lead-free" before the survey, but bulk samples contained as much as 17% lead. The authors speculated that lead driers, rather than pigments, may have accounted for some of the lead in the paint. In another ship overhaul, oxyacetylene torches were used to cut old lead-painted iron plates. Air lead levels were much lower when exhaust ventilation was used (mean 91 $\mu g/m^3$) compared to when it was not (mean 547 $\mu g/m^3$). The mean blood lead level in the shipfitters was 38 $\mu g/dl$ (range 25 to 53 $\mu g/dl$).[55]

G. FOULING ONE'S OWN NEST: LEAD EXPOSURE AMONG CHILDREN OF LEAD WORKERS

In 1977, Baker and colleagues reported a 42% prevalence of blood lead levels above 30 $\mu g/dl$ among children of workers at a lead smelter.[56] Blood lead levels correlated with levels of housedust lead contamination, presumably from lead dust brought home on workers' clothing. Elevated blood lead levels have also been reported in children of battery workers.[57] To address this hazard, the OSHA lead standard requires employers to provide showers, changing facilities and laundered work clothes and to ensure that workers shower and change out of work clothes before leaving the workplace.[31] Changing clothes without showering and shampooing may not adequately protect children,[58] as lead dust may be carried home on the skin or hair.

Small shops pose a special risk to children of lead workers. They are often located in or near the home and are exempted from the OSHA requirement to provide changing and showering facilities. The small scale battery reclaiming operation described earlier[44] caused lead encephalopathy in a 3-year-old girl. Home lead exposure was aggravated by using discarded battery casings for fuel in a wood stove. Another lead reclamation worker[44] melted lead on his kitchen stove, and his 16-month-old child had a blood lead level of 63. Children of the home cutlery workers described earlier were also found to have increased lead absorption (PbB range 15 to 35 $\mu g/dl$).[45]

H. OCCUPATIONAL LEAD EXPOSURE IN DEVELOPING COUNTRIES

Lead exposure in developing countries is of special concern because such countries often lack the industrial hygiene expertise, regulation, and worker education[33] that have been used to reduce occupational lead exposure in traditional lead-related industries in developed countries. Furthermore, lead is widely used in industry and relatively easily monitored, so occupational lead exposure may be an index of the general level of occupational health in a nation.[30]

Although limited data are available, recent studies suggest that current lead exposure at battery manufacturers in developing countries is similar to that prevailing in developed countries prior to recent improvements. At a large-scale Korean manufacturer, 38% of workers had blood lead levels of 60 $\mu g/dl$ and above, and department-mean air lead levels ranged from 70 to 380 $\mu g/m^3$.[59] At a smaller factory in the Sudan, 95% of workers had

PbB of 40 µg/dl and above, and department-mean air lead levels ranged from 1800 to 2200 µg/m³.[60] In China, where occupational health regulations are, by developing country standards, relatively strong, the mean PbB among workers at a battery plant was 62 µg/dl, and the mean air lead level was 578 µg/m³.[61] Two of the authors (Matte and Baker) recently had the opportunity to study three small battery manufacturers (5 to 30 production workers) in Jamaica, and found that 90% of air lead samples exceeded 50 µg/m³. Of the workers, 28% had blood lead levels above 60 µg/dl.[62] While some hygienic measures, such as separate lunch rooms, showers, and changing out of work clothes at the end of a shift, had been adopted voluntarily by manufacturers, engineering controls and respiratory protection were inadequate or nonexistent.

In developing countries, the importance of small-scale industries,[63] often located at or near homes, also increases the potential for occupational lead exposure. Self-employed "battery chargers" in Nigeria were found to have a 63% prevalence of blood lead levels 40 µg/dl and above.[64] We also studied lead exposure among workers at battery repair shops in Jamaica and found that about two thirds had blood lead levels above 60 µg/dl. The poor hygienic conditions found at the shops also caused high-level lead contamination of yard soil and house dust at residences sharing premises with these yards, and a high risk (43%) of blood lead levels above 70 µg/dl among children less than 12 years old living at these residences.[65] In Barbados, Koplan and colleagues[66] found a mean blood lead level of 49 µg/dl among potters. The pottery shops adjoined residences where elevated blood lead levels (mean 35 µg/dl) were also found among household members not involved in pottery work. Papier-mâché workers in India were found to have a mean blood lead of 69 µg/dl, presumably from white lead pigment used in decorating various molded objects.[67] Of note is that workers studied were as young as 10 years of age. Silver jewelry makers in the same country refined impure silver by heating it with lead and were found to have blood lead levels ranging from 40 to 210 µg/dl.[68]

III. HEALTH EFFECTS IN LEAD WORKERS

Since this subject is discussed in detail elsewhere in this monograph as well as in a recent review,[69] it will not be treated fully here. The earliest manifestations of lead toxicity occur as a result of metabolic disruption through enzyme inhibition, specifically effecting heme biosynthesis. Such effects occur at relatively low blood lead concentrations. As inhibition becomes increasingly severe, with rising blood lead level concentrations, hemoglobin concentrations fall and overt anemia may develop. While acute lead encephalopathy is rare in occupationally-exposed adults today, milder forms of central nervous system (CNS) dysfunction, characterized by mood disturbances (particularly depression and irritability), psychomotor impairment, and short-term memory impairment have been frequently reported. Reversibility of certain CNS changes has been demonstrated upon removal from occupational lead exposure.[70] Acute effects are typically limited to these hematopoietic and neurologic effects.

More prolonged exposures are associated with increase in the severity and irreversibility of CNS manifestations as well as development of peripheral nervous system (PNS) dysfunction, renal disorders, and other effects. PNS disorders may be subclinical (i.e., slowed nerve conduction times) or overt (wrist-drop). Renal toxicity is poorly understood but may progress to end-stage renal disease. Other effects may include chromosomal changes, carcinogenicity, and reproductive disorders.

Standardized psychological testing may detect subtle CNS dysfunction at blood lead levels below which symptoms or signs of lead toxicity are evident on a routine history and physical exam.

IV. CASE MANAGEMENT

Proper management of lead poisoning in adult workers requires both attention to the individual and to his or her work environment. Unlike most occupational diseases, lead poisoning can be treated. Individuals with significantly elevated blood lead concentrations and evidence of functional impairment attributable to lead exposure, are successfully treated with chelating drugs, most notably EDTA (ethylene diamine tetraacetic acid) and penicillamine.[32] In the past, chelating drugs such as these have been used inappropriately in a prophylactic manner. This practice has been appropriately condemned since it subjects the worker to the dual toxicity of lead and the chelating drug while failing to remove the fundamental cause responsible for the worker's illness. Recent studies have shown that modest doses of EDTA can successfully chelate appreciable quantities of lead and, as a result, successfully treat workers with overt evidence of lead toxicity. Chelation therapy should not be used to "treat the blood lead concentration". Workers undergoing chelation therapy should be removed from lead exposure during the treatment course and until clear resolution of functional disturbances is seen.

V. PREVENTION

Primary prevention of occupational lead poisoning can be accomplished by routine monitoring of workplace air-borne-lead concentrations coupled with careful engineering control of sources of lead exposure in the workplace. Although of secondary importance, the use of respirators to provide further worker protection is viewed by NIOSH[71] as an acceptable adjunct to the development of engineering controls. Other important preventive measures include hand washing before meals to reduce the ingestion hazard, banning smoking from the work area to reduce the incremental lead absorption previously associated with this practice, and showering and proper use of work clothing to protect both the worker and his family.

Secondary prevention occurs through the detection of individuals with elevated blood lead concentrations followed by appropriate administrative action designed to reduce worker exposure and thereby facilitate lowering of the body lead burden through normal excretory processes. In the OSHA lead standard,[31] specific definitions are given, which specify blood lead concentrations at which workers must be removed from exposure and levels at which return to work may occur. Currently, the average blood lead concentration can not legally exceed 50 µg/dl. The OSHA standard further stipulates that workers, removed from exposure for medical reasons due to these provisions, should not suffer loss of salary or other benefits. This "medical removal protection" provision of the standard enhances the likelihood that employees will be willing to accept job transfer for reasons which derive from health concerns.

As in the case of other occupational diseases, one of the physician's most important roles is recognition of the work-relatedness of the individual case. The process of making this determination has been described elsewhere in detail[72] and relies heavily on the collection of a detailed occupational history.[73] In the case of lead exposure, the occupational history should include a clear statement of the occupation of the case and a description of the type of industry in which the worker is employed. Further, a description of the job tasks during which potential lead exposure occurs is useful. After obtaining a complete medical and occupational history, laboratory tests are performed to evaluate the extent of lead absorption and organ system functional status.

The measurement of lead concentration in whole blood is essential to the appropriate monitoring and management of lead-exposed workers and workers with lead toxicity. Quantification of lead concentration in blood should be performed only by those laboratories that have achieved proficiency in the proficiency testing program administered by the Centers for Disease Control, which is accepted by OSHA as a standard for determining lab acceptability.

Once a case of lead poisoning is identified by a physician, it should be reported to the appropriate state or federal agency. Many states have occupational disease reporting laws that require physicians to report suspect occupational disease cases to an agency of state government that has the responsibility for followup. Recently, to assist in the reestablishment of the capacity of state government to respond to such reports, the NIOSH has proposed and funded demonstration projects in ten states. This approach, the Sentinel Event Notification System for Occupational Risks, relies on physicians to recognize and report occupational diseases, such as lead poisoning, to a responsible state agency, which then performs appropriate followup and investigation.[74]

In some states where adequate resources do not exist in state government, referral of the suspect case to OSHA may be appropriate. Further, employers covered under the record keeping requirements of OSHA may be required to list the case on a log of occupational illness and injury.

In summary, prevention of occupational lead poisoning relies upon a combination of the classic tools of primary and secondary prevention in the workplace. Since lead is easily measured in a variety of media, the efficacy of control programs can be readily assessed. In view of the progress that has been made in recent years, occupational lead poisoning is an imminently preventable disease.

REFERENCES

1. **Woolley, C. L.**, Ur Excavations, 2, British Museum and the Museum of The University of Pennsylvania, London, 1934.
2. **Aitchison, L.**, *A History of Metals*, MacDonald and Evans, London, 1960.
3. **Hunter, Sir D.**, *The Diseases of Occupations*, 4th ed., Little, Brown, and Company, Boston, 1969.
4. **Waldron, H. A.**, Lead poisoning in the ancient world, *Med. Hist. (Suppl.)*, 17, 391, 1973.
5. **Murozumi, M., Chow, T. J., and Patterson, C.**, Chemical concentrations of pollutant lead aerosols, terrestrial dusts, and sea salts in Greenland and Antarctic Snow Strata, *Geochim. Cosmochim. Acta*, 33, 1247, 1969.
6. **Patterson, C. C.**, Contaminated and natural lead environments of man, *Arch. Environ. Health*, 11, 344, 1965.
7. **Patterson, C. C., Chow, T. J., and Murozumi, M.**, The possibility of measuring variations in the intensity of world-wide lead smelting during medieval times, in *Scientific Methods in Medieval Archaeology*, University of California Press, Berkeley, 1970.
8. U.S. Department of the Interior, Bureau of Mines Minerals Yearbook, Superintendent of Documents, Washington, D.C., 1986.
9. U.S. Department of the Interior, Lead, in Mineral Facts and Problems, Bureau of Mines, Bulletin 675, Superintendent of Documents, Washington, D.C., 1985.
10. **Ramazzini, B.**, *de Morbis Artificum Diatriba*, Transl. Wright, W. C. (1940), University of Chicago Press, Chicago, 1713.
11. **Thackrah, C. T.**, *The Effects of Arts, Trades, and Professions, and of Civic States, and Habits of Living on Health and Longevity with Suggestions for the Removal of Many of the Agents which Produce Disease, and Shorten the Duration of Life*, 2nd ed., Longman, Rees, Orme, Brown, Green, and Longman, 1832.
12. **Schilling, R. S. F.**, *Occupational Health Practice*, Butterworths, London, 1973, chap. 1.
13. **Legge, Sir T.**, *Industrial Maladies*, Oxford University Press, London, 1934.
14. **Oliver, Sir T.**, *Lead Poisoning from the Industrial, Medical and Social Points of View*, H.K. Lewis, London, 1914.
15. Her Majesty's Chief Inspector of Factories, Annual Report, Her Majesty's Stationery Office, London, 1973.
16. **McCord, C. P.**, Lead and lead poisoning in early American lead mines and lead poisoning, *Ind. Med. Surg.*, 22, 394, 1953.
17. **Hamilton, A.**, *Exploring the Dangerous Trades*, Little, Brown, Boston, 1943.

18. **Hamilton, A.**, Lead poisoning in the United States, *Am. J. Public Health,* 4:477, 1914.
19. **Hamilton, A.**, Lead Poisoning in the Manufacture of Storage Batteries, U.S. Bureau of Labor Statistics Bulletin 165, U.S. Government Printing Office, Washington, D.C., 1915.
20. **Hamilton, A. and Verrill, C. H.**, Hygiene of the Printing Trades, U.S. Bureau of Labor Statistics, Bulletin 200, U.S. Government Printing Office, Washington, D.C., 1917.
21. **Hamilton, A.**, Lead poisoning in American industry, *J. Ind. Hyg.,* 1:8, 1919.
22. **Hamilton, A.**, *Industrial Toxicology,* Haper & Brothers, Publishers, New York, 1934.
23. **Stokinger, H. E.**, Recent history of lead exposure in U.S. industry 1935—1965, in Symposium on Environmental Lead Contamination, U.S. Public Health Service Publication 1440, U.S. Department of Health, Education, and Welfare, Washington, D.C., 1965.
24. World Health Organization, Environmental Health Criteria 3, Lead, World Health Organization, Geneva, 1977.
25. Centers for Disease Control, Results of blood lead determinations among workers potentially exposed to lead — United States, *MMWR,* 32, 216, 1983.
26. **Hernberg, S., Dodson, V. N., and Zenz, C.**, Lead and its compounds, in *Occupational Medicine,* 2nd ed., Zenz, C., Ed., Year Book Medical Publishers, Chicago, 1988, chap. 36.
27. **Gartside, P. S., Buncher, C. R., and Lerner, S.**, Relationship of air lead and blood lead for workers at an automobile battery factory, *Int. Arch. Occup. Environ. Health,* 50, 1, 1982.
28. **Froines, J. R., Wen-Chen, V. L., Hinds, W. C., and Wegman, D. H.**, Effect of aerosol size on the blood lead distribution of industrial workers, *Am. J. Ind. Med.,* 9, 227, 1986.
29. **Chavalitnitikul, C., Levin, L., and Chen, L. C.**, Study and models of total lead exposures of battery workers, *Am. Ind. Hyg. Assoc. J.,* 45, 802, 1984.
30. **Hernberg, S. and Tola, S.**, The battle against occupational lead poisoning in Finland, experiences during the 15-year period 1964-1978, *Scand. J. Work Environ. Health,* 5, 336, 1979.
31. U.S. Department of Labor, Occupational Safety and Health Administration, Standard for occupational exposure to lead, 29 CFR 1910.1025, 1978.
32. **Cullen, M. R., Robins, J. M., and Eskenzi, B.**, Adult inorganic lead intoxication: Presentation of 31 new cases and a review of recent advances in the literature, *Medicine,* 62, 221, 1983.
33. **Michaels, D., Barrera, C., and Gacharna, M. G.**, Economic development and occupational health in Latin America: new directions for public health in less developed countries, *Am. J. Publ. Health,* 75, 536, 1985.
34. **Fischbein, A.**, Environmental and occupational lead exposure, in *Environmental and Occupational Medicine,* Rom, W. N., *Ed.*, Little, Brown, Boston, 1983, chap. 38.
35. Agency for Toxic Substances and Disease Registry, The Nature and Extent of Lead Poisoning in Children in the United States: A Report to Congress, U.S. Department of Health and Human Services, Atlanta, 1988.
36. **Burgess, W. A.**, *Recognition of Health Hazards in Industry,* John Wiley & Sons, New York, 1981.
37. **Simonson, A. V., Mecham, C. C., and Allcott, G. A.**, Evaluation of pavement cleaning methods and equipment for exposure control in secondary lead smelters, Final technical report, Demonstration project 14, National Institute for Occupational Safety and Health, Cincinnati, 1983.
38. **Lilis, R., Fischbein, A., Eisinger, J., et al.**, Prevalence of lead disease among secondary lead smelter workers and biological indicators of lead exposure, *Environ. Res.,* 14, 255, 1977.
39. **Lilis, R., Valciukas, J. A., Kon, S., Sarkosi, L., Campbell, C., and Selikoff, I. J.**, Assessment of lead health hazards in a body shop of an automobile assembly plant, *Am. J. Ind. Med.,* 3, 33, 1982.
40. **Goldberg, R., Garabrant, D. H., Peters, J. M., and Simonowitz, J. A.**, Excessive lead absorption resulting from exposure to lead naphthenate, *J. Occup. Med.,* 29, 750, 1987.
41. **Alkes, I., Teitelbaum, D., Kadushin, F., et al.**, Lead poisoning in a capacitor and resistor plant, *Morbid. Mortal. Wkly. Rep.,* 34, 384, 1985.
42. **Seligman, P. J. and Halperin, W.**, Targeting workplace inspections for lead, *Am. J. Ind. Med.,* 1991, in press.
43. **Goldman, R. H., Baker, E. L., Hannan, M., and Kamerow, D. B.**, Lead poisoning in automobile radiator mechanics, *N. Engl. J. Med.,* 317, 214, 1987.
44. **Dolcourt, J. L., Finch, C., Coleman, G. D., Klimas, A. J., and Milar, C. R.**, Hazard of lead exposure in the home from recycled automobile storage batteries, *Pediatrics,* 68, 225, 1981.
45. **Kawai, M., Toriumi, H., Katagiri, Y., and Maruyama, Y.**, Home lead-work as a potential source of lead exposure for children, *Int. Arch. Occup. Environ. Health,* 53, 37, 1983.
46. **Landrigan, P. J., Tamblyn, P. B., Nelson, M., Kerndt, P., Kronoveter, K. J., and Zack, M. M.**, Lead exposure in stained glass workers, *Am. J. Ind. Med.,* 1, 177, 1980.
47. **Novotny, T., Cook, M., Hughes, J., and Lee, S. A.**, Lead exposure in a firing range, *Am. J. Public Health,* 77, 1225, 1987.
48. **Feldman, R. G.**, Urban lead mining: lead intoxication among deleaders, *N. Engl. J. Med.,* 298, 1143, 1978.

49. Centers for Disease Control, Potential increased demand for lead testing as a result of recent HUD regulations, *MMWR*, 36, 325, 1987.
50. Fischbein, A., Leeds, M., and Solomon, S., Lead exposure among iron workers in New York City, *N.Y. State J. Med.*, 84, 445, 1984.
51. Spee, T. and Zwennis, W. C. M., Lead exposure during demolition of a steel structure coated with lead-based paints, *Scand. J. Work Environ. Health*, 13, 52, 1987.
52. Landrigan, P. J., Baker, E. L., Himmelstein, J. S., et al., Exposure to lead from the mystic river bridge: the dilemma of deleading, *N. Engl. J. Med.*, 306, 673, 1982.
53. Pollock, C. A. and Ibels, L. S., Lead intoxication in paint removal workers on the Sydney Harbor Bridge, *Med. J. Aust.*, 145, 635, 1986.
54. Booher, L. E., Lead exposure in a ship overhaul facility during paint removal, *Am. Ind. Hyg. Assoc. J.*, 49, 121, 1988.
55. Landrigan, P. J. and Straub, W. E., Occupational lead exposure aboard a tall ship, *Am. J. Ind. Med.*, 8, 233, 1985.
56. Baker, E. L., Folland, D. S., Taylor, T. A., et al., Lead poisoning in children of lead workers, *N. Engl. J. Med.*, 296, 260, 1977.
57. Watson, W. N., Witherell, L. E., and Giguere, G. C., Increased lead absorption in children of workers in a lead storage battery plant, *J. Occup. Med.*, 20, 759, 1978.
58. Morton, D. E., Saah, A. J., Silberg, S. L., et al., Lead absorption in children of employees in a lead-related industry, *Am. J. Epidemiol.*, 115, 549, 1982.
59. Lee, B. K., Occupational lead exposure of storage battery workers in Korea, *Br. J. Ind. Med.*, 39, 283, 1982.
60. Mohamed, A., Hamed, A. S., Elhaimi, Y. A. A., Osman, Y., Effects of exposure to lead among lead-acid battery factory workers in Sudan, *Arch. Environ. Health*, 41, 261, 1986.
61. Yi-Lan, W., Pei-Kun, L., Zi-qiang, C., et al., Effects of occupational lead exposure, *Scand. J. Work Environ. Health*, 11, 20, 1985.
62. Matte, T. D., Figueroa, J. P., Burr, G., Flesch, J. P., Keenleyside, R. A., Baker, E. L., Lead exposure among lead-acid battery workers in Jamaica, *Am. J. Ind. Med.*, 16, 167, 1989.
63. Phoon, W. Recent developments in occupational health in tropical countries, *Trop. Dis. Bull.*, 79, 653, 1982.
64. Asogwa, S. E., The risk of lead poisoning in battery chargers and the possible hazard of their occupation on the environment, *Nig. Med. J.*, 9, 189, 1979.
65. Matte, T. D., Figueroa, J. P., Ostrowski, S. R., Burr, G., Jackson-Hunt, L., Keenlyside, R. A., and Baker, E. L., Lead poisoning among household members exposed to lead-acid battery repair shops in Jamaica, *Int. J. Epidemiol.*, 18, 874, 1989.
66. Koplan, J. P., Wells, A. V., Diggory, H. J. P., et al., Lead absorption in a community of potters in Barbados, *Int. J. Epidemiol.*, 6, 225, 1977.
67. Kaul, P. S. and Kaul, B., Blood lead and erythrocyte protoporphyrin levels among papier-mache workers in Kashmir, *Mt. Sinai J. Med.*, 53, 145, 1986.
68. Behari, J. R., Singh, S., and Tandon, S. K., Lead poisoning among indian silver jewelry makers, *Ann. Occup. Hyg.*, 27, 107, 1983.
69. Landrigan, P. J., Froines, J. R., and Mahaffey, K. R., Body lead burden: summary of epidemiological data on its relation to environmental sources and toxic effects in dietary and environmental lead, in *Human Health Effects*, Mahaffey, K. R., Ed., Elsevier Science, Amsterdam, 1985, Chap. 2.
70. Baker, E. L., White, R. F., Pothier, L. J., Berkey, C. S., Dinse, C. E., Travers, P. H., Harley, J. P., and Feldman, R. G., Occupational lead neurotoxicity: improvement in behavioral effects after removal from exposure, *Br. J. Ind. Med.*, 42:507, 1985.
71. Schulte, H. F., Personal protective devices, in The Industrial Environment-its Evaluation and Control, National Institute for Occupational Safety and Health, U.S. Department of Health and Human Services, 1973.
72. Kusnetz, S. and Hutchinson, M., A guide to the work relatedness of disease, rev. ed., U.S. Department of Health and Human Services, Department of Health, Education and Welfare (NIOSH) Publication No. 79-116, 1979.
73. Goldman, R. H. and Peters, J. M., The occupational health history, *JAMA*, 246:2831-6, 1981.
74. Baker, E. L., Sentinel Event Notification System for Occupational Risks (SENSOR): The Concept, *Am. J. Public Health*, 79s, 18, 1989.

Chapter 10

LEAD, THE KIDNEY, AND HYPERTENSION

Richard P. Wedeen

TABLE OF CONTENTS

I.	Introduction	170
II.	Acute Lead Nephropathy in Children	170
	A. The Fanconi Syndrome	170
	B. Queensland Nephritis	172
III.	Lead Nephropathy in Adults	173
	A. Symptomatic Lead Poisoning	173
	B. Cumulative Lead Absorption	173
	C. Lead Nephropathy in Asymptomatic Adults	176
	D. Gout Nephropathy	177
IV.	Lead and Hypertension	179
	A. Low-Level Lead Absorption and Blood Pressure	179
	B. Lead and Hypertensive Cardiovascular Disease	180
	C. Mortality Studies	181
V.	Pathophysiology of Lead-Induced Hypertension	181
	A. Cation Transport Systems	181
	1. Calcium	181
	2. Sodium	182
	B. Hormonal Mediation	183
	1. The Renin-Angiotensin-Aldosterone System	183
	2. Vasoactive Hormones	183
VI.	Implications	184
References		185

I. INTRODUCTION

Kidney disease due to inorganic lead was first described in 1862.[1] Late 19th century authors regularly included lead as a cause of kidney disease in adults but sometimes failed to differentiate lead-induced tubulointerstitial nephritis (meager proteinuria) from glomerulonephritis (heavy proteinuria). The confusion is not surprising since Bright's Disease was defined by proteinuria, and urinary protein excretion progressively rises as tubulointerstitial disease advances. Histologically, tubulointerstitial nephritis is distinguished by tubular atrophy, interstitial fibroblast proliferation, a paucity of round cell infiltrates, and relative preservation of glomerular morphology until renal failure is advanced (Figure 1).

In the twentieth century, chronic interstitial nephritis due to lead has been confused with the transient renal dysfunction associated with acute lead poisoning and the renal disease associated with hypertension; nephrosclerosis. Although these entities overlap, they are treated separately in the medical literature. Three clinical forms of renal disease due to lead are currently recognized:

1. Acute lead nephropathy — A reversible proximal tubule reabsorptive defect (Fanconi syndrome) following acute massive exposure in childhood and experimental animal models.
2. Chronic lead nephropathy — Chronic tubulointerstitial nephritis in adults, following occupational or sporadic environmental exposure (often associated with gout and hypertension).
3. Hypertension — Exposure to ambient environmental lead resulting in hypertension and subclinical renal dysfunction.

These entities are associated with differences in the intensity and duration of exposure. Exposure intensity is highest in the acute form, intermediate in occupational exposure and lowest in ambient exposure. Duration is longest in ambient exposure, and shortest in acute episodes.

Despite widespread recognition of lead nephropathy,[2-6] prevention of this progressive form of renal disease remains a low priority in public health circles. Lead nephropathy remains controversial because of incomplete epidemiologic data, diagnostic misconceptions, and the absence of a compelling theory of pathogenesis. Nevertheless, there are 1.4 million lead workers in the U.S., and over 2 million Americans may have unrecognized excessive lead exposure.[7]

II. ACUTE LEAD NEPHROPATHY IN CHILDREN

A. THE FANCONI SYNDROME

In 1926, McKhann and Vogt[8] showed that acute lead poisoning resulting from childhood pica was accompanied by glycosuria in the absence of hyperglycemia (renal glycosuria). Increased urinary excretion of amino acids and phosphate was subsequently noted.[1] Since these organic substances are normally reabsorbed by the proximal tubule and appear in the urine in minute quantities, acute lead intoxication appeared to impair proximal tubule reabsorption, and produce an acquired "Fanconi syndrome". Disordered calcium metabolism due to phosphaturia sometimes resulted in renal rickets.

In acute lead poisoning, renal biopsies show acid-fast intranuclear inclusion bodies in proximal tubule epithelial cells (Figure 2). Intranuclear inclusions are also found in nonrenal tissues including the liver, osteoclasts, anterior horn cells of the spinal cord and in neurons of the substantia nigra.[9] The characteristic lead- and protein-containing inclusions are regularly found in acutely lead-poisoned experimental animals,[10,11] but are uncommon in the chronic lead nephropathy of adults in the absence of acute intoxication.[12,13]

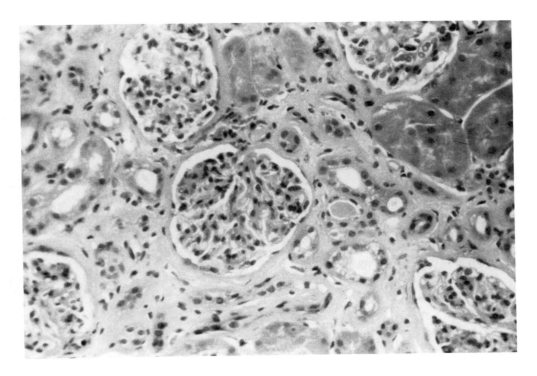

FIGURE 1. Focal interstitial nephritis in a patient with lead nephropathy. (From Wedeen, R. P., in *Current Nephrology*, Vol. 11, Year Book Publishers, 1988, chap. 3. With permission.)

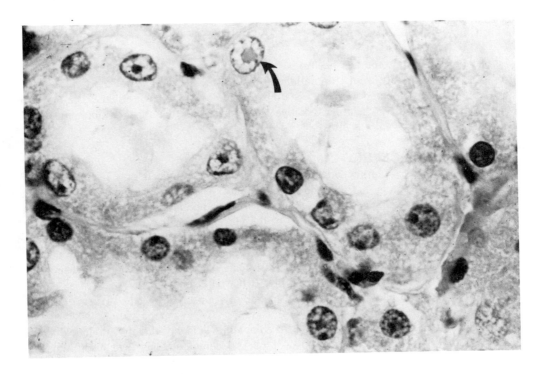

FIGURE 2. Characteristic acid fast intranuclear inclusion body (arrow) in proximal tubule. Postmortem tissue from adult who died from lead encephalopathy.

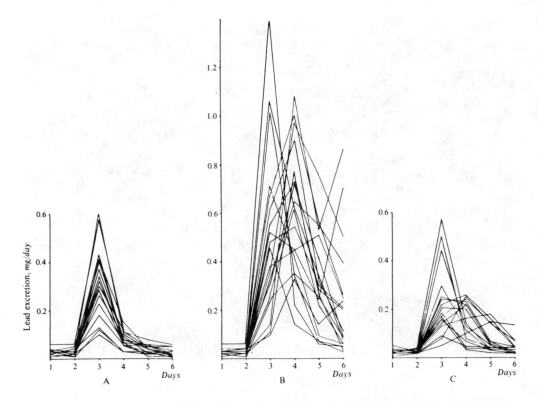

FIGURE 3. CaEDTA lead mobilization tests in: (A) normal control subjects; (B) chronic lead nephropathy; (C) chronic renal failure of nonlead etiology. Emmerson performed these EDTA tests by collecting 24 h urines for 6 consecutive days; 1 g EDTA was given i.v. in 250 ml saline over 1 h on day 3. (From Emmerson, B. T., *Kidney Int.*, 4, 1, 1973. With permission.)

The Fanconi syndrome has been observed only at extremely high blood lead levels, usually greater than 150 µg/dl.[14] in childhood plumbism, the tubular transport defect is of minor importance compared to the devastating central nervous system effects. The Fanconi syndrome is reversed by chelation therapy and is unusual in the chronic tubulointerstitial nephritis that develops later in life.[10-16]

B. QUEENSLAND NEPHRITIS

As early as 1893, chronic interstitial nephritis was recognized as a late sequelae of childhood lead poisoning in Australia.[1] The disease, originally known as "Queensland nephritis", afflicted young adults brought up in Brisbane and arose from a form of pica that differed from that seen in the inner cities of the U.S. Rather than eating paint chips, the Brisbane children ingested weathered paint dust from the enclosed outdoor verandas in Queensland. Neurologic symptoms were similar to those in the U.S., but chelation therapy was not available. In a followup study of 401 Queensland children who had been lead poisoned in childhood, Henderson[17] found that 101 of 187 survivors had renal disease. The records of 108 of 165 who had died under the age of 40 also showed the presence of renal disease.

Studies of Queensland nephritis provided compelling evidence that the ethylene diamine tetraacetic acid (EDTA) lead mobilization test assesses cumulative past lead absorption even in the presence of renal failure (Figure 3).[18] Australian investigators showed that the bone lead concentration correlates with past lead absorption.[18,19] No comparable followup study of untreated children has been reported in the U.S.

III. LEAD NEPHROPATHY IN ADULTS

A. SYMPTOMATIC LEAD POISONING

Chronic tubulointerstitial nephritis due to lead is most readily recognized in association with symptomatic lead poisoning. Descriptions of lead nephropathy in adults generally originate from occupational settings and are found in "moonshine" consumers in the southeastern U.S.[1] Lead nephropathy in moonshiners was frequently accompanied by gout and hypertension.[21,22] While renal failure is universally associated with hyperuricemia, gout is distinctly rare in patients with renal failure except in the renal disease due to lead.

Lead nephropathy has only recently been diagnosed in the absence of overt intoxication. Early detection of tubulointerstitial nephritis is difficult because of the absence of a simple urine test to detect nonglomerular renal dysfunction. In early lead nephropathy, identification of characteristic excretion patterns in the urine of low molecular-weight proteins, such as β-2-microglobulin, lysozyme or N-acetyl glucoseaminidase, (so called "tubular proteinuria"), has proven unreliable.[3,13,16,23,24] At postmortem examination, the histologic features of lead nephropathy are easily confused with pyelonephritis or hypertensive nephrolsclerosis.[1]

Failure to diagnose lead nephropathy was fostered by the mistaken notion that adverse renal effects do not occur at blood lead levels below 80 μg/dl (Table 1).[25] Current data indicates that hypertension and renal disease occur at blood lead concentrations below 40 μg/dl (Figure 4). Blood lead measurements are both convenient and accurate. In pediatric practice and epidemiologic studies, they have proven of practical value. The blood concentration, however, primarily reflects recent absorption. The blood lead concentration progressively falls following the termination of exposure. Since over 90% of the body-lead burden resides in bone with a biological half-life approximating a decade,[26] the bone should contain the record of past lead absorption.[27]

B. CUMULATIVE LEAD ABSORPTION

The EDTA calcium disodium lead-mobilization test offers the most satisfactory method presently available for the assessment of cumulative past lead absorption. The chelation test has, therefore, become the "gold standard" for the diagnosis of chronic lead nephropathy in adults. Because lead exposure is ubiquitous, it is naive to speak of "normal" body lead burdens; "unexposed" control groups are difficult to find.[28-30]

Among armed service veterans in New Jersey, "essential" hypertensives with renal failure had a mean blood lead of 19 μg/dl while those without nephropathy had a mean blood level of 18 μg/dl.[31] Excessive body-lead burdens in hypertensive patients with renal failure was demonstrated by the CaEDTA lead-mobilization test. Patients with renal disease of nonlead etiology had a mean blood lead of 15 μg/dl. The small difference in blood levels between these groups indicates that blood lead concentration is an inadequate measure of cumulative past lead absorption.[32]

Chelatable lead is relatively stable in adults who have no unusual exposure. This is relatively constant over CaEDTA doses ranging from 0.5 to 2.8 g (Table 2).[33] The upper limit of chelatable lead in controls is about 650 μg in 24 h. Urine collections are extended to 3 or 4 d in the presence of renal failure (serum creatinine [Scr] >1.5 mg/dl), but this upper limit of chelatable lead varies only slightly.[13] Control levels of chelatable lead are similar in children even though the EDTA dose is adjusted to body weight. The upper limit of normal used by various authors ranges from about 400 to 1500 μg/d/1.73 m² body surface area.[34-36]

The validity of the EDTA lead-mobilization test as a measure of body lead stores has been demonstrated by direct comparisons between chelatable lead and bone lead concentration. The correlation coefficient (r) obtained from 35 measurements of chelatable lead and bone lead in iliac biopsies was 0.87 (Figure 5).[37] In this study, transiliac biopsies from 11

TABLE 1
Representation of the Relationship of Blood Lead Levels to the Clinical Manifestations of Lead Poisoning in 1976[a]

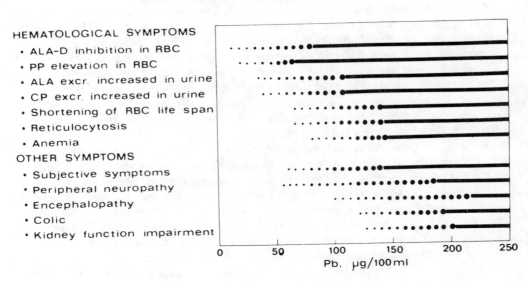

[a] Kidney function impairment was not suspected when the blood lead concentration was less than 125 μg/dl.

(From Hernberg, S., in *Effects and Dose-Response Relationships of Toxic Metals*, Nordberg, G. F., Ed., Elsevier, New York, 1976, chap. 815. With permission.)

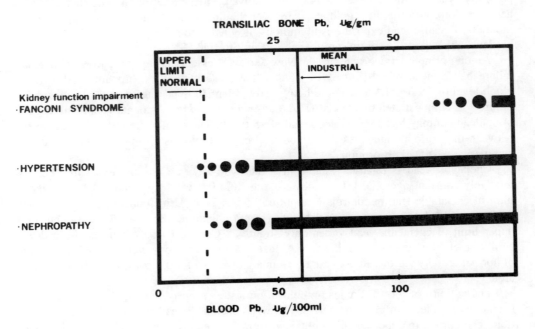

FIGURE 4. A current view of the relationship of blood lead levels to renal effects. Hypertension and nephropathy occur at blood levels below 25 μg/dl.

TABLE 2
Tests in Control Adult Subjects

Reference[a]	No.[a]	N	CaEDTA Route/dose (g)	Urinary Pb chelate[b] (μg) Mean ± SD	Range
Leckie et al. (1958)	29	8	i.v./2	415 ± 152	250—650
Teisinger et al. (1959)	30	50	i.v./2.8	—	58—350
Albahary et al. (1961)	21	20	i.v./0.5 or 1	—	85—465
Emmerson (1963)	27	19	i.v./1	—	90—640
Hernberg et al. (1963)	31	25	i.v./2	157 ± 97	—
Araki (1973)	32	45	i.v./20 mg/kg	—	22—172
Batuman et al. (1981)	3	10	i.m./2	424 ± 72	—

[a] From Wedeen, R. P., in *Current Nephrology,* Vol. 11, Year Book Publishers, 1988, chap. 3. With permission.

[b] Urine collections vary from 1 to 4 d.

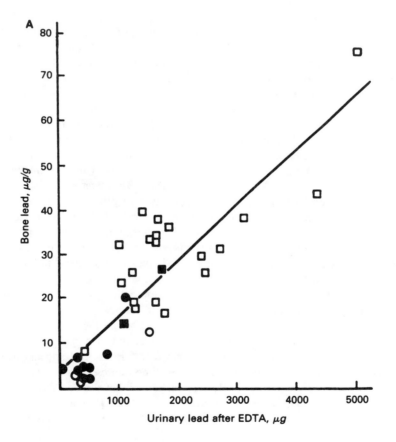

FIGURE 5. Urinary lead excretion after the EDTA test versus transiliac bone lead concentration in 35 subjects. ○ = nonlead exposed, normal renal function; ● = nonlead exposed, renal failure; □ = lead worker, normal renal function; ■ = lead worker, renal failure. (From Van de Vyver, F. L., O'Haese, P. C., Visser, W. J., Elseviers, M. M., Knippenberg, L. J., Lamberts, L. V., Wedeen, R. P., and De Broe, M. E., *Kidney Int.,* 33, 601, 1988. With permission.)

TABLE 3
Transiliac Bone Lead Concentrations and Lead to Calcium Concentration Ratios Determined by Atomic Adsorption Spectroscopy of Bone Biopsies

Group	N	Lead (μg/g)	Lead/calcium ratio × 10^6
Dialysis	153	5.5 ± 4.6	53 ± 49
Dialysis (top 5%)	8	20.6 ± 2.7	190 ± 42
Dialysis due to analgesic nephropathy	10	3.7 ± 2.1	39 ± 19
Cadavers without lead exposure	11	6.3 ± 1.1	65 ± 19
Lead workers	22	30.1 ± 13.8	285 ± 179[a]

[a] Measured in 12 lead workers

(From Van de Vyver, F. L., D'Haese, P. C., Visser, W. J., Elseviers, M. M., Knippenberg, L. J., Lamberts, L. V., Wedeen, R. P., and De Broe, M. E., *Kidney Int.*, 33, 601, 1988. With permission.)

nonlead-exposed cadavers had a mean lead concentration of 6.3 μg/g wet weight while 22 lead workers (averaging 1.9 mg chelatable lead) had a mean lead concentration of 30.1 μg/g bone (Table 3). Schutz et al.[38] found a mean lead concentration of 29 μg/g wet weight in 27 vertebral bone biopsies from active Swedish lead workers. Since iliac biopsies from 153 patients with end-stage renal disease had a mean lead concentration of 5.5 μg/g, it is evident that renal failure per se does not increase bone lead stores.[37]

Renal failure decreases the rate of elimination of lead from the blood,[26] but does not alter the renal clearance of lead which ranges from about 1 to 3 ml/min.[39] Similarly, EDTA testing in patients with renal disease from causes other than lead shows chelatable lead is not increased by renal failure.[31,40-42] In 10 Belgian patients with documented analgesic nephropathy, the mean iliac lead was 4 μg/g (see Table 3) indicating no selective increase in bone lead in tubulointerstitial nephritis as compared to glomerulonephritis.[37] Immunoreactive parathormone has been found to be both increased,[42] and decreased,[43] in lead nephropathy compared to other renal diseases. Parathormone does not appear to explain selective increases in chelatable lead in patients with lead nephropathy. It is, therefore, reasonable to conclude that elevated chelatable lead in patients with renal failure is the cause, rather than the consequence, of the renal disease when no other etiology can be identified. Unfortunately, the EDTA lead mobilization test is impractical for large scale use because it requires both injections and repetitive 24-h urine collections.

The bone lead concentration may prove to be a better biological index of cumulative lead absorption than the EDTA test. An ill-defined fraction of bone lead exchanges with blood. Exchangeable lead is excreted in urine and feces. Bone lead appears to follow the metabolic pathways of bone calcium and, at the same time, represents a source of endogenous exposure. *In vivo* X-ray fluorescence (XRF) is a technique that can be used to measure bone lead without removing bone samples for analysis. XRF is a rapid, safe, accurate, and transportable method for determining body lead stores in large populations.[43-47]

C. LEAD NEPHROPATHY IN ASYMPTOMATIC ADULTS

The EDTA mobilization test has proven of value in identifying lead nephropathy despite normal blood lead concentrations and the absence of symptoms of acute poisoning. Wedeen et al.[48] screened 140 lead workers without previously recognized renal dysfunction, with the EDTA lead mobilization test; 1 g CaNa$_2$EDTA, i.m., twice with procaine, 8 to 12 h apart. Glomerular filtration rates ([^{125}I]-iothalamate clearance) were measured in 57 of the

113 lead workers who excreted 1000 μg/d or more of lead chelate. In 21 workers, glomerular filtration rates were less than the lower limit of normal (<90 ml/min/1.73 m² body surface area) indicating the presence of unexpected early renal disease. Six of 12 renal biopsies revealed focal tubulointerstitial nephritis without intranuclear inclusion bodies. After excluding workers who had experienced symptomatic lead intoxication, were over age 55, and had gout, kidney stones, hypertension, diabetes, or other recognized causes of renal disease, 12 workers were identified with subclinical lead nephropathy.

Eight of these men with occupational lead nephropathy were given chelation therapy consisting of 1 g EDTA with procaine, i.m., three times weekly until the repeated EDTA lead mobilization test (at least 1 week after withholding chelation therapy) returned to less than 800 μg/d.[48] Chelation therapy for 6 to 50 months resulted in a consistent increase in glomerular filtration rate of 20% or more in four of the eight patients. The therapy demonstrated a reversible component of subclinical chronic lead nephropathy. The reversibility of the renal dysfunction in these men supports the diagnosis of lead nephropathy but should not be interpreted as indicating that more advanced renal failure (Scr >3 mg/dl) is similarly reversible. Clinical renal failure indicates destruction of more than two thirds of the kidneys. No evidence that such extensive tubulointerstitial disease is reversible has been presented. Rather, both published[49] and unpublished anecdotes indicate reversibility is unlikely after the Scr exceeds 3 mg/dl. The reversible component of renal dysfunction observed in the four lead workers given long-term, low-dose chelation therapy probably represents a prerenal effect of lead. Chelation presumably ameliorated chronic hypovolemia or renal vasomotor hyperactivity. Chronic volume depletion could be explained by renal salt wasting which has been demonstrated in acutely lead-poisoned rats.[50,51]

The effect of chelation therapy in more advanced renal failure is unknown and, therefore, should be undertaken only with considerable caution. If treatment is attempted, a clear clinical end-point for therapy should be identified in advance. Appropriate end-points for chelation therapy could include restoration of organ function. Chelation therapy should probably not be continued after chelatable lead has returned to the presently accepted upper limit of normal; 650 μg. (Testing is done at least 1 week after cessation of chelation therapy.) It should also be recognized that removal from lead exposure will result in a continuous, albeit slow, net loss of lead from the body at a rate approximating 100 μg/d. Thus, natural deleading must be weighed against the unknown risks of chelation therapy. The safety of the EDTA *test* in renal failure,[52] does not guarantee the safety of long-term EDTA *therapy* in renal failure. The cumulative effect of multiple injections of EDTA over a prolonged period in renal failure is unknown. Oral chelation therapy with dimercaptosuccinic acid (DMSA, succimer) has proven effective and safe in patients without renal disease.[53,54] While these agents are, therefore, promising for long-term chelation, as well as, for chelation testing, their efficacy and toxicity in chronic renal failure is also unknown.

D. GOUT NEPHROPATHY

Gout has been associated with lead poisoning since the 18th century.[1] Renal disease, however, was not identified as a common complication of gout until the description of the role of uric acid in gouty arthritis by Alfred Baring Garrod in 1859. Of Garrod's patients, one third had previously experienced acute lead poisoning. Reluctance to accept the role of lead in gout nephropathy stems from knowledge that other factors including diet (e.g., alcohol, uric acid precursors, etc.), body mass, and genetic predisposition play contributory etiologic roles in this disease.

Clearly not all gout, nor all renal disease associated with gout, is due to lead. Renal disease secondary to uric acid kidney stones is unrelated to lead. Lead tends to diminish, rather than increase, urinary uric acid excretion.[55] The acute renal failure induced by intraluminal uric acid deposition in the presence of massive hyperuricemia (serum uric acid >25

FIGURE 6. Intrarenal urate precipitate in freeze-dried kidney section from rat given 5% oxonic acid (a uricase inhibitor) in chow for 12 d followed by an intravenous lithium urate infusion (1.7 mmol over 1 h). Intrarenal urate crystals are seen in tubular lumens in the cortex and medulla.

mg/dl) is, similarly, unrelated to lead. In the presence of severe hyperuricemia, retention of intraluminal crystals may cause tubular atrophy and interstitial fibrosis (Figure 6).[4,56] When the intrarenal crystals are resorbed, differentiation from tubulointerstitial nephritis due to lead may be difficult using morphologic criteria alone.

The slowly progressive tubulointerstitial nephritis observed in gout patients in the absence of nephrolithiasis or intraluminal crystal deposition sometimes appears to be due to lead nephropathy. More than half of the patients with Queensland nephritis or moonshine-induced lead nephropathy had gout. The role of lead in gout nephropathy was substantiated by the observation that the mean chelatable lead in 22 male armed service veterans in New Jersey with gout and renal failure (Scr >1.5 mg/dl) was 806 μg/3 d while chelatable lead averaged only 470 μg/3 d in 22 gouty veterans with normal renal function.[40] Patient recollections of prior exposure to lead was not different between the two groups, mean blood lead levels were comparable and none of the patients had ever sustained symptomatic lead poisoning. Since renal failure per se does not increase chelatable lead, the EDTA test results indicate that insidious renal failure in gout patients may be due to unrecognized excessive past lead absorption. The coexistence of gout and renal failure is now considered a marker of prior lead poisoning,[3,4,44,57] notwithstanding dissent based on unusual statistical analyses.[58]

The study by Batuman et al.[40] has recently been confirmed by chelation testing in Italian patients with gout and renal failure.[42] Mean total chelatable lead after, 1.5 g CaEDTA, i.m., was 505 μg/2 d in 12 patients compared to a mean total of 180 μg/2 d in 12 controls with comparable renal failure due to glomerulonephritis. It is more difficult to compare a recent German study of chelatable lead in gout and renal failure because of the minimal lead exposure in the Germans.[41,57] Chelatable lead in control patients averaged only 64 μg/d/ 1.73 m². As judged by the lead mobilization tests, the history of excessive lead exposure in the German patients may have been inaccurate. Behringer et al.[41] consider a total chelatable lead greater than 250 μg/4 d indicative of excessive lead absorption although 650 μg is usually accepted as the upper limit of normal.

In the absence of renal failure, chelation testing does not appear to separate gouty from nongouty hypertensives.[59,60] Since blood lead predicts blood pressure at all levels of both measurements (described subsequently), it would be anticipated that some gouty hypertensives would have excessive lead stores. Chelation testing, however, is more predictive when renal disease complicates gout or hypertension. The failure to identify excess mobilizable lead in hypertensives with gout may arise from the current criteria used to define abnormal EDTA tests and renal function. Renal failure is not usually detected until more than two thirds of renal function is lost. Since the 650 μg standard for chelatable lead has been determined by the presence of clinical renal failure, this standard may be too high to detect lead-induced gout and hypertension associated with preclinical renal injury. With current methodology, a reduced glomerular filtration rate appears to be the key to selecting appropriate candidates for chelation testing.

The frequency of lead poisoning among gout patients may depend on demographic characteristics of the population studied. The progressive disappearance of renal disease in gout since Garrod's time corresponds to the lowering of overall exposure to lead in the general population.[61] Similarly, the progressive reduction in cardiovascular complications in gout patients, including renal disease, may reflect a progressive reduction in lead exposure.[62]

IV. LEAD AND HYPERTENSION

A. LOW-LEVEL LEAD ABSORPTION AND BLOOD PRESSURE

Renal disease and hypertension have been intertwined since the first measurements of blood pressure were undertaken in the 19th century.[1] It has long been recognized that renal disease can cause hypertension and hypertension can cause renal disease. Causality is difficult to prove when both these conditions are present. There is, nevertheless, considerable evidence supporting the view that low-level lead absorption contributes to the hypertension that occurs in more than 25% of the adult population. This evidence has been presented in epidemiologic studies showing a correlation between blood lead and blood pressure and in clinical studies showing the contribution of lead to hypertensive cardiovascular disease.

Although blood lead is of limited value for assessing cumulative lead stores, it can be a useful epidemiologic tool. The Second National Health and Nutrition Examination Survey (NHANES II) performed between 1976 and 1980 in over 20,000 noninstitutionalized Americans, 6 months to 74 years of age,[30] included blood pressure and venous blood lead measurements in almost 10,000 individuals.[63] Multiple regression analysis revealed a significant linear correlation between blood pressure and the natural log of blood lead even when both measures were within the accepted normal range (Figure 7).[64] An increase in blood lead from 14 to 30 μg/dl resulted in an increase in systolic pressure of 7 mmHg and diastolic pressure of 3 mmHg.[65] The effect of lead persisted after adjusting for 34 covariates. The predictive effect of blood lead on blood pressure was most evident in black males. The relationship was not significant in men and women over 56 years of age. In a separate analysis of the NHANES II data limited to white males 40 to 59 years of age, a significant correlation was found between blood lead and blood pressure after adjusting for 87 cofactors.[66] This study was designed to minimize the effects of sex, race, and age on blood pressure. The relationship of blood lead to blood pressure was strongest at low levels of blood lead and for systolic pressures. But the association was also evident for diastolic pressures over 90 mmHg.

The NHANES II analyses apply to the lowest level lead exposure arising from ambient atmospheric lead. Although these analyses have been challenged,[67] the conclusions are

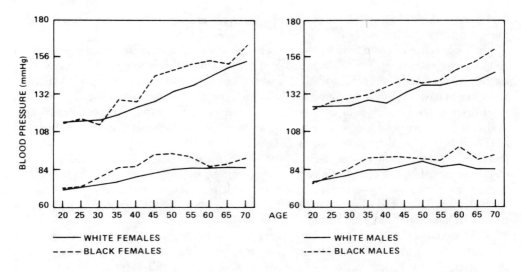

FIGURE 7. Relationship between blood lead and blood pressure for black and white females and males. (From Harlan, W. R. et al., *Environ. Health Perspect.*, 78, 9, 1988. With permission.)

reinforced by a British investigation of blood lead and blood pressure in 7371 middle-aged men.[68] A number of smaller studies using both occupationally exposed,[69,70] and nonoccupationally exposed cohorts,[71-78] support the relationship between lead exposure and hypertensive cardiovascular disease. Absence of a significant correlation between blood lead and blood pressure has been reported in other small studies.[79,80] The statistical power to detect differences in blood pressure as a function of blood lead in small cohorts is considerably less than that of the large American and British reports, but the dose-response relationship demonstrated in some of these studies,[73,75] nevertheless, supports the lead association.

Confirmatory evidence that "low-level" lead exposure increases blood pressure has been obtained in rats.[81] Long Evans rats given 0.1 to 0.5 ppm lead in drinking water for up to 18 months showed consistent increases in systolic blood pressure of about 14 mmHg compared to controls. In these studies, control rats received low lead diets with a mineral mix containing low-calcium supplements. Systolic pressure averaged 110 mmHg in anesthetized and 120 mmHg in unanesthetized control animals. After 12 months, 37% of the rats receiving lead had pressures over 140 mmHg but no controls showed this effect. Perry et al.[81] believe these uncomplicated blood pressure elevations are analogous to essential hypertension in humans. The model is important because it may reflect the preclinical effects of ambient lead exposure on blood pressure.

B. LEAD AND HYPERTENSIVE CARDIOVASCULAR DISEASE

Hypertension increases the risk of atherosclerosis, coronary artery disease, heart disease, stroke, and end-stage renal disease.[69,82] Although lead has long been suspected of inducing hypertension, conflicting results were reported from early uncontrolled clinical investigations.[1] The renal histologic changes of hypertensive nephrosclerosis may be seen in lead nephropathy in the absence of clinical hypertension.[12,13,19,83,84] This suggests that the chronic kidney damage induced by lead may be initiated in the renal vascular endothelium rather than in the proximal tubule as has sometimes been suggested.[85] Because clinical assessment of renal function (Scr) is insensitive to minor reductions in glomerular filtration rate, early renal injury may not be detected until after hypertension is manifest. Hypertension has frequently been produced in experimental animals by lead.[86] However, variation in methodology and response makes these models difficult to relate to human exposure.

The role of lead in the induction of hypertensive renal disease was suggested by EDTA tests in hypertensive armed service veterans with elevated serum-creatinine levels (Scr >1.5 mg/dl) and in controls with no renal dysfunction.[31] The mean chelatable lead was 860 μg Pb/3 d in 27 hypertensive men with renal failure (mean Scr = 3.2 mg/dl) compared to 340 μg/3 d in 21 hypertensives with normal renal function. Control patients with renal failure of known nonlead causes (mean Scr = 3.4 mg/dl) excreted 440 μg Pb/3 d. The duration of hypertension did not appear to contribute to the renal disease in these veterans since those with renal failure had had hypertension for a mean of 7 years while those without renal failure had a 14-year history of high blood pressure. All of these patients were classified as having "essential hypertension" prior to chelation testing. Demographic differences might well produce different results in cohorts of hypertensives who have lower exposures to environmental lead.

C. MORTALITY STUDIES

The cardiovascular complications of excessive lead absorption are evident in a number of mortality studies performed on lead workers. Death certificates are a notoriously poor source of etiologic information.[87,88] Lead nephropathy is not included in the coding of causes of end-stage renal disease. The International Classification of Diseases offers such ambiguous terms as, "other hypertensive disease," "chronic nephritis," "vascular lesions, central nervous system," or "ill-defined conditions," for death certificates.[89]

Despite these caveats, renal disease, hypertension, and cerebrovascular disease are frequently reported as significant causes of excess mortality in lead workers.[85,89-93] In an examination of 2215 death certificates from American lead workers from 1947 and 1980, Cooper et al.[89] found more than double the number of expected deaths attributed to "hypertensive heart disease," "other hypertensive disease," and "chronic or unspecified nephritis". Similar results were obtained from death certificates of 1,987 secondary smelter workers near the Bunker Hill Mine in Idaho.[85] The incidence and prevalence of end-stage renal disease in this area was five- to sixfold greater than that in surrounding Idaho communities,[3] among the highest recorded in the U.S.[94] In a study of 1,898 retired British lead workers, a significant excess of deaths from cerebrovascular accidents as well as from renal disease was found.[92] In a case-control analysis of 867 workers, only cerebrovascular disease appeared as a cause of excessive deaths.[91]

V. PATHOPHYSIOLOGY OF LEAD-INDUCED HYPERTENSION

A. CATION TRANSPORT SYSTEMS

1. Calcium

Calcium may mediate the effects of lead on blood pressure and the kidney at both whole body and subcellular levels. Increased dietary calcium protects against high blood pressure in humans[95] and spontaneously hypertensive rats.[96] Although the mechanism of this effect has been questioned,[97-99] the observation that blood pressure varies inversely with calcium intake has not been challenged. The paradox that increased calcium intake lowers, while increased intracellular calcium increases blood pressure could, in part, be explained by calcium/lead interactions.[100,101] Low calcium diets increase lead absorption. Lead-induced increases in cellular permeability to calcium would increase vasomotor activity causing hypertension. Lead could be one of the circulating factors in essential hypertension which raise platelet cytosolic free calcium.[102] Platelet dysfunction could contribute to the intrarenal vascular disease associated with hypertension by interfering with normal endothelial-cell function causing changes in vascular smooth muscle contractility.[103] Lead interactions with vascular smooth muscle are more likely to be responsible for hypertension and tubulointerstitial nephritis than is the effect of lead on the proximal renal tubule.

The response of smooth muscle to vasoconstrictors such as angiotensin and adrenaline, and to vasodilators, such as kallikrine, is mediated by the concentration of free intracellular calcium. Intracellular calcium also controls aldosterone and renin secretion; increased intracellular calcium reduces renin secretion.[104] Increased cellular permeability to calcium is believed to be responsible for the increased vasopressor response associated with lead-induced hypertension.[105] Cytosolic free calcium is modulated by Na/K ATPase, calcium channels, and a variety of cation exchange processes. Inhibition of Na/K ATPase by lead would permit both calcium and sodium ions to enter the cell to trigger vascular smooth muscle contraction and hypertension.

2. Sodium

A role for sodium in hypertension is firmly established. As the major cation of extracellular fluid, intravascular volume expansion is accompanied by increased circulating sodium. In the absence of an expanded vascular bed, increased intravascular volume produces hypertension. In addition, chronic increases in sodium chloride intake induce systemic vasoconstriction.[103] However, it is equally clear that increased sodium is not the sole cause of high blood pressure. Hypertensives are often classified as salt sensitive or salt insensitive.

An increased capacity to excrete a saline load is characteristic of hypertensives. Hypertensives excrete more of a saline load in the urine more quickly than do nonhypertensives. But this enhanced sodium excretion could be an effect, rather than a cause, of hypertension. It could also represent an epiphenomenon rather than a primary cellular transport defect. Nevertheless, the ouabain-like activity of the hypothesized natriuretic hormone could mediate cellular sodium, vascular tone, and salt wasting by the kidney.

Evidence that blood vessels in hypertensives contain more sodium than in nonhypertensives suggests a role for cellular sodium transport in hypertension. Unfortunately, cellular functions are difficult to study in blood vessels *in vivo*. It is more convenient to examine membrane transport in the readily accessible red blood cell.[106,107] There is evidence that increased intracellular sodium causes increased intracellular calcium mediated by a Na/K exchange mechanism.[108] Lead interactions with transmembrane sodium transport, therefore, may have important implications for vasomotor tone.

The evidence that cellular sodium transport is abnormal in essential hypertension remains controversial. Abnormal sodium transport may be a genetic marker of hypertension or an acquired characteristic of hypertension. Sodium transport defects have rarely been examined in light of environmental triggers (such as lead) which might interact with genetic factors to produce hypertensive cardiovascular disease. Evidence that lead induces defects in cellular sodium transport comparable to those found in essential hypertensives is beginning to emerge.

Unbound cytosolic sodium is the net result of sodium entry and efflux. Ouabain inhibitable pumps (Na/K ATPase) remove sodium from the cell while both passive permeability and protein-mediated processes (furosemide and *p*-chloromercuribenzene sulfonate inhibitable Na/K cotransport) modulate bidirectional fluxes. In addition, sodium may exchange with other cations such as lithium, calcium, or hydrogen across the cell membrane (Na/Li countertransport).[109]

Vascular reactivity is in part controlled by intracellular electrolyte composition, which in turn is modulated by membrane bound Na/K ATPase. In the kidney, inhibition of Na/K ATPase is followed by sodium wasting. Saline loading decreases Na/K ATPase in association with enhancement of sodium excretion.[106] Inhibition of Na/K ATPase has been noted in the red cells of lead workers,[110,111] as well as, following a single oral dose of lead in rats.[112,113] Chronic inhibition of membrane bound ATPase would be expected to increase blood pressure. A close correlation has been found between membrane bound lead (but not blood lead) and Na/K ATPase in humans.[114,115] Batuman[116] has recently demonstrated that lead increases red cell Na/Li countertransport *in vitro*. This finding indicates an effect of lead on red-cell sodium transport similar to that found in essential hypertension and in hyperlipidemia. However, Moreau et al.[73] failed to find any correlation between Na/Li countertransport and blood pressure in red cells from 129 Paris police officers. None of these policemen had unusual exposure to lead or hypertension. The absence of a correlation, therefore, may have no bearing on the role of excessive lead absorption in hypertension because of the narrow range of blood leads and blood pressures examined. Lead inhibition of Na/K ATPase and stimulation of red-cell Na/Li countertransport *in vitro* supports the notion that genetic and environmental factors interact at the cellular level in modulating cellular cation content, cytosolic calcium, and vasomotor responsiveness.

Genetic differences in membrane cation transport may account for differing vulnerability to lead toxicity. Na/Li countertransport has been reported to be reduced in nonhypertensive blacks as compared to white subjects.[117,118] These cell membrane transport systems, may, therefore, help to identify mechanisms of increased susceptibility to lead-induced hypertension and renal disease in blacks.

B. HORMONAL MEDIATION
1. The Renin-Angiotensin-Aldosterone System

Angiotensin II is a potent endogenous vasoconstrictor derived from the kidney which plays a pivotal role in renovascular hypertension. Because the control of blood pressure is multifactorial, it has been difficult to define the contribution of angiotensin II in other forms of hypertension. Patients with lead-induced hypertensive nephrosclerosis due to moonshine consumption showed depression of angiotensin activity and diminished aldosterone production in response to salt depletion.[119] The abnormal response was reversed by chelation therapy,[120,121] an effect compatible with the correction of prerenal renal dysfunction by chelation therapy in early lead nephropathy.[48] Lead-induced hyporeninemic hypoaldosteronism could also account for the fluorocortisone resistant hyperkalemia and distal renal tubular acidosis observed in lead nephropathy.[121,122] Studies of occupationally exposed subjects, however, have shown variable renin-angiotensin responses.[123-125] Renin profiling in hypertensives designated "high, normal, or low renin" by the Laragh group at New York Hospital-Cornell Medical Center showed a tendency to higher blood lead levels (mean 16.8 μg/dl, N = 5) in high renin hypertensives.[126] These preliminary observations conflict with the expectations engendered by studies in moonshiners,[120-122] and should not be considered conclusive. None of these patients had either high exposure to lead or hypertension with renal failure.

In contrast to the diminished angiotensin activity demonstrated in chronic lead nephropathy in humans, an increased renin response to salt depletion has been demonstrated in occupationally lead-exposed normotensive men,[124,125] as well as, in acutely lead-poisoned experimental animals.[127] Although, increased renin-angiotensin activity is usually seen in acutely lead-intoxicated animals, chronic "low dose" lead administration (100 ppm in drinking water) reduces plasma renin activity and concentration in association with hypertension in rats.[126] The biphasic response is believed to be consistent with increased angiotensin production prior to the advent of hypertension followed by reduced angiotensin activity once sustained hypertension supervenes. Boscolo et al.[128] report similar biphasic effects in occupationally exposed workers but the small size of this study and the marked clinical heterogeneity makes it unclear whether the groups were clinically defined prior to data analysis.

2. Vasoactive Hormones

Hypertension and the cardiotoxic effects of lead could be explained in part by either increased release of endogenous catecholamines or increased cellular responsiveness to normal levels. Increased adrenergic activity has been reported in lead-poisoned humans,[129,130] and rodents.[127,131] Lead enhances cardiac[130] and vascular reactivity to adrenergic agents both *in vivo* and *in vitro* in experimental animals.[132,133] It has been postulated that lead desensitizes β-adrenergic receptors resulting in the loss of vasomotor control in acute lead-poisoning.[129,130] β-adrenergic reactivity could also be enhanced by increased intracellular free calcium mediated by lead interactions with protein kinase C.[132]

Kallikrine is a potent renal vasodilator which has been reported to be reduced in lead-poisoned men.[133] The hypothesis that lead depletion of renal kallikrine induces hypertension has not, however, been substantiated by evidence of specificity.

VI. IMPLICATIONS

Hypertension and lead nephropathy appear to be important consequences of lead exposure in adults. Assigning a causal role to lead is difficult for a number of reasons: the damage is delayed for many years, the renal failure is multifactorial, and there are complex interactions with genetic background, diet, and systemic disease. Interpretation of epidemiologic data on body lead stores in the general population is made more difficult by reliance on blood lead concentrations. In an era when physicians have become familiar with risk ratios as indicators of disease causation, risk ratios are not available for lead nephropathy. XRF measurements of bone lead stores may provide more definitive measurement of the risk of end-stage renal disease from environmental lead exposure.

Multiple bone lead measurements made in Marc E. DeBroe's laboratory in Antwerp, Belgium provide an answer to the question: How much lead is too much? Transiliac biopsies show that unselected elderly males have about 5 μg Pb/g bone wet weight compared to occupationally exposed workers who have more than 20 μg/g (see Table 3). A lifetime of exposure thus results in a bone lead content about one fourth of that found in occupationally exposed individuals. Five μg/g ilium corresponds to about 200 mg total body lead burden expected in a 60- to 70-year-old man.[134] Since over 90% of the body burden is stored in bone, lead workers have over 800 mg of lead in bone, i.e., at least four times the level in unexposed individuals.

Available data also allows estimation of the time it would take to convert bone lead stores from normal levels to the levels found in lead workers, i.e., exposure levels at high risk for kidney disease and/or hypertension. Current Federal Food and Drug Administration standards, for example, permit lead-glazed imported ceramics to be used if they release up to 5 ppm lead into a standard acidic solution (4% acetic acid). If we assume that an individual consumes a liter of such liquid a day, up to 5 mg lead would be ingested per day. This allowable intake comes to 1.82 g lead per year, more than 10 times the World Health Organization's "recommended provisional tolerable intake for adults of 3 mg per week."[134] Since lead absorption in an adult approximates 10% of the lead ingested,[135] about 500 μg/d will be added to the body lead burden. This amounts to 182 mg/year of absorbed lead which approaches the total body burden expected in a lifetime of ambient exposure. In less than five years an adult absorbing 0.5 mg lead/d from improperly glazed ceramics will have the bone lead stores found in lead workers. That person would be subject to the incidence of kidney disease and hypertension anticipated in occupational lead exposure.

Such absorption levels are not rare. About 1% of the adult American population have elevated blood lead levels (>30 μg/dl) according to NHANES II.[30] This represents about 2 million adults with excessive lead absorption within whom the source of lead exposure is unrecognized. Clinical symptoms of acute lead intoxication are absent and the diagnosis is overlooked because the source is not identified. Their risk of developing renal disease can be estimated to be at least five times that of the general population since about 5% of end-stage renal disease patients have elevated bone lead stores,[37] compared to 1% of the general population.

REFERENCES

1. **Wedeen, R. P.,** *Poison in the Pot: The Legacy of Lead,* Southern Illinois University Press, Carbondale, Il, 1984.
2. **Pinto de Almeida, A. R., Carvalho, F. M., Spinola, A. G., and Rocha, H.,** Renal dysfunction in Brazilian lead workers, *Am. J. Nephrol.,* 7, 455, 1987.
3. **Bennett, W. M.,** Lead nephropathy, *Kidney Int.,* 28, 212, 1985.
4. **Beck, L. H.,** Requiem of gouty nephropathy, *Kidney Int.,* 30, 280, 1986.
5. **Ibels, L. S. and Pollock, C. A.,** Lead intoxication, *Med. Toxicol.,* 1, 387, 1986.
6. **Wedeen, R. P.,** Heavy metals, in *Diseases of the Kidney,* 14th ed. Schrier, R. W. and Gottschalk, C. W., Eds., Little, Brown, Boston, 1988, chap. 46.
7. Editor, Leading work-related diseases and injuries — United States, *Morbidity Mortality Wkly Rep.,* 34, 537, 1985.
8. **McKhann, C. F. and Vogt, E. C.,** Lead poisoning in children: with notes on therapy, *Am. J. Dis. Child.,* 32, 386, 1926.
9. **Osheroff, M. R., Uno, H., and Bowman, R. E.,** Lead inclusion bodies in the anterior horn cells and neurons of the substantia nigra in the adult rhesus monkey, *Toxicol. Appl. Pharmacol.,* 64, 570, 1982.
10. **Goyer, R. A., Leonard, D. L., Moore, J. F., Rhyme, B., and Krigman, M. R.,** Lead dosage and the role of the intranuclear inclusion body: an experimental study, *Lab. Invest.,* 22, 245, 1970.
11. **Goyer, R. A., and Wilson, M. H.,** Lead-induced intranuclear inclusion bodies: results of ethylenediaminetetra acetic acid treatment, *Lab. Invest.,* 32, 149, 1975.
12. **Cramer, K., Goyer, R. A., Jagenberg, R., and Wilson, M. H.,** Renal ultrastructure, renal function, and parameters of lead toxicity in workers with different periods of lead exposure, *Br. J. Ind. Med.,* 31, 113, 1974.
13. **Wedeen, R. P., Maesaka, J. K., Weiner, B., Lipat, G. A., Lyons, M. M., Vitale, L. F., and Joselow, M. M.,** Occupational lead nephropathy, *Am. J. Med.,* 59, 630, 1975.
14. Committee on Biologic Effects of Atmospheric Pollutants, Airborne Lead in Perspective, National Academy of Sciences, Washington, D.C., 1972, p. 91.
15. **Clarkson, T. W. and Kench, J. E.,** Urinary excretion of amino acids by men absorbing metals, *Biochem. J.,* 62, 361, 1956.
16. **Hong, C. D., Hanenson, I. G., Lerner, S., Hammond, P. B., Pesce, A. J., and Pollak, V. E.,** Occupational exposure to lead; effects on renal function *Kidney Int.,* 18, 489, 1980.
17. **Henderson, D. A.,** A follow-up of cases of plumbism in children, *Australas. Ann. Med.,* 3, 219, 1954.
18. **Emmerson, B. T.,** Chronic lead nephropathy, *Kidney Int.,* 4, 1, 1973.
19. **Inglis, J. A., Henderson, D. A., and Emmerson, B. T.,** The pathology and pathogenesis of chronic lead nephropathy occurring in Queensland, *J. Pathol.,* 124, 65, 1978.
20. **Moel, D. I, Sachs, H. K., Cohn, R. A., and Drayton, M. A.,** Renal function 9 to 17 years after childhood lead poisoning, *J. Pediatr.,* 106, 729, 1985.
21. **Ball, D. J. and Sorenson, L. B.,** Pathogenesis of hyperuricemia in saturnine gout, *N. Engl. J. Med.,* 280, 1199, 1969.
22. **Morgan, J. A., Hartley, M. W., and Miller, R. E.,** Nephropathy in chronic lead poisoning, *Arch. Intern. Med.,* 118, 17, 1966.
23. **Buchet, J. P., Roels, H., Bernard, A., Jr., and Lauwerys, R.,** Assessment of renal function of workers exposed to inorganic lead, cadmium or mercury vapor, *J. Occ. Med.,* 22, 741, 1980.
24. **Meyer, B. R., Fischbein, A., Rosenman, K., Lerman, Y., Dreyer, D. E., and Reidenberg, M. M.,** Increased urinary enzyme excretion in workers exposed to nephrotoxic chemicals, *Am. J. Med.,* 76, 989, 1984.
25. **Hernberg, S.,** Biochemical, subclinical, and clinical responses to lead and their relation to different exposure levels, as indicated by the concentration of lead in the blood, in *Effects and Dose-Response Relationships of Toxic Metals,* Nordberg, G. F., Ed., Elsevier, New York, 1976, chap. B15.
26. **Hryhorczuk, D. O., Rabinowitz, M. B., Hessl, S. M., Hoffman, D., Hogan, M. M., Mallin, K., Finch, H., Orris, P., and Berman, E.,** Elimination kinetics of blood lead in workers with chronic lead intoxication, *Am. J. Ind. Med.,* 8, 33, 1985.
27. **Landrigan, P. J., Froines, J. R., and Mahaffey, K. R.,** Body lead burden: summary of epidemiological data on its relation to environmental sources and toxic effects, in *Dietary and Environmental Lead: Human Health Effects,* Mahaffey, K. R., Ed., Elsevier, Amsterdam, 1985, chap. 14.
28. **Piomelli, S., Seaman, C., Zullow, D., Curran, A., and Davidoff, B.,** Threshold for lead damage to heme synthesis in urban children, *Proc. Natl. Acad. Sci. U.S.A.,* 79, 3335, 1982.
29. **Wedeen, R. P.,** Bone lead, hypertension, and lead nephropathy, *Environ. Health Perspect.,* 78, 57, 1988.
30. **Mahaffey, K. R., Annest, J. L., Roberts, J., and Murphy, R. S.,** National estimates of blood lead levels, 1976—1980: association with selected demographic and socioeconomic factors, *N. Engl. J. Med.,* 307, 573, 1982.

31. **Batuman, V., Landy, E., Maesaka, J. K., and Wedeen, R. P.,** Contribution of lead to hypertension with renal failure, *N. Eng. J. Med.*, 309, 17, 1983.
32. **Vitale, L. F., Joselow, M. M., Wedeen, R. P., and Pawlow, M.,** Blood lead — an inadequate measure of occupational exposure, *J. Occup. Med.*, 17, 155, 1975.
33. **Wedeen, R. P.,** Occupational and environmental renal diseases, in *Current Nephrology,* Vol. 11, Year Book Medical Publishers, 1988, chap. 3.
34. **Chisolm, J. J., Mellits, E. D., and Barrett, M. B.,** Interrelationships among blood lead concentration, quantitative daily ALA-U and urinary lead output following calcium EDTA, in *Effects and Dose-Response Relationships of Toxic Metals,* Nordberg, G. F., Ed., Elsevier, New York, 1976, chap. B6.
35. **Markovitz, M. E. and Rosen, J. F.,** Assessment of lead stores in children: validation of an 8 hour $CaNa_2EDTA$ provocative test, *J. Pediatr.*, 104, 337, 1984.
36. **Rosen, J. F., Markowtiz, M. E., Bijur, P. E., Jenks, S. T., Wieloposki, L., Kalef-Ezra, J. A., and Slatkin, D. N.** L-line x-ray fluorescence of cortical bone lead compared with $CaNa_2EDTA$ test in lead-toxic children: public health implications, *Proc. Natl. Acad. Sci. U.S.A.*, 86, 685, 1989.
37. **Van de Vyver, F. L., D'Haese, P. C., Visser, W. J., Elseviers, M. M., Knippenberg, L. J., Lamberts, L. V., Wedeen, R. P., and De Broe, M. E.,** Bone lead in dialysis patients, *Kidney Int.*, 33, 601, 1988.
38. **Schutz, A., Skerfving, S. Mattson, S. Christoffersson, J. O., and Ahlgren, L.,** Lead in vertebral bone biopsies from active and retired lead workers, *Arch. Environ. Health,* 42, 340, 1987.
39. **Campbell, B. C., Elliott, H. L., and Meredith, P. A.,** Lead exposure and renal failure: does renal insufficiency influence lead kinetics?, *Toxicol. Lett.*, 9, 121, 1981.
40. **Batuman, V., Maesaka, J. K., Haddad, B., Tepper, E., Landy, E., and Wedeen, R. P.** The role of lead in gout nephropathy, *N. Engl. J. Med.*, 304, 520, 1981.
41. **Behringer, D., Craswell, P., Mohl, C., Stoeppler, M., and Ritz, E.,** Urinary lead excretion in uremic patients, *Nephron,* 42, 323, 1986.
42. **Colleoni, N. and D'Amico, G.,** Chronic lead accumulation as a possible cause of renal failure in gouty patients, *Nephron,* 44, 32, 1986.
43. **Craswell, P. W., Price, J., Boyle, P. D., Heazelwood, V. J., Baddeley, H., Lloyd, H. M., Thomas, B. J., Thomas, B. W., and Williams, G. M.,** Chronic lead nephropathy in Queensland: alternative methods of diagnosis, *Aust. N.Z. J. Med.*, 16, 11, 1986.
44. **Craswell, P. W., Price, J., Boyle, P. D., Heazlewood, V. J., Baddeley, H., Lloye, H. M., Thomas, B. J., and Thomas, B. W.,** Chronic renal failure with gout: a marker of chronic lead poisoning, *Kidney Int.,* 26, 319, 1984.
45. **Greenberg, A., Parkinson, D. K., Fetterolf, D. E., Ellis, K. J., Wielopolski, L, Vaswani, A. N., Cohn, S. H., Landrigan, P. J., and Puschett, J. B.,** Effects of elevated lead and cadmium burdens on renal function and calcium metabolism, *Arch. Envir. Health,* 41, 69, 1986.
46. **Somervaille, L. J., Chettle, D. R., and Scott, M. C.,** In vivo measurement of lead in bone using x-ray fluorescence, *Phys. Med. Biol.*, 30, 929, 1985.
47. **Wedeen, R. P., Batuman, V., Quinless, F., Williams, F. H., Jr., Bogden, J., Schidlovsky, G., and Jones, K. W.,** In vivo x-ray fluorescence (XRF) for assessing body lead stores, in *In Vivo Body Composition Studies.,* Ellis K. J., Yasumura, S., and Morgen, W., Eds., The Institute of Physical Sciences in Medicine, London, 1987, chap. 56.
48. **Wedeen, R. P., Mallik, D. K., and Batuman, V.,** Detection and treatment of occupational lead nephropathy, *Arch. Int. Med.*, 139, 53, 1979.
49. **Germain, M. J., Braden, G. L., and Fitzgibbons, J. P.,** Failure of chelation therapy in lead nephropathy, *Arch. Intern. Med.*, 144, 2419, 1984.
50. **Fleischer, N., Mouw, D. R., and Vander, A. J.,** Chronic effects of lead on renin and renal sodium excretion, *J. Lab. Clin. Med.*, 95, 759, 1980.
51. **Mouw, D. R., Vander, A. J., Cox, J., and Fleischer, N.,** Acute effects of lead on renal electrolyte excretion and plasma renin activity, *Toxicol. Applied Pharmacol!.*, 46, 435, 1978.
52. **Wedeen, R. P., Batuman, V., and Landy, E.,** The safety of the EDTA lead-mobilization test, *Environ. Res.*, 30, 58, 1983.
53. **Aposhian, H. V.,** DMSA and DMPS-water soluble antidotes for heavy metal poisoning, *Ann. Rev. Pharmacol. Toxicol.*, 23, 193, 1983.
54. **Graziano, J. H.,** Role of 2,3-dimercaptosuccinic acid in the treatment of heavy metal poisoning, *Med. Toxicol.*, 1, 155, 1986.
55. **Emmerson, B. T., Mirosch, W., and Douglas, J. B.,** The relative contributions of tubular reabsorption and secretion to urate excretion in lead nephropathy, *Aust. N.Z. Med.*, 4, 353, 1971.
56. **Linnane, J. W., Barry, A. F., and Emmerson, B. T.,** Urate deposits in the renal medulla, *Nephron,* 29, 216, 1981.
57. **Ritz, E., Mann, J., and Stoeppler, M.,** Lead and the kidney, *Adv. Nephrol.*, 17, 241, 1988.
58. **Reynolds, P. P., Knapp, M. J., Baraf, H. S. B., and Holmes, E. W.,** Moonshine and lead. Relationship to the pathogenesis of hyperuricemia in gout, *Arthritis Rheum.*, 26, 1057, 1983.

59. **Peitzman, S. J., Bodison, W., and Ellis, I.,** Moonshine drinking among hypertensive veterans in Philadelphia, *Arch. Intern. Med.,* 145, 632, 1985.
60. **Wright, L. F., Saylor, R. P., and Cecere, F. A.,** Occult lead intoxication in patients with gout and kidney disease, *J. Rheumatol.,* 11, 517, 1984.
61. **Wedeen, R. P.,** Lead and the gouty kidney, *Am. J. Kid. Dis.,* 2, 559, 1983.
62. **Yu, T. F. and Berger, L.,** Impaired renal function in gout: its association with hypertensive vascular disease and intrinsic renal disease, *Am. J. Med.,* 72, 95, 1982.
63. **Harlan, W. R., Landis, J. R., Schmouder, R. L., Goldstein, N. G., and Harlan, L. C.,** Blood lead and blood pressure. Relationship in the adolescent and adult U.S. population, *JAMA,* 253, 530, 1985.
64. **Harlan, W. R., Hull, A. L., Schmouder, R. L., Landis, J. R. Larkin, F. A., and Thompson, F. E.,** High blood pressure in older Americans. The first national health and nutrition examination survey, *Hypertension,* 6, 802, 1984.
65. **Harlan, W. R.,** The relationship of blood lead levels to blood pressure in the U.S. population, *Environ. Health Perspect.,* 78, 9, 1988.
66. **Pirkle, J. L., Schwartz, J., Landis, J. R., and Harlan, W. R.,** The relationship between blood lead levels and blood pressure and its cardiovascular risk implications, *Am. J. Epidemiol.,* 121, 246, 1985.
67. **Gartside, P. S.,** The relationship of blood lead levels and blood pressure in NHANES II: additional calculations, *Environ. Health Perspect.,* 78, 31, 1988.
68. **Pocock, S. J., Shaper, A. G., Ashby, D., Delves, H. T., and Clayton, B. E.,** The relationship between blood lead, blood pressure, stroke, and heart attacks in middle-aged British men, *Environ. Health Perspect.,* 78, 23, 1988.
69. **Kirkby, H. and Gyntelberg, F.,** Blood pressure and other cardiovascular risk factors of long term exposure to lead, *Scand. J. Work Environ. Health,* 11, 15, 1985.
70. **de Kort, W. L. A. M., Vershoor, M. A., Wibowo, A. E. E., van Hemmen, J. J.,** Occupational exposure to lead and blood pressure, *Am. J. Indust. Med.,* 11, 145, 1987.
71. **Kromhout, D., Wibowo, A. A. E., Herber, R. F. M., Dalderup, L. M., Heerink, H., de Lezenne Coulander, C., and Zielhuis, R. L.,** Trace metals and coronary heart disease risk indicators in 162 elderly men (the Zutohen study), *Am. J. Epidemiol.,* 122, 378, 1985.
72. **Moreau, T., Orssaud, G., Juguet, B., and Busquet, G.,** Lettre-a-l'editeur: plombemie et pression arterielle: premier resultats d'une enguete transversale de 431 sujet de sexe masculin, *Rev. d'Epidemiol. Sante Publique,* 30, 395, 1982.
73. **Moreau, T., Hannaert, P., Orssaud, G., Huel, G., Garay, R. P., Claude, J. R., Juguet, B., Festy, B., and Lellouch, J.,** Influence of membrane sodium transport upon the relation between blood lead and blood pressure in a general male population, *Environ. Health Perspect.,* 78, 47, 1988.
74. **Neri, L. C., Hewitt, D., and Orser, B.,** Blood lead and blood pressure: analysis of cross-sectional and longitudinal data from Canada, *Environ. Health Perspect.,* 78, 123, 1988.
75. **Orssaud, G., Claude, J. R., Moreau, T., Lellough, J., Juguet, B., and Festy, B.,** Blood lead concentration and blood pressure, *Br. Med. J.,* 290, 244, 1985.
76. **Sharp, D. S., Becker, C. E., and Smith, A. H.,** Chronic low-level lead exposure its role in the pathogenesis of hypertension, *Med. Toxicol.,* 2, 210, 1987.
77. **Sharp, D. S., Osterloh, J., Becker, C. E., Bernard, B., Smith, A. H., Fisher, J. M., Syme, S. L., Holman, B. L., and Johnson, T.,** Blood pressure and blood lead concentration in bus drivers, *Environ. Health Perspect.,* 78, 131, 1988.
78. **Weiss, S. T., Munoz, A., Stein, A., Sparrow, D., and Speizer, F. E.,** The relationship of blood lead to blood pressure in a longitudinal study of working men, *Am. J. Epidemiol.,* 123, 800, 1986.
79. **Elwood, P. C., Davey-Smith, G., Oldham, P. D., and Toothill, C.,** Two Welsh surveys of blood lead and blood pressure, *Environ. Health Perspect.,* 78, 119, 1988.
80. **Parkinson, D. K., Hodgson, M. J., Bromet, E. J., Dew, M. A., and Connell, M. M.,** Occupational lead exposure and blood pressure, *Br. J. Indust. Med.,* 44, 744, 1987.
81. **Perry, H. M., Jr., Erlanger, M. W., and Perry, E. F.,** Increase in the blood pressure of rats chronically fed low levels of lead, *Environ. Health Perspect.,* 78, 107, 1988.
82. **Tarugi, P., Calandra, S., Borella, P., and Vivoli, G. F.,** Heavy metals and experimental atherosclerosis, *Atherosclerosis,* 45, 221, 1982.
83. **Morgan, J. A.,** Chelation therapy in chronic lead nephropathy, *South. Med. J.,* 68, 1001, 1975.
84. **Morgan, J. A.,** Hyperkalemia and acidosis in chronic lead nephropathy, *South. Med. J.,* 69, 881, 1976.
85. **Selevan, S. G., Landrigan, P. J., Stern, F. B., and Jones, J. H.,** Mortality of lead smelter workers, *Am. J. Epidemiol.,* 122, 673, 1985.
86. **Victery, W.,** Evidence for effects of chronic lead exposure on blood pressure in experimental animals: an overview, *Environ. Health Perspect.,* 78, 71, 1988.
87. **Feinstein, A. R.,** Fraud, distortion, delusion, and consensus: the problems of human and natural deception in epidemiologic science, *Am. J. Med.,* 84, 474, 1988.

88. **Schade, W. J. and Swanson, G. M.**, Comparison of death certificate occupation and industry data with lifetime occupational histories obtained by interviews: variations in the accuracy of death certificate entries, *Am. J. Ind. Med.*, 14, 121, 1988.
89. **Cooper, W. C., Wong, O., and Kheifets, L.**, Mortality among employees of lead battery plants and lead producing plants, 1947—1980, *Scand. J. Work Environ. Health*, 11, 331, 1985.
90. **Davies, J. M.**, Long term mortality study of chromate pigment workers who suffered lead poisoning, *Br. J. Ind. Med.*, 41, 170, 1984.
91. **Fanning, D. A.**, Mortality study of lead workers, 1926—1985, *Arch. Environ. Health*, 43, 247, 1988.
92. **Malcolm, D. and Barnett, H. A. R.**, A mortality study of lead workers 1925—76, *Br. J. Ind. Med.*, 39, 404, 1982.
93. **McMichael, A. J. and Johnson, H. M.**, Long-term mortality profile of heavily-exposed lead smelter workers, *J. Occup. Med.*, 24, 375, 1982.
94. **Sugimoto, T. and Rosansky, S. J.**, The incidence of treated end stage renal disease in the eastern United States: 1973—1979, *Am. J. Public Health*, 74, 14, 1984.
95. **McCarron, D. A. and Morris, C. D.**, Blood pressure response to oral calcium in persons with mild to moderate hypertension, *Ann. Intern. Med.*, 103, 825, 1985.
96. **Hatton, C., Muntzel, M., McCarron, D. A., Pressley, M., and Bukoski, R. D.**, Early effects of dietary calcium on blood pressure, plasma volume, and vascular reactivity, *Kidney Int.*, 34 (Suppl. 25), S-16, 1988.
97. **Kaplan, N. M. and Meese, R. B.**, The calcium deficiency hypothesis of hypertension: a critique, *Ann. Inter. Med.*, 105, 947, 1986.
98. **McCarron, D. A. and Morris, C. D.**, The calcium deficiency hypothesis of hypertension, *Ann. Intern. Med.*, 107, 919, 1987.
99. **Reed, D., McGee, D., Yano, K., and Hankin, J.**, Diet, blood pressure, and multicollinearity, *Hypertension*, 7, 405, 1985.
100. **Pounds, J. G.**, Effect of lead intoxication on calcium homeostasis and calcium-mediated cell function: a review, *Neurotoxicology*, 5, 295, 1984.
101. **Wedeen, R. P.**, Blood lead levels, dietary calcium, and hypertension, *Ann. Intern. Med.*, 102, 403, 1985.
102. **Lindner, A., Kenny, M., and Meacham, A. J.**, Effects of a circulating factor in patients with essential hypertension on intracellular free calcium in normal platelets, *N. Engl. J. Med.*, 316, 509, 1987.
103. **Van Houte, P. M.**, The endothelium-modulator of vascular smooth-muscle, *N. Engl. J. Med.*, 319, 513, 1988.
104. **Meredith, P. A. and Campbell, B. C.**, The influence of lead exposure on the renin-angiotensin system in man, *Brit. J. Pharmacol.*, 19, 602P, 1983.
105. **Piccinini, F., Favalli, L., and Chiari, M. C.**, Experimental investigations on the contraction induced by lead in arterial smooth muscle, *Toxicology*, 8, 43, 1977.
106. **Haddy, F. J.**, Ionic control of vascular smooth muscle cells, *Kidney Int.*, 34 (Suppl 25), S1, 1988.
107. **Hilton, P. J.**, Cellular sodium transport in essential hypertension, *N. Engl. J. Med.*, 313, 222, 1986.
108. **Blaustein, M. P.**, Sodium ions, calcium ions, blood pressure regulation and hypertension: a reassessment and a hypothesis, *Am. J. Physiol.*, 232, C165, 1977.
109. **Canessa, M., Brugnara, C., and Escobales, N.**, The $Li+-Na+$ exchange and $Na+-K+-Cl-$ cotransport systems in essential hypertension, *Hypertension*, 10 (Suppl. I), I-4, 1987.
110. **Hernberg, S., Vihko, V., and Hasan, J.**, Red cell membrane ATPases in workers exposed to inorganic lead, *Arch. Environ. Health*, 14, 319, 1967.
111. **Hasan, J., Vihko, V., and Hernberg, S.**, Deficient red cell membrane $/Na++ K+/-$ ATPase in lead poisoning, *Arch. Environ. Health*, 14, 313, 1967.
112. **Suketa, Y., Hasegawa, S., and Yamamoto, T.**, Changes in sodium and potassium in urine and serum of lead intoxicated rats, *Toxicol. Appl. Pharmacol.*, 47, 203, 1979.
113. **Suketa, Y., Ban, K., and Yamamoto, T.**, Effects of ethanol and lead ingestion on urinary sodium excretion and related enzyme activity in rat kidney, *Biochem. Pharmacol.*, 30, 2293, 1981.
114. **Raghavan, S. R. V., Culver, B. D., and Gonick, H. C.**, Erythrocyte lead-binding protein after occupational exposure. II. Influence on lead inhibition of membrane $Na+$, $K+$ adenosinetriphosphatase, *J. Toxicol. Environ. Health*, 7, 561, 1981.
115. **Weiler, E., Khalil-Manesh, F., and Gonick, H.**, Effects of lead and natriuretic hormone on kinetics of sodium-potassium-activated adenosine triphosphatase: possible relevance to hypertension, *Environ. Health Perspect.*, 78, 113, 1988.
116. **Batuman, V., Dreisbach, A., Chun, E., and Naumoff, M.**, Lead increases the red cell sodium-lithium countertransport, in press.
117. **M'Buyamba-Kabangu, J. R., Lijnen, P., Groeseneken, D., Staessen, J., Lissens, W., Goossens, W., Fagard, R., and Amery, A.**, Racial differences in intracellular concentration and transmembrane fluxes of sodium and potassium in erythrocytes of normal male subjects, *J. Hypertension*, 2, 647, 1984.

118. **Trevisan, M., Ostrow, D., Cooper, R. S., Sempos, C., and Stamler, J.,** Sex and race differences in sodium-lithium countertransport and red cell sodium concentration, *Am. J. Epidemiol.,* 120, 537, 1984.
119. **Sandstead, H. H., Michelakis, A. M., and Temple, T. E.,** Lead intoxication: its effect on the renin-aldosterone response to sodium deprivation, *Arch. Environ. Health,* 20, 356, 1970.
120. **McAllister, R. G., Michelakis, A. M., and Sandstead, H. H.,** Plasma renin activity in chronic plumbism: effect of treatment, *Arch. Intern. Med.,* 127, 919, 1971.
121. **Gonzalez, J. J., Werk, E. E., Thrasher, K., Behar, R., and Loadholdt, C. B.,** Renin aldosterone system and potassium levels in chronic lead intoxication, *South. Med. J.,* 72, 433, 1979.
122. **Ashouri, O. S.,** Hyperkalemic distal tubular acidosis and selective aldosterone deficiency: combination in a patient with lead nephropathy, *Arch. Inter. Med.,* 145, 1306, 1985.
123. **Campbell, B. C., Beattie, A. D., Elliott, H. L., Goldberg, A., Moore, M. R., Beevers, D. G., and Tree, M.,** Occupational lead exposure and renin release, *Arch. Environ. Health,* 34, 439, 1979.
124. **Campbell, B. C., Meredith, P. A., and Scott, J. J. C.,** Lead exposure and changes in the renin-angiotensin-aldosterone system in man, *Toxicol. Lett.,* 55, 25, 1985.
125. **Boscolo, P., Galli, G., Iannaccone, A., Martino, F., Porcelli, G., and Troncone, L.,** Renin activity and urinary kallikrein excretion in lead-exposed workers as related to hypertension and nephropathy, *Life Sci.,* 28, 175, 1981.
126. **Vander, A. J.,** Chronic effects of lead on the renin-angiotensin system, *Environ. Health Perspect.,* 78, 77, 1988.
127. **Goldman, J. M., Vander, A. J., Mouw, D. R., Keiser, J., and Nicholls, M. G.,** Multiple short-term effects of lead on the renin-angiotensin system, *Lab. Clin. Med.,* 97, 251, 1981.
128. **Boscolo, P., Porcelli, G., Cecchetti, G., Salimei, E., and Iannaccone, A.,** Urinary kallikrein activity of workers exposed to lead, *Br. J. Ind. Med.,* 35, 226, 1978.
129. **Bertel, H., Kneip, T. J., and Ott, J.,** Lead-induced hypertension: blunted beta adrenoreceptor-mediated response, *Br. Med. J.,* 1, 551, 1978.
130. **Kopp, S. J., Barron, J. T., and Tow, J. P.,** Cardiovascular actions of lead and relationship to hypertension: a review, *Scand. J. Work Environ. Health,* 11, 15, 1985.
131. **Webb, R. C., Winquist, R. J., Victery, W., and Vander, A. J.,** *In vivo* and *in vitro* effects of lead on vascular reactivity in rats, *Am. J. Physiol.,* 241, H211, 1981.
132. **Chai, S. and Webb, R. C.,** Effects of lead on vascular reactivity, *Environ. Health Perspect.,* 78, 85, 1988.
133. **Boscolo P. and Carmignani, M.,** Neurohumoral blood pressure regulation in lead exposure, *Environ. Health Perspect.,* 78, 101, 1988.
134. **World Health Organization,** Ceramic foodware safety, Critical review of sampling, analysis, and limits for lead and cadmium release, Report of a Meeting, World Health Organization, Geneva, 12—14 November 1979, HCS/79.7.
135. **Chamberlain, A. C.,** Prediction of response of blood lead to airborne and dietary lead from volunteer experiments with lead isotopes, *Proc. R. Soc. Lond.,* B244, 149, 1985.

Chapter 11

NEURODEVELOPMENTAL EFFECTS OF LOW-LEVEL LEAD EXPOSURE IN CHILDREN

David Bellinger and Herbert L. Needleman

TABLE OF CONTENTS

I.	Introduction	192
II.	The Nature of the Deficit	193
III.	Level of Exposure and Severity of Effect	194
	A. Assessment of Exposure	194
	B. Assessment of Effects	195
	C. Consistency Among Studies in Exposure-Effect Relationships	195
IV.	Effect Modification	198
	A. Socioeconomic Status	198
	B. Sex	198
	C. Other Bases of Response Variability	198
	1. Inter-individual Variability	199
	2. Intra-individual Variability	199
V.	Reversibility/Persistence	201
	A. Measurement and Conceptual Issues	201
	B. Animal and Human Studies	203
VI.	Concluding Remarks	204
	Acknowledgments	205
	References	205

I. INTRODUCTION

More effort has been devoted to characterizing the developmental neurotoxicity of lead than any other environmental pollutant. For most toxicants, the distribution of human exposure is largely unknown, and inferences about health risks are drawn by applying somewhat arbitrary uncertainty factors to data collected *in vitro* or on animals receiving doses of questionable relevance to human exposure.[1] In contrast, nationally-representative data are available on population exposure to lead,[2] and several dozen human studies have been conducted on the reproductive and developmental correlates of exposure levels prevalent among contemporary urban children. Moreover, a large body of animal studies, many involving nonhuman primates with blood lead levels well within the range of current human exposure are available to clarify the interpretation of human studies (see Chapter 8). Free of some of the biases to which observational studies are subject, these animal studies provide opportunities to investigate experimentally specific hypotheses that, for ethical or practical reasons, cannot be addressed through human studies.

Despite an impressive research effort over the past two decades, lead's neurobehavioral toxicity in children remains controversial although the intensity of the debate has lessened somewhat in recent years. Some argue that the evidence of adverse neuropsychologic effects at lower exposures, although incomplete, is sufficiently persuasive that additional reductions in the medical and environmental action levels are warranted, as well as more vigorous primary prevention programs. Others focus on the dangers of drawing causal inferences from epidemiologic studies. They argue that the scientific data base remains insufficient, or the demonstrated effect is too weak, to justify costly environmental clean-up programs or changes in industrial practices. As Darwin wrote in the concluding chapter of the second edition of *On the Origin of Species,* "Anyone whose disposition leads him to attach more weight to unexplained difficulties than to the explanation of a certain number of facts will certainly reject the theory [of evolution]."[3]

Why has lead continued to be a major focus of pediatric environmental health research? While undue exposure to other pollutants with adverse developmental sequelae may occur as the result of an accident[4] or residence in a particular locale,[5] virtually all children in industrialized nations are chronically exposed to lead. Many carry lead burdens that are disturbingly close to those at which health is compromised.[6] If lead exposures at the upper end of the "normal" range produce only a slight increase in the risk of developmental dysfunction, the public health impact would be great due to the large number of children at risk.[7] The identification of a factor that contributes even modestly to the pandemic problem of learning disability would be a major achievement, producing substantial reductions in medical, educational, and emotional costs. Finally, for excess lead exposure, unlike many conditions thought to increase developmental morbidity, simple, low-technology solutions are readily available. Although these solutions would be costly in the short-term, the long-term costs of failing to act will certainly be greater.

In this chapter, we discuss the developmental and neuropsychologic expressions of lead toxicity, focusing on recent studies that investigated exposures below those presently considered to be hazardous. Several components of a risk assessment are discussed: (1) the nature of the deficits identified in exposed children; (2) relationship between level of exposure and severity of effect; (3) factors that increase a child's vulnerability; and (4) the reversibility of deficits. Although it is impossible to address these issues without touching on some of the many epidemiological and biostatistical pitfalls of such research (e.g., sampling bias, laboratory analysis, adjustment for confounding bias), the major focus is substantive findings, their interpretation, and remaining gaps in knowledge.

II. THE NATURE OF THE DEFICIT

Recognition of lead's neuropsychological impact on children has evolved rapidly over the past five decades. Childhood lead poisoning was first described in 1893.[8] For 40 years, considerable doubt persisted as to whether clinical poisoning has any durable central nervous system (CNS) effects. The previous view was clearly expressed by McKhann[9] in 1932:

The neurologic manifestations of lead poisoning usually subside without serious consequences if the ingestion of lead is stopped and the removal of lead from the circulation and its deposition in inert form in the bones can be hastened . . .

This was challenged by the findings of Byers and Lord[10] who studied 20 lead-poisoned children. Although only half had signs of encephalopathy, 19 of the 20 had reading, learning, sensorimotor or behavior disorders. The mean IQ of 19 children tested was 92 ± 10.5. Byers and Lord offered a developmental hypothesis that presaged the beginning of the modern era of lead studies. They argued that the toxicity of lead may not be limited to brain hemorrhage and cell death but may include interference with the normal growth and development of the cerebral cortex. This was a new concept. They also speculated that only a small percentage of affected children were correctly diagnosed as lead poisoned. This study initiated a tradition of investigations, continuing today, on the question of whether exposures within the range considered "normal" increase a child's risk of cognitive impairment. Of course the criterion for a "normal" blood lead level has changed dramatically over the decades.

Epidemiologic studies of lead at low dose began to proliferate in greater numbers after 1970. The serious methodological deficiencies of many of these studies have been identified.[11-13] No uniform battery of outcome measures was used. Most studies used a conventional measure of psychometric intelligence, although some used screening instruments such as the Denver Developmental Screening Test or group tests such as the British Eleven-plus examination. In some studies, reading skills, language competence, visual-motor integration, or reaction time were also evaluated. In several of them, children's behavior was assessed using checklists or rating scales. More recently, efforts have been made to employ standard "core" batteries of assessments in order to increase study comparability.[14-15]

It is difficult to summarize succinctly current knowledge about lead's neurobehavioral toxicity. A recent meta-analysis of the IQ results of many of the major studies yielded a joint p-value of .0001 for the "lead effects" hypothesis in studies based on blood lead, and a joint p-value of .004 in studies based on tooth lead.[15a] The magnitude of the difference between the mean scores of the groups considered "exposed" and "not exposed" is typically between four and seven points. In reviewing the recent literature, the U.S. Environmental Protection Agency (EPA) concluded that blood lead levels exceeding 50 µg/dl are associated with a five point decline in IQ, levels of 30 to 50 µg/dl with a four point decline, and levels of 15 to 30 µg/dl with a decline of perhaps one to two points.[16] EPA concluded that peripheral nervous system dysfunction, manifested as slowed nerve conduction velocity or EEG and evoked potential abnormalities, occurs at blood lead levels within this lower range.

IQ differences of the magnitude reported are sometimes dismissed as unimportant. Because of the sigmoid shape of a normal cumulative frequency distribution, however, a shift of four to seven points in the mean can represent a fourfold difference in the percentage of children in the extreme tails of the distribution.[7,17-18]

The specific nature of the cognitive deficits associated with increased lead exposure is not well understood. In some studies, children's verbal and auditory skills appear to be most affected.[19-20] Several studies have reported an inverse relationship between reading scores and lead.[21-23] In other studies the deficits of children with higher exposures have appeared to lie in the domain of visuo-spatial and visual-motor integration skills (e.g., figure reproduction, jigsaw puzzle completion, eye-hand coordination).[24-27]

One factor that may contribute to these inconsistencies among studies is differences among cohorts in the timing of exposure. In a study of clinically poisoned children, Shaheen[28] observed that exposure prior to age two was associated with linguistic deficits, whereas exposure in the 2 to 3 year range was associated with spatial deficits. Differences among cohorts in sociodemographic characteristics may also produce different patterns of deficit. For instance, linguistic skills may be less vulnerable to lead insult in children from families who place great value on academic achievement and who provide supports sufficient to "buffer" lead's adverse impact in this domain.

One of the most consistent findings across studies is an inverse association between children's exposure and their ability to pay attention. Clinicians have long noted that children who recover from lead poisoning have short attention spans. Recent evidence suggests that children with exposures that are not sufficient to produce clinical intoxication nevertheless perform worse than less exposed peers on simple reaction time under varying intervals of delay,[19,29-30] serial choice reaction tasks[30-31] and the Continuous Performance Test.[26]

These deficiencies on laboratory tasks are mirrored in teachers' ratings of children's classroom behavior. Using the same 11-item forced-choice questionnaire, several investigators have reported dose-dependent increases in the frequency of negative ratings on attention-related dimensions of behavior in this setting (e.g., distractibility, organization, impulsivity, frustration tolerance, ability to follow directions).[19,32-33] Similar findings have been reported in studies using the Rutter B2 scale,[23,32,34] the Connors teacher rating scale,[32] and a composite of both.[35] Other studies using these instruments have failed to find associations between lead exposure and classroom behavior.[36-38]

III. LEVEL OF EXPOSURE AND SEVERITY OF EFFECT

The exposure limits that underlie regulatory policy are based on dose-response relationships. An inaccurate characterization of this relationship may result in standards that are either unnecessarily restrictive or insufficiently protective. The assessment of both exposure and effect (or response) pose special challenges for efforts to define the relationship between lead exposure and children's cognitive function.

A. ASSESSMENT OF EXPOSURE

The dose sustained by the target organ, the brain, cannot be measured directly. The amount of lead in another body tissue must be accepted as a proxy measure. The possibilities include the concentration of lead in blood, a short-term storage site, or in bone, a long-term storage site. Shed deciduous tooth has been the bone sampled for this purpose, although methods for *in vivo* measurement of bone lead are currently being developed.[39] An alternative measure of exposure is free erythrocyte protoporphyrin (FEP), an index of biologic response to lead. This has become a less attractive option because of recent evidence that an FEP level is less informative about a child's blood lead level than once believed.[40-41]

Because alternative exposure indices reflect different aspects of lead biokinetics (and are measured in different units), quantitative features of the dose-response relationships linking them to children's cognitive function are not comparable. Furthermore, studies that rely on tooth lead concentration as the index of exposure may differ in terms of the portion of tooth anatomy sampled. Because the regional distribution of lead in tooth is not homogeneous,[42] and the concentration varies by tooth type and location in the jaw,[43] comparability of "tooth lead" values reported in different studies should not be assumed unless identical protocols are used in dissecting teeth to obtain specimens for analysis and "adjusting" concentrations for covariables.

B. ASSESSMENT OF EFFECTS

Risk assessment analysis has been developed largely in the context of carcinogensis where tumor development, a relatively discrete event, is the "response" of interest. At issue is the incidence of "response" at different levels of toxicant exposure. In neurotoxicity risk assessment, "response" is typically less well-defined, consisting of a continuously graded functional deficit. The dose-response curve may convey information about the "severity" of a response as well as its incidence.[43a] An accurate characterization of response gradation involves the array of psychometric issues that attend all such measurements.[44]

With respect to studies of lead toxicity, the goal is to estimate a child's "cognitive function" and to identify the aspect(s) most sensitive to lead toxicity. Recent research in cognitive psychology has amply demonstrated that cognitive function is not monolithic but the result of complex, context-dependent interplay among numerous aspects of information processing.[45] A global measure, such as an intelligence test, may not be the most valid or sensitive measure of the quality, efficiency, or flexibility of a child's cognition or of any effects lead may have on it. Its use is likely to bias toward zero the estimate of the slope of the dose-response relationship.

Intelligence tests, however, have played an important role in neurotoxicological assessment. Their administration and scoring are standardized and familiar, facilitating comparison of results from different studies. One should bear in mind that small changes in integrative indices such as IQ scores may be markers of substantial changes in more basic cognitive competences. These considerations have motivated efforts to employ measures thought to assess more discrete aspects of children's nervous system function, such as vigilance,[29] evoked potentials,[46] and postural disequilibrium.[47] Besides meeting the practical objective of providing an outcome potentially more sensitive to lead toxicity, use of more focused measures provides the data needed to conduct an analysis of lead's impact that is more firmly rooted in neuropsychological theory. Many endpoints assessed in animal studies involve "process" measures such as rate of skill acquisition rather than simply the performance of a well-established skill. Traditional psychometric tests for children tend to focus on the latter. The adaptation for children of paradigms developed on animals is a promising development in assessment strategy.[48]

C. CONSISTENCY AMONG STUDIES IN EXPOSURE-EFFECT RELATIONSHIPS

Recognizing these uncertainties in the measurement of both exposure and effect, we evaluated the extent to which major studies converge on a common dose-response relationship for lead and cognitive function. This analysis was limited to studies that report the mean full-scale IQ or verbal IQ scores of children by exposure category. Thus, the focus is the "severity" dimension of the response curve, specifically the consistency among studies in the mean deficit associated with low-level lead exposure.

For some studies in the literature, the only measure of association reported is the partial regression coefficient assigned to lead. This constrains the lead-IQ relationship to be linear over the range of exposures represented in the sample, an assumption that may impede identification of "lowest observed effect level" and other features germane to risk assessment. Limiting the studies to those that present children's mean scores by exposure category should not introduce bias because the manner in which an investigator chooses to report data is unlikely to be systematically related to the magnitude of the association of interest. Indeed, some of the investigators who conducted studies excluded from our analysis interpreted their data as evidence of significant lead effects[49-50] while others did not.[51-52]

Because blood lead and tooth lead levels are not strictly comparable as markers of exposure, neither are the dose-response relationships based on them. Therefore, the relationship between lead level and IQ is presented separately for these indices (Figure 1 for

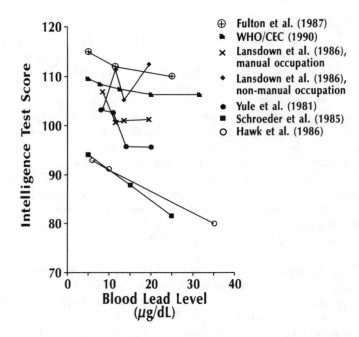

FIGURE 1. Blood lead levels versus intelligence test scores. Data are presented from cross-sectional and retrospective cohort studies that relied on blood lead level as the index of children's exposure. Whether or not a study is included in the figure depends on the form in which the data were reported. For a study to be included, the investigators had to present either mean IQ scores for children by blood lead strata or sufficient information about the regression of IQ on blood lead to specify the regression line (i.e., the coefficient for blood lead and the intercept of the regression line, or a figure from which these statistics could be determined). Except where noted, scores are adjusted for confounding. The source of information provided for each study depicted is as follows: **Yule et al.:** mean full-scale WISC-R IQ scores for children in blood lead quartiles (see Table 7, p.571 in Reference 21). **WHO/CEC studies, Winneke et al.:** WISC scores based on four subscales: Vocabulary, Comprehension, Picture Completion, Block Design (see Table 5, p.558 in Reference 53). These scores appear not to be adjusted for confounding. **Fulton et al.:**[22] regression of British Ability Scales combined score on blood lead: derived by Grant and Davis using method described in a footnote to Figure 2 in their review paper (see p.95 in Reference 92) **Lansdown et al.:** mean WISC-R IQ scores are not presented for the complete cohort — only for children stratified by parental occupation (manual versus nonmanual; see Table 9, p.232 in Reference 54). These scores are not adjusted for confounding. **Hawk et al.:** regression of Stanford-Binet IQ scores on blood lead level [derived from Figure 1 (p.180) in Reference 55]. Data can be considered adjusted because no control or interaction variables contributed significantly to the regression. **Schroeder et al.:**[55a] regression of Stanford-Binet IQ scores on blood lead level [derived from Figure 5 (p.151) in Reference 55a]. These data represent the (apparently unadjusted) regression of IQ on contemporary blood lead level among 6 to 12 year olds. Data from previous assessments of this cohort are also presented in this chapter, but these follow-up data were selected for inclusion here because they are most similar to those from a traditional cross-sectional study.

blood lead studies[21-22,53-55,55a] and Figure 2 for tooth lead studies.[19,26,56-59] In both figures, straight lines are drawn between the symbols for scores of adjacent exposure groups within a given study. These are included only to help the reader identify observations from the same study and should not be interpreted to mean that the lead-IQ association is linear over the range of exposures spanning adjacent groups. In most cases, a point represents the mean score of children in an exposure category adjusted for the variables considered to be potential confounders by the investigators. With some exceptions (e.g., maternal IQ, social class, sex), these variables tend to be study-specific. Error bars representing the standard errors of the estimated values are omitted to prevent the figures from becoming too cluttered.

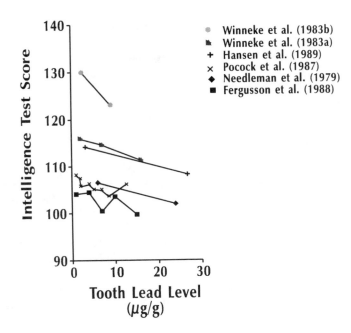

FIGURE 2. Tooth lead levels versus intelligence test scores. Data are presented from retrospective cohort studies that relied on the concentration of lead in some portion of tooth as the index of children's exposure. The data reporting requirements for inclusion of a study in the figure are the same as described for Figure 1. The source of information provided for each study depicted is as follows: **Winneke et al.:** verbal IQ scores from the German adaptation of the WISC (see p.240 in Reference 56); full-scale IQ scores for children in different tooth lead strata were not provided. **Winneke et al.:** full-scale IQ scores from the German adaptation of the WISC from Table 3 (p.177) in Reference 57. These are scores for matched groups. **Pocock et al.:** WISC-R scores from Figure 4 (p.64) in Reference 58. **Needleman et al.:** full-scale WISC-R scores from Table 7 (p.693) in Reference 19. **Hansen et al.:**[26] full-scale scores on the Danish adaptation of the WISC; these are scores for matched groups. **Fergusson et al.:** unadjusted full-scale WISC-R scores at age 8 years from Table 2 (p.800) in Reference 59. Tooth lead values displayed are midpoints of the ranges. No range is provided for the highest tooth lead strata (12+), therefore, 15 μg/g was chosen. A similar pattern was evident for WISC-R scores at 9 years of age (not shown). Adjusted scores are not provided.

Rutter[60] noted that it may be difficult to compare quantitative aspects of dose-response relationships derived from different studies. Studies differ in many respects, including the intelligence test used and its appropriateness to the population sampled, the demographic and psychosocial characteristics of the population, the age of the children at exposure and/or testing, etc. Indeed, Figures 1 and 2 clearly show that the intercepts of regression lines fit to the observations within individual studies would differ dramatically.

In all studies, regardless of exposure index, children's IQ scores declined with increasing lead level, usually in a monotonic fashion. Moreover, the rate of decline is qualitatively similar, despite the fact that the range of exposures represented in the different samples varies considerably. The trend toward lower scores with increasing exposure was statistically significant in some studies, but not in others. This, too, is not surprising given that a p-value depends not only on effect size but also on sample size and precision of measurement.

Another striking feature of both figures is the apparent absence of a clear threshold of effect in most studies. The abscissa of Figure 1 could be truncated so that only the scores of children with blood lead values less than 15 μg/dl are displayed, yet the general features of the dose-response relationship would not differ substantially. Similarly, the abscissa of Figure 2 could be truncated at approximately 10 ppm without much loss of information.

IV. EFFECT MODIFICATION

Several aspects of "variability" in response to lead exposure can be distinguished. One type, often called "effect modification", concerns characteristics of an individual that influence the likelihood of response or the severity of effect. In epidemiology, response rates are most frequently stratified by social class, sex, age, and race. We discuss the evidence available on differences in lead toxicity across strata of social class and sex.

A. SOCIOECONOMIC STATUS

A child already struggling with medical or psychosocial stresses appears to be more vulnerable to the developmental adversities associated with increased lead exposure. In a sample of West German schoolchildren, Winneke and Kraemer[61] noted a significant association between tooth lead level and children's scores on visual-motor integration and reaction time tasks, but only among lower-class children. Among British preschool children, increased blood lead was associated with lower IQ scores only for those whose parents were manual workers.[51] Whereas increased lead exposure was significantly associated with poorer cognitive function in British schoolchildren who participated in the first London study,[21] it was not in the second London study.[54] Lansdown et al.[54] speculated that the higher overall socioeconomic status of the cohort assessed in the second study may have contributed to the discrepancy.

Two of the prospective studies have reported data consistent with the hypothesis that the magnitude of lead's cognitive impact differs across social strata. In the Cincinnati cohort, infants in the lower half of the SES distribution displayed a decrement of 16 points in Mental Development Index (MDI) score at age 6 months across the 1 to 22 $\mu g/dl$ range of 10-d blood lead levels.[49] In the Boston cohort, infants in the lower half of the SES distribution displayed MDI deficits in the second year of life at lower levels of prenatal exposure than did infants in the upper half of the distribution.[62] Specifically, while the performance of lower-class infants appeared to suffer when cord blood lead levels were 6 to 7 $\mu g/dl$, the decline in the scores of upper class infants was apparent only when cord blood lead levels exceeded 10 $\mu g/dl$. At these higher levels of exposure, however, the performance of children in the two social class groups was equivalent. The class-related factors that confer developmental advantage on higher SES children were overcome at higher doses of lead.

B. SEX

The greater vulnerability of males to a variety of biologic insults may extend to lead as well. In the cohort of 6-year olds assessed in the Institute of Child Health/Southampton study, the tooth lead-IQ relationship had a significant negative slope for boys (a decline of three points in IQ for each log unit increase in tooth lead), but a nonsignificant positive slope for girls.[58] Dietrich et al.[49] reported a significant interaction between infant gender and prenatal lead exposure in terms of infants' MDI scores at age 6 months. The impact of a given lead level was greater on boys, who showed a decrement of nearly 23 points over the maternal blood lead range of 1 to 27 $\mu g/dl$.

In some studies, sex and/or SES have not proved to be significant modifiers of the association between lead and children's cognition.[52,58,62a] The likelihood of detecting synergistic associations may depend on specific exposure parameters within a cohort or the distributions on sociodemographic factors.

C. OTHER BASES OF RESPONSE VARIABILITY

Even within groups that appear to be relatively homogeneous, individuals differ in their response to lead. Because no sociodemographic or clinical correlate may be apparent, this is presumed to reflect biological variation. This type of variability contributes to within-group outcome variance. Often viewed strictly as "noise",[63] it increases the standard error of the mean response in an exposure group, reducing the likelihood that a significant difference will be detected in the mean response of groups experiencing different levels of exposure.[64] Yet, an increase in variability of response among individuals or within-individuals over time may be viewed as another potentially exposure-related dimension of outcome rather than solely as a source of error variance.[65] Investigators using animal models have taken the lead in developing analytic methods for such endpoints.[66] The failure of investigators conducting human studies to follow their lead is puzzling in view of the requirement that regulatory policy be formulated in such a way that the most vulnerable members of the population are protected.[67]

In our longitudinal study of infant development, we addressed the issue of inter- and intra-individual response variability as a function of exposure level.

1. Inter-individual Variability

At all four ages that the Bayley Scales were administered, the children with "high" prenatal exposure (10 to 25 µg/dl in umbilical cord blood), performed significantly worse than infants with either "medium" (6 to 7 µg/dl) or "low" prenatal exposure (<3 µg/dl). The mean deficit of these children was four to eight points, which can be interpreted as their mean "response" to a high prenatal lead dose. Our goal was to characterize the nature of the underlying distribution of "responses" among high lead children. This concerns the "incidence" dimension of the response curve, as described in the previous section. For instance, it might be severely skewed, or even bimodal, with a small subset of high lead children displaying very poor development, but most children performing at a level comparable to that of children with lower exposures. If the prevalence of "responders" is in fact low, the most effective strategy for reducing lead-associated developmental morbidity would be to increase "host resistance" of the subset of vulnerable children by, for instance, providing nutritional supplements or instructing parents in ways to increase the cognitive stimulation available to their children. Alternatively, most children with high lead levels might display a performance deficit that varies normally around the mean value. This finding would suggest that while improving nutritional status and psychosocial supports may be beneficial for many reasons, they serve as useful adjuncts to but not substitutes for, efforts to prevent lead exposure.

We plotted the distribution of MDI "residual" scores at 12 months of age for infants classified into four groups based on umbilical cord blood lead level: <5, 5 to 9.9, 10 to 14.9, and 15 to 20 µg/dl (Figure 3). These scores represent the difference between a child's observed MDI score and the score "expected" on the basis of his or her values on important covariates such as maternal IQ, family social class, birthorder, quality of home environment, etc. (Statistically, this score is $Y - \hat{Y}$, where Y is observed MDI score and $\hat{Y}$ is the fitted value calculated from the equation for the ordinary least-squares regression line.) At this age, infants with cord lead values of 15 to 20 µg/dl achieved MDI scores that were, on average, 7.2 points lower than expected. The distribution of residual scores in this group (as well as all others) does not deviate significantly from normal suggesting that the mean "response" of the children is clearly not the product of an underlying bimodal distribution involving a small number of extreme "responders" and a majority of "nonresponders". Rather the response and, presumably the susceptibility, of a group of children to a given prenatal lead dose follows a Gaussian distribution, with random variation about a central value that can be accepted as a meaningful estimate of the population parameter.

2. Intra-individual Variability

The prevalence of deficient performance among children with high cord lead levels can also be assessed by examining the consistency over time in a child's test scores. We attempted

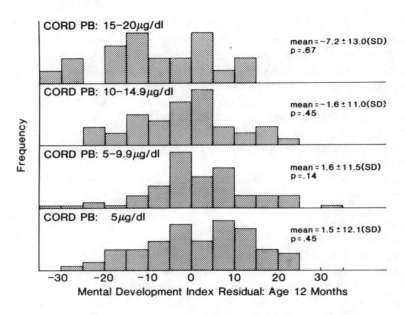

FIGURE 3. Distributions of "residual" scores at age 12 months for four groups of infants classified by the concentration of lead in umbilical cord blood. A child's residual score was calculated by subtracting the fitted Mental Development Index (MDI) score (Bayley Scales of Infant Development) from the observed score. The fitted score was derived using the equation describing the regression of MDI score against 12 potential confounders [maternal IQ, age, education, family social class, ethnicity, alcohol consumption during pregnancy, smoking history, Home Observation for Measurement of the Environment (HOME) score, gestational age, birth weight, birth order, and sex]. p-Values are the probabilities that the distributions of residual scores for the exposure groups do not deviate from normality. (From Bellinger, D., Prenatal/early postnatal exposure to lead and risk of developmental impairment, in *Research in Infant Assessment*, Paul, N., Ed., March of Dimes Birth Defects Foundation, White Plains, NY, BD:OAS, 25(6), 88, 1989. With permission.)

to distinguish among three alternative explanations of the fact that the children in this group achieved significantly lower scores at each age: (1) most children consistently achieved lower scores than expected; (2) the same small set of children consistently achieved lower scores than expected; (3) a different set of children in this group manifest poor development at any given age.

For each child in the cohort, we determined the number of ages at which the MDI score achieved exceeded the MDI score expected (i.e., $Y - \hat{Y} > 0$). The maximum score of four, the number of times the Bayley Scales were administered, means that on each occasion, the child's MDI score was higher than estimated on the basis of such factors as maternal IQ, family social class, birth weight, and birth order. Table 1 presents the distributions of summary scores for infants classified by umbilical cord blood lead concentration. Nearly one in three (30%) children with levels of 15 μg/dl or more (but less than 25) did not achieve their expected level of performance on any of the four occasions. Among infants with levels below 10 μg/dl, the corresponding frequency was approximately 7% (9/121). The percentage of children who scored higher than expected on, at most, one of the four occasions increased monotonically across exposure groups: 23.7, 25.8, 45.7, and 65.0%, respectively. The modal score was three occasions in the lowest exposure group, two occasions in the two intermediate groups, and one occasion in the highest exposure group. Interestingly, the percentage of children who scored higher than expected on all four occasions was unrelated to exposure.

TABLE 1
Classification of Infants According to Umbilical Cord Blood Lead Level and Number of Occasions at Which Observed Mental Development Index (MDI) Score Exceeded Expected Score

Umbilical cord blood lead level (μg/dl)	Number of occasions observed MDI exceeded expected MDI				
	0	1	2	3	4
0—5	5 (8.5)[a]	9 (15.3)	18 (30.5)	20 (33.9)	7 (11.9)
5—10	4 (6.5)	12 (19.4)	22 (35.5)	17 (27.4)	7 (11.3)
10—15	7 (20.0)	9 (25.7)	11 (31.4)	5 (14.3)	3 (8.6)
15—20	6 (30.0)	7 (35.0)	2 (10.0)	3 (15.0)	2 (10.0)

[a] Row percentages.

The analyses of inter-individual variability of performance suggested that at a given age most infants "respond" to higher lead doses with performance that is below the expected level. The analyses of intra-individual variability of performance suggest that the expression of this "response" is consistent over time for most children with high lead levels. We conclude that a child who manifests a deficit at one age is likely to do so at the other ages as well, at least within the first 2 years of life.

V. REVERSIBILITY/PERSISTENCE

A. MEASUREMENT AND CONCEPTUAL ISSUES

The concept of reversibility/persistence is deceptively complicated. At issue is whether a lead-associated cognitive deficit attenuates once a child's exposure ends, or whether function is thereafter limited even if subsequent lead exposure is low.

The effort to select among alternative hypotheses is impeded by a complex of conceptual and measurement issues. To conclude that a deficit is irreversible, an investigator must observe continuity of impairment in a process that, as part of normal development, changes over time. For instance, in infancy the best index of a child's visual-motor integration skills is coordination in visually-directed reaching. Subsequently, the child's ability to complete formboards and, later still, skill in reproducing complex line figures serve this function. To accurately describe the time course of a child's performance within a cognitive domain requires the use at different ages of instruments and procedures that are equivalent in discriminating power (i.e., difficulty and reliability). Although tests appropriate to different ages may focus on the same general domain of functioning, it is inevitable that they assess slightly different aspects of the domain and vary in psychometric properties. This is a recurring problem for investigators trying to distinguish specific and generalized deficits.[68] In toxicological studies, this may lead to an inaccurate characterization of the natural history of a deficit.

These measurement issues are overlaid on a conceptual controversy regarding the continuity of normal development. A lack of continuity, it is claimed, accounts for the failure of tests of infant sensory-motor abilities to predict accurately the IQ scores of school-aged children. Others view this more as a measurement artifact, attributable to the focus of traditional infant tests on sensory-motor skills rather than on information-processing capacities, which appear to be more stable.[69-70]

A cognitive deficit may appear to have disappeared unless a thorough neuropsychological evaluation, including "limit testing", is conducted. Bright children often learn to "work around" a mild specific deficit and, using compensatory strategies, function adequately under ordinary circumstances. Producing satisfactory work may be more "effortful" for them, however[71] and performance may deteriorate when task demands are increased (e.g., processing load increased, distractions or time constraints introduced) or when the child is otherwise already stressed (e.g., tired or sick). The approach required to detect toxic effects manifested as limited "reserve capacity" or lack of resilience[72] has been applied more often in studies using animal models than in human studies. For instance, the administration of lead may have effects not apparent until an animal is challenged pharmacologically or stressed in some way. As adults, rats exposed neonatally to lead appear normal but manifest greater lithium-induced polydipsia.[73] A lead-associated developmental delay in the distress vocalization of neonatal rats is not apparent until they are subjected to cold stress.[74] Lead-exposed monkeys perform as well as controls on simple discrimination but not in the formation of complex pattern discrimination learning sets[75] and perform relatively worse when irrelevant cues are introduced[76] or on reversal trials after the acquisition of discrimination learning sets.[77]

The time-course of a child's recovery from a deficit may vary considerably depending on the specific nature of the deficit and its particular ramifications. Children with learning disabilities tend to present with social and emotional disorders, often secondary to the cognitive difficulties.[78] Even if a cognitive deficit is successfully remediated through educational interventions, the adverse impact that academic difficulties have on a child's self-esteem, peer relations, motivation, and attitudes toward learning is often considerably more resistant. No longer coupled to the academic problems in which they originated, these psychosocial difficulties are perpetuated by maladaptive interactional patterns or unfavorable self-concept.

Just as susceptibility to lead toxicity may vary among cognitive functions, so too may the reversibility of any deficits. The mechanism of lead's impact might not be the same for its myriad effects, in some cases representing an acute pharmacologic effect that attenuates once exposure has ended[79] and in others a more permanent modification of neuronal structure or physiology.[80-81] Thus, for some cognitive functions, current exposure may be the most predictive exposure index while for others, an index of past exposure is the most predictive. For instance, Winneke et al.[31] reported that the performance of school-age children on a visual-motor integration task was significantly related to lead exposure early in life. In contrast, reaction time was more strongly associated with current exposure. David et al.[82] reported that the administration of penicillamine to children with blood lead levels of 25 to 55 µg/dl was followed by reductions in the frequency of hyperactive behaviors as reported by parents and teachers.

The extent to which a persisting underlying deficit is expressed may change with development. For instance, although lead exposure during infancy may affect brain structures or processes involved in language, a performance deficit might not be observed for several years. There are several possible explanations. The instruments or procedures used to assess primitive language skills (e.g., appearance of first words or first two-word sentences) might not be sufficiently sensitive to identify deficit, early language skills may be relatively resistant to lead insult, or the expression of toxicity may depend on the maturation of lead-sensitive, language-related brain structures. The late appearance of deficit after certain types of brain injury has been reported in both animal and human studies.[83] What appears to be a late effect may be the overt expression of an irreversible but heretofore latent (or at least undetected) deficit.

B. ANIMAL AND HUMAN STUDIES

The extent to which any adverse cognitive effects of lead in children are reversible is poorly understood. This is not due to a lack of interest but to the relative unavailability of samples of children whose exposure histories have the necessary temporal characteristics. If exposure is chronic, acute pharmacologic effects attributable to current exposure may not be distinguishable from residual effects of past exposure (unless the expression of cognitive toxicity is highly specific to age at exposure). Continuing exposure would produce adverse effects that, for practical purposes, would be permanent.

Less ambiguous data on reversibility have been obtained using animal models because temporal features of exposure can be manipulated experimentally. These studies generally support the view that years after the cessation of low-level lead exposure, nonhuman primates display significant performance deficits. For instance, monkeys with only very early or low lifetime exposures display deficits in performance on various schedules of reinforcement at age 3,[65,84] the formation of spatial cue reversal learning sets at age 4,[85] delayed spatial alternation and Hamilton Search Task performance at age 5—6,[86] delayed alternation at age 7—8,[87] and spatial discrimination reversal learning at age 9—10.[76]

Several follow-up studies of human cohorts have been reported. In most cases, the relative strength of the association between children's performance at time two and their lead exposure at time one and time two is evaluated in order to identify ages at which exposure is most deleterious. Statistically controlling for exposure at time two when evaluating the contribution of exposure at time one to performance at time two would address this problem to some extent. The more highly correlated the children's exposures at times one and two, however, the more difficult it is to draw inferences about reversibility or persistence, even when similar deficits are observed on the two occasions. For instance, Schroeder et al.[55a] reported that children's blood lead levels were associated with concurrent IQ, but not with IQ measured 5 years later (after controlling for blood lead level at the time of the 5 year follow-up). They concluded that any adverse impact of early exposure had been reversed. Because the correlation between initial and follow-up blood lead levels was 0.74, including follow-up lead level as a covariate most likely led to underestimation of the contribution of early lead exposure to later performance.

Winneke and colleagues[27,88] conducted 3 and 6 year follow-up evaluations of schoolchildren in Nordenham and Stolberg (FRG), respectively. For some outcomes (delayed reaction time, errors on the Wiener reaction device), performance was significantly associated with the index of earlier exposure, in some cases more strongly than to the index of current exposure. For other outcomes (e.g., WISC, visual-motor integration), the patterns were less clear. In these studies, like that of Schroeder et al.,[55a] the stability in children's lead exposure over time limits the inferences that can be drawn about the reversibility of deficit. For children in the Nordenham sample, the correlation between blood lead levels measured in 1982 and 1985 was 0.82. The correlation between tooth lead level and 1982 blood lead level was 0.57; and 0.63 for 1985 blood lead level.

In the cohort assembled by Needleman et al.,[19] children with elevated dentine lead levels displayed performance deficits in the first and second grades, and throughout their elementary years, higher rates of grade retention and referral for remedial academic aid.[38] Follow-up blood lead levels could not be obtained, however, leaving open the possibility that differences in concurrent exposure were responsible for what appeared to be residual deficits among children with higher dentine lead levels. Higher concurrent lead exposure could be rejected as an explanation for the higher rates of reading disability and failure to complete high school when these children were 17 to 19 years old,[89] since blood lead levels in adulthood were all below 7 μg/dl. Having higher dentine lead levels in 1979 were associated with a sevenfold increased risk of high school failure, and a sixfold risk of reading disabilities. High dentine lead was also associated with increased absenteeism in the last year of high school, lower class standing, poorer vocabulary scores, slower reaction times, and inferior eye-hand coordination.

The cohort of infants that we have followed since their birth have exposure patterns particularly well-suited to an assessment of reversibility. Although the infants in the three cord blood lead categories differed considerably in terms of prenatal exposure, their postnatal exposures were comparable.[90] We concluded that the performance deficits observed through two years of age among the children with higher prenatal exposures are not attributable to group differences in ongoing exposure.

Follow-up assessments of these children suggest that between ages 2 and 5, the association between prenatal exposure and development attenuated.[25] Additional analyses revealed that the extent of "recovery" or "compensation" over this period by the children with high prenatal exposure was not uniform. The performance of boys, children of lower socioeconomic standing, and those with higher postnatal exposure failed to improve and, indeed, appeared to deteriorate somewhat.[90]

VI. CONCLUDING REMARKS

An impressive volume of research has been conducted in the last decade on the neurobehavioral toxicity of lead in children.[92] The "lowest observed effect level" and the maximum acceptable blood lead level continue to be topics of spirited debate, suggesting that the final chapter in the history of regulatory policy for lead exposure is yet to be written.

While the weight of evidence supports the basic hypothesis that low-level exposure is associated with adverse neuropsychological function,[6,16] remarkably little insight has been gained into fundamental aspects of these neurotoxic expressions (including the specific nature of the deficits, their severity, persistence, and implications for adaptive capacities). More speculative yet are their neurochemical and neuropathological bases (see Chapter 5).

Part of the explanation for lack of progress may be that this problem has generally been addressed by investigators trained in medicine and public health rather than neuropsychology and neurobiology. Rarely has lead been selected as a model exposure for studying the response of the developing nervous system to insult.[28] The motivation to undertake complex and costly human epidemiologic studies has most often been a desire to answer the question "Can *any* differences be identified in the neuropsychological performance of children experiencing higher levels of lead exposure?" Thus, methodological issues pertaining to study validity have been the major foci of discussion (e.g., evaluation of sampling bias, the choice of exposure index and analytical methods, adjustment for confounding bias). Furthermore, the need to insure communicability of study results has more or less obligated investigators to construct neuropsychological assessment protocols around such familiar (but by no means universally accepted) measures as standardized IQ-type tests. In general, these protocols have tended not to be structured like those used in clinical evaluations, in which the goal is to evaluate the strengths and weaknesses of a child's information processing style. This is accomplished by characterizing problem-solving strategies rather than simply recording success or failure on an item, as done in the administration of most IQ tests. Practical constraints of cost and participant compliance and the desire to maximize statistical power have also contributed to the heavy reliance on IQ-type tests, which yield a little information about a broad range of cognitive functions in a relatively brief period.

Recent research has provided increasingly persuasive evidence of an association between low-level lead exposure and neuropsychological performance deficits, as well as on the exposure levels at which such deficits can be detected. In this chapter, we have attempted to identify important aspects of this association that remain poorly characterized.

ACKNOWLEDGMENTS

Research in this chapter was supported by grants HD08945, HD17407, and P30-HD18655 (Mental Retardation Center Grant) from the National Institute of Child Health and Human Development, and grants ES04095 and ES00138 (Research Career Development Award) from the National Institute of Environmental Health Sciences.

REFERENCES

1. **McMillan, D. E.**, Risk assessment for neurobehavioral toxicity, *Environ. Health Perspect.*, 76, 155, 1987.
2. **Mahaffey, K., Annest, J., Roberts, J., and Murphy, R.**, National estimates of the blood lead levels: United States, 1976—1980: association with selected demographic and socioeconomic factors, *N. Engl. J. Med.*, 307, 573, 1982.
3. **Darwin, C.**, Recapitulation and conclusion (from The Origin of Species, 2nd ed.) in *The Sacred Beetle and Other Great Essays in Science*, Gardner, M., Ed., New American Library, New York, 1984, 5.
4. **Rogan, W. J., Gladen, B. C., Hung, K.-L., Koong, B.-L., Shih, L.-Y., Taylor, J. S., Wu, Y.-C., Yang, D., Ragan, N. B., and Hsu, C.-C.**, Congenital poisoning by polychlorinated biphenyls and their contaminants in Taiwan, *Science*, 241, 334, 1988.
5. **Jacobson, S. W., Fein, G. G., Jacobson, J. L., Schwartz, P. M., and Dowler, J. K.**, The effect of intrauterine PCB exposure on visual recognition memory, *Child Dev.*, 56, 853, 1985.
6. Agency for Toxic Substances and Disease Registry, The Nature and Extent of Lead Poisoning in Children in the United States: A Report to Congress, U.S. Department of Health and Human Services, Public Health Service, Atlanta, GA, 30333.
7. **Weiss, B.**, Neurobehavioral toxicity as a basis for risk assessment, *Trends Pharmacol. Sci.*, 9, 59, 1988.
8. **Gibson, J., Love, W., Hardine, D., Bancroft, P., and Turner, A.**, Notes on lead-poisoning as observed among children in Brisbane, in *Intercolonial Medical Congress of Australasia: Transactions of the 3rd Session*, Huxtable, L., Ed., Charles Potter, Sydney, 1893, 76.
9. **McKhann, C.**, Lead poisoning in children: the cerebral manifestations, *Arch. Neurol. Psychiatr.*, 27, 294, 1932.
10. **Byers, R. and Lord, E.**, Late effects of lead poisoning on mental development, *Am. J. Dis. Child.*, 66, 471, 1943.
11. **Rutter, M.**, Raised lead levels and impaired cognitive/behavioral functioning: a review, *Dev. Med. Child Neurol.*, Suppl. 47, 1980.
12. **Bornschein, R., Pearson, D., and Reiter, L.**, Behavioral effects of moderate lead exposure in children and animal models. I. Clinical studies, *CRC Rev. Toxicol.*, 8, 43, 1980.
13. **Bellinger, D. and Needleman, H.**, Low level lead exposure and psychological deficit in children, in *Advances in Developmental and Behavioral Pediatrics*, Vol. 3, Wolraich, M. and Routh, D., Eds., JAI Press, Greenwich, CT, 1982, 1.
14. World Health Organization, Regional Office for Europe, Epidemiological Study Protocol on Biological Indicators of Lead Neurotoxicity in Children, March, 1983.
15. **Bornschein, R. and Rabinowitz, M.**, *Environ. Res.*, 38, 1, 1985.
15a. **Needleman, H. and Gatsonis, C. A.**, 1990.
16. U.S. Environmental Protection Agency, Air Quality Criteria for Lead, (EPA Report No. EPA/600/8-83/028aF-dF) Office of Health and Environmental Assessment, Research Triangle Park, NC, 1986.
17. **Needleman, H., Leviton, A., and Bellinger, D.**, Lead-associated intellectual deficit, *N. Engl. J. Med.*, 306, 367, 1982.
18. **Bellinger, D.**, Recent studies of the developmental and neuropsychologic effects of low-level lead exposure, in *Proc. Natl. Conf. Childhood Lead Poisoning: Current Perspectives*, 1987, 32.
19. **Needleman, H. L., Gunnoe, C. E., Leviton, A., Reed, R., Peresie, H., Maher, C., and Barrett, P.**, Deficits in psychologic and classroom performance of children with elevated dentine lead levels, *N. Engl. J. Med.*, 300, 689, 1979.
20. **Mayfield, S.**, Language and speech behaviors of children with undue lead absorption: a review of the literature, *J. Speech Hear. Res.*, 26, 362, 1983.
21. **Yule, W., Lansdown, R., Millar, I., and Urbanowicz, M.**, The relationship between blood lead concentrations, intelligence, and attainment in a school population: a pilot study, *Dev. Med. Child Neurol.*, 23, 567, 1981.

22. **Fulton, M., Raab, G., Thomson, G., Laxen, D., Hunter, R., and Hepburn, W.,** Influence of blood lead on the ability and attainment of children in Edinburgh, *Lancet*, 1, 1221, 1987.
23. **Silva, P., Hughes, P., Williams, S., and Faed, J.,** Blood lead, intelligence, reading attainment, and behavior in eleven year old children in Dunedin, New Zealand. *J. Child Psychol. Psychiatr.*, 29, 43, 1988.
24. **McBride, W., Black, B., and English, B.,** Blood lead levels and behaviour of 400 preschool children, *Med. J. Aust.*, 2, 26, 1982.
25. **Bellinger, D., Sloman, J., Leviton, A., Waternaux, C., Needleman, H., and Rabinowitz, M.,** Low-level lead exposure and child development: assessment at age 5 of a cohort followed from birth, in *Proc. 6th Int. Conf. Heavy Metals in the Environment*, Lindberg, S. and Hutchinson, T., Eds., CEP Consultants, Edinburgh, 1987, 49; *Pediatrics*, 219, 1991.
26. **Hansen, O., Trillingsgaard, A., Beese, I., Lyngbye, T., and Grandjean, P.,** A neuropsychological study of children with elevated dentine lead level, in *Proc. 6th Int. Conf. Heavy Metals in the Environment*, Lindberg, S. and Hutchinson, T., Eds., CEP Consultants, Edinburgh, 1987, 54.
27. **Winneke, G., Collet, W., Kraemer, U., Brockhaus, A., Ewert, T., and Krause, C.,** Three- and six-year follow-up studies in lead-exposed children, in *Proc. 6th Int. Conf. Heavy Metals in the Environment*, Lindberg, S. and Hutchinson, T., Eds., CEP Consultants, Edinburgh, 1987, 60.
28. **Shaheen, S.,** Neuromaturation and behavior development: The case of childhood lead poisoning, *Dev. Psychol.*, 20, 542, 1984.
29. **Hunter, J., Urbanowicz, M.-A., Yule, W., and Lansdown, R.,** Automated testing of reaction time and its association with lead in children, *Int. Arch. Occup. Environ. Health*, 57, 27, 1985.
30. **Hatzakis, A., Kokkevi, A., Katsouyanni, K., Maravelias, K., Salaminios, F., Kalandidi, A., Koutselinis, A., Stefanis, K., and Trichopoulus, D.,** Psychometric intelligence and attentional performance deficits in lead-exposed children, in *Proc. 6th Int. Conf. Heavy Metals in the Environment*, Lindberg, S. and Hutchinson, T., Eds., CEP Consultants, Edinburgh, 1987, 204.
31. **Winneke, G., Beginn, U., Ewert, T., Havestadt, C., Kraemer, U., Krause, C., Thron, H. L., and Wagner, H. M.,** Comparing the effects of perinatal and later childhood lead exposure on neuropsychological outcome, *Environ. Res.*, 38, 155, 1985.
32. **Yule, W., Urbanowicz, M., Lansdown, R., and Millar, I.,** Teachers' ratings of children's behavior in relation to blood lead levels, *Br. J. Dev. Psychol.*, 2, 295, 1984.
33. **Hatzakis, A., Salaminios, F., Kokevi, A., Katsouyanni, K., Maravelias, K., Kalandidi, A., Koutselinis, A., Stefanis, K., and Trichopoulos, D.,** Blood lead and classroom behavior in two communities with different degree of lead exposure: evidence of a dose-response effect?, in *Proc. 5th Int. Conf. Heavy Metals in the Environment*, Lekkas, T., Ed., CEP Consultants, Edinburgh, 1985, 47.
34. **Raab, G., Fulton, M., Thomson, G., Hunter, R., Hepburn, W., and Laxen, D.,** The influence of blood lead levels on school attainment, mental abilities, and behavior — results from the Edinburgh Lead Study, in *Proc. 6th Int. Conf. Heavy Metals in the Environment*, Lindberg, S. and Hutchinson, T., Eds., CEP Consultants, Edinburgh, 1987, 213.
35. **Fergusson, D., Fergusson, J., Horwood, L., and Kinzett, N.,** A longitudinal study of dentine lead levels, intelligence, school performance and behavior. III. Dentine lead levels and inattention/activity, *J. Child Psychol. Psychiatr.*, 29, 811, 1988.
36. **Ernhart, C., Landa, B., and Schell, N.,** Subclinical levels of lead and developmental deficit: a multivariate follow-up reassessment, *Pediatrics*, 67, 911, 1981.
37. **Smith, M., Delves, T., Lansdown, R., Clayton, B., and Graham, P.,** The effects of lead exposure on urban children: the Institute of Child Health/Southampton Study, *Dev. Med. Child Neurol.*, Suppl. 47, 1983.
38. **Bellinger, D. C., Needleman, H. L., Bromfield, R., and Mintz, M.,** A follow-up study of the academic attainment and classroom behavior of children with elevated dentine lead levels, *Biol. Trace Element Res.*, 6, 207, 1984.
39. **Rosen, J.,** The toxicological importance of lead in bone: the evaluation and potential uses of bone lead measurements by X-ray fluorescence to evaluate treatment outcomes in moderately lead toxic children, in *Biological Monitoring of Toxic Metals*, Clarkson, T., Friberg, G., Nordberg, S., and Sager, P., Eds., Plenum Press, New York, 1988, 603.
40. **Mahaffey, K. and Annest, J.,** Association of erythrocyte protoporphyrin with blood lead level and iron status in the second National Health and Nutrition Examination Survey, 1976—1980, *Environ. Res.*, 41, 327, 1986.
41. **Marcus, A. H. and Schwartz, J.,** Dose-response curves for erythrocyte protoporphyrin versus blood lead: effects of iron status, *Environ. Res.*, 44, 221, 1987.
42. **Purchase, N. and Fergusson, J.,** Lead in teeth: the influence of the tooth type and the sample within a tooth on lead levels, *Sci. Total Environ.* 52, 239, 1986.
43. **Paterson, L. J., Raab, G. M., Hunter, R., Laxen, D. P. H., Fulton, M., Fell, G. S., Halls, D. J., and Sutcliffe, P.,** Factors influencing lead concentrations in shed deciduous teeth, *Sci. Total Environ.*, 74, 219, 1988.

43a. **Wyzga, R.,** personal communication.
44. **Messick, S.,** Assessment of children, in *Handbook of Child Psychology: History, Theory, and Methods,* Vol. 1, 4th ed., Kessen, W., Ed., John Wiley & Sons, New York, 1983, 497.
45. **Gazzaniga, M., Ed.,** *Handbook of Cognitive Neuroscience,* Plenum Press, New York, 1984.
46. **Otto, D., Hudnekk, K., Boyes, W., Janssen, R., and Dyer, R.,** Electrophysiological measures of visual and auditory function as indices of neurotoxicity, *Toxicology,* 49, 205, 1988.
47. **Bhattacharya, A., Shukla, R., Bornschein, R., Dietrich, K., and Kopke, J.,** Postural disequilibrium quantification in children with chronic lead exposure: a pilot study, *Neurotoxicology,* 9, 327, 1988.
48. **Paule, M., Cranmer, J., Wilkins, J., Stern, H., and Hoffman, E.,** Quantitation of complex brain function in children: preliminary evaluation using a nonhuman primate behavioral test battery, *Neurotoxicology,* 9, 367, 1988.
49. **Dietrich, K. N., Krafft, K. M., Bornschein, R. L., Hammond, P. B., Berger, O., Succop, P. A., and Bier, M.,** Low-level fetal lead exposure effect on neurobehavioral development in early infancy, *Pediatrics,* 80, 721, 1987.
50. **Hawk, B. A., Schroeder, S. R., Robinson, G., Otto, D., Mushak, P., Kleinbaum, D., and Dawson, G.,** Relation of lead and social factors to IQ of low-SES children: a partial replication, *Am. J. Mental Defic.,* 91, 178, 1986.
51. **Harvey, P. G., Hamlin, M. W., and Kumar, R.,** Blood lead, behaviour and intelligence test performance in preschool children, *Sci. Total Environ.,* 40, 45, 1984.
52. **Ernhart, C. B., Morrow-Tlucak, M., Marler, M. R., and Wolf, A. W.,** Low level lead exposure in the prenatal and early preschool periods: early preschool development, *Neurobehav. Toxicol. Teratol.,* 9, 259, 1987.
53. **Winneke, G., Brockhaus, A., Ewers, U., Kraemer, U., and Neuf, M.,** Results from the European Multicenter Study on lead neurotoxicity in children: implications for risk assessment, *Neurotoxicol. Teratol.,* 12, 553, 1990.
54. **Lansdown, R., Yule, W., Urbanowicz, M.-A., and Hunter, J.,** The relationship between blood-lead concentrations, intelligence, attainment and behavior in a school population: the second London study, *Int. Arch. Occup. Environ. Health,* 57, 225, 1986.
55. **Hawk, B., Schroeder, S., Robinson, G., Otto, D., Mushak, P., Kleinbaum, D., and Dawson, G.,** Relation of lead and social factors to IQ of low-SES children: a partial replication, *Am. J. Ment. Defic.,* 91, 178, 1986.
55a. **Schroeder, S., Hawk, B., Otto, D., Mushak, P., and Hicks, R.,** Separating the effects of lead and social factors on IQ, *Environ. Res.,* 38, 144, 1985.
56. **Winneke, G., Kraemer, U., Brockhaus, A., Ewers, U., Kujanek, G., Lechner, H., and Janke, W.,** Neuropsychological studies in children with elevated tooth-lead concentrations. II. Extended Study, *Int. Arch. Occup. Environ. Health,* 51, 231, 1983.
57. **Winneke, G., Hrdina, K.-G., and Brockhaus, A.,** Neuropsychological studies in children with elevated tooth-lead concentrations, *Int. Arch. Occup. Environ. Health,* 51, 169, 1983.
58. **Pocock, S. J., Ashby, D., and Smith, M. A.,** Lead exposure and children's intellectual performance, *Int. J. Epidemiol.,* 16, 57, 1987.
59. **Fergusson, D. M., Fergusson, J. E., Horwood, L. J., and Kinzett, N. G.,** A longitudinal study of dentine lead levels, intelligence, school performance, and behaviour. II. Dentine lead and cognitive ability, *J. Child Psychol. Psychiatr.,* 29, 793, 1988.
60. **Rutter, M.,** Low level lead exposure: sources, effects, and implications, in *Lead Versus Health,* Rutter, M. and Jones, R., Eds., John Wiley & Sons, New York, 1983, 333.
61. **Winneke, G. and Kraemer, U.,** Neuropsychological effects of lead in children: interactions with social background variables, *Neuropsychobiology,* 11, 195, 1984.
62. **Bellinger, D., Leviton, A., Needleman, H., Waternaux, C., and Rabinowitz, M.,** Low-level lead exposure, social class and infant development, *Neurotoxicol. Teratol.*
62a. **McMichael, A. J., Baghurst, P. A., Wigg, N. R., Vimpani, G. V., Robertson, E. F., and Roberts, R. J.,** Port Pirie Cohort Study: environmental exposure to lead and children's abilities at the age of four years, *N. Engl. J. Med.,* 319, 468, 1988.
63. **Guidotti, T. L.,** Exposure to hazard and individual risk: when occupational medicine gets personal, *J. Occup. Med.,* 30, 570, 1988.
64. **Weiss, B.,** Quantitative perspectives on behavioral toxicology, *Toxicol. Lett.,* 43, 285, 1988.
65. **Rice, D.,** Schedule-controlled behavior in infant and juvenile monkeys exposed to lead from birth, *Neurotoxicology,* 9, 75, 1988.
66. **Cox, C., and Cory-Slechta, D.,** Analysis of longitudinal 'time series' data in toxicology, *Fundam. Appl. Toxicol.,* 8, 159, 1987.
67. **Cory-Slechta, D. A.,** The lessons of lead for behavioral toxicology, in *Lead Exposure and Child Development: An International Assessment,* Smith, M., Grant, L., and Sors, A., Eds., Kluwer Press, Lancaster, UK, in press.

68. **Strauss, M. E. and Allred, L. J.,** Methodological issues in detecting specific long-term consequences of perinatal drug exposure, *Neurobehav. Toxicol. Teratol.,* 8, 369, 1986.
69. **Bornstein, M. H. and Sigman, M. D.,** Continuity in mental development from infancy, *Child Dev.,* 57, 251, 1986.
70. **Vietze, P. and Coates, D.,** Information-processing approaches to early identification of mental retardation, *Ann. N.Y. Acad. Sci.,* 477, 266, 1986.
71. **Zubrick, S. R., Macartney, H., and Stanley, F.,** Hidden handicap in school-age children who received neonatal intensive care, *Dev. Med. Child Neurol.,* 30, 145, 1988.
72. **Newman, L. M. and Johnson, E. M.,** Teratogen-induced decrements of postnatal functional capacity, *J. Am. Coll. Toxicol.,* 5, 517, 1986.
73. **Mailman, R. and DeHaven, D.,** Responses of neurotransmitter systems to toxicant exposure, in *Cellular and Molecular Neurotoxicity,* Narahashi, T., Ed., Raven Press, New York, 1984, 207.
74. **Davis, M.,** Ethological approaches to behavioral toxicology, in *Nervous System Toxicology,* Mitchell, C., Ed., Raven Press, New York, 1982, 29.
75. **Lilienthal, H., Winneke, G., Brockhaus, A., and Molik, B.,** Pre- and postnatal lead-exposure in monkeys: effects on activity and learning set formation, *Neurobehav. Toxicol. Teratol.,* 8, 265, 1986.
76. **Gilbert, S. G. and Rice, D. C.,** Low-level lifetime lead exposure produces behavioral toxicity (spatial discrimination reversal) in adult monkeys, *Toxicol. Appl. Pharmacol.,* 91, 484, 1987.
77. **Bushnell, P. J. and Bowman, R. E.,** Reversal learning deficits in young monkeys exposed to lead, *Pharmacol. Biochem. Behav.,* 10, 733, 1979.
78. **Bruck, M.,** Social and emotional adjustments of learning-disabled children: a review of the issues, in *Handbook of Cognitive, Social, and Neuropsychological Aspects of Learning Disabilities,* Vol. 1, Ceci, S., Ed., Lawrence Erlbaum Associates, Hillsdale, NJ, 1986, 361.
79. **Coria, F. and Monton, F.,** Recovery of the early cellular changes induced by lead in rat peripheral nerves after withdrawal of the toxin, *J. Neuropathol. Exp. Neurol.,* 47, 282, 1988.
80. **Audesirk, G.,** Effects of lead exposure on the physiology of neurons, *Prog. Neurobiol.,* 24, 199, 1985.
81. **Altman, J.,** Morphological and behavioral markers of environmentally induced retardation of brain development: an animal model, *Environ. Health Perspect.,* 74, 153, 1987.
82. **David, O., Hoffman, S., Clark, J., Grad, G., and Sverd, J.,** Penicillamine in the treatment of hyperactive children with moderately elevated lead levels, in *Lead Versus Health: Sources and Effects of Low Level Lead Exposure,* Rutter, M. and Jones, R., Eds., John Wiley & Sons, New York, 1983, 297.
83. **Rodier, P. M.,** Exogenous sources of malformations in development: CNS malformations and developmental repair processes, in *Malformations of Development: Biological and Psychological Sources and Consequences,* Gollin, E., Ed., Academic Press, New York, 1984, 287.
84. **Rice, D. C. and Gilbert, S. G.,** Low lead exposure from birth produces behavioral toxicity (DRL) in monkeys, *Toxicol. Appl. Pharmacol.,* 80, 421, 1985.
85. **Bushnell, P. J. and Bowman, R. E.,** Persistence of impaired reversal learning in young monkeys exposed to low levels of dietary lead, *J. Toxicol. Environ. Health,* 5, 1015, 1979.
86. **Levin, E. D. and Bowman, R. E.,** The effect of pre- or postnatal lead exposure on Hamilton Search Task in monkeys, *Neurobehav. Toxicol. Teratol.,* 5, 391, 1983.
87. **Rice, D. C. and Karpinski, K. F.,** Lifetime low-level lead exposure produces deficits in delayed alternation in adult monkeys, *Neurotoxicol. Teratol.,* 10, 207, 1988.
88. **Winneke, G., Collet, W., and Lilienthal, H.,** The effects of lead in laboratory animals and environmentally-exposed children, *Toxicology,* 49, 291, 1988.
89. **Needleman, H., Schell, A., and Bellinger, D.,** unpublished data, 1989.
90. **Rabinowitz, M., Leviton, A., and Needleman, H. L.,** Variability of blood lead concentrations during infancy, *Arch. Environ. Health,* 39, 74, 1984.
91. **Bellinger, D., Leviton, A., and Sloman, J.,** Antecedents and correlates of improved cognitive performance in children exposed *in utero* to low levels of lead, *Environ. Health Perspect.,* in press.
92. **Smith, M., Grant, L., and Sors, A., Eds.,** *Lead Exposure and Child Development: An International Assessment,* Kluwer Publishers, Lancaster, UK, in press.

Chapter 12

LOW-LEVEL LEAD EXPOSURE: EFFECT ON QUANTITATIVE ELECTROENCEPHALOGRAPHY AND CORRELATION WITH NEUROPSYCHOLOGIC MEASURES

James L. Burchfiel, Frank H. Duffy, Peter H. Bartels, and Herbert L. Needleman

TABLE OF CONTENTS

I.	Introduction	210
II.	Methods	211
	A. Subjects	211
	B. EEG Evaluation	211
	C. Psychologic Data	212
	D. Multivariate Statistical Analysis	212
	1. Reduction of the Dimensions of the Data Set	212
	2. Tests of Separation of High- and Low-Lead Populations and Retrospective Classification of Subjects	213
	3. Prospective Classification of Subjects	213
III.	Results	214
	A. EEG Evaluation	214
	B. Discrimination Between High- and Low-Lead Children Based on EEG and Psychologic Variables	215
	C. Prospective Classification of Subjects into High- or Low-Lead Group Based on EEG and Psychologic Variables	217
IV.	Discussion	218
	References	221

I. INTRODUCTION

Do levels of lead exposure below those causing overt signs of poisoning have a significant, detrimental effect on the cerebral nervous system (CNS)? What are the CNS consequences of levels of lead exposure which a normal child accumulates in the course of everyday life? In order to answer these key questions, we used quantitative electroencephalography to assess the effect of low-level lead exposure on a group of asymptomatic children.

The electroencephalogram (EEG) is an appropriate measure of CNS pathology, and high levels of lead exposure produce clearly demonstrable EEG alterations.[1] Lower levels of lead exposure would be expected to have less dramatic EEG effects, and these would be difficult to document in a conventional EEG tracing. Conventional EEG is quite sensitive to the detection of paroxysmal events such as an epileptic discharge, but the sensitivity with which a human interpreter can reliably detect subtle alterations of background activity in a tracing is relatively low.

Substantially greater sensitivity is achieved by applying quantitative techniques to EEG analysis. The most widely used technique is spectral (or Fourier) analysis. This technique mathematically breaks down the complex voltage oscillations of the EEG within a given time epoch into their component sinusoidal frequencies. It transforms the EEG signal into a frequency spectrum — a quantitative representation of the signal which defines the range of frequencies in the signal and the numerical magnitude of each frequency.

Spectral analysis is potentially a major addition to the armamentarium of neurotoxicology. Even though the technique has been utilized by neurophysiologists for over 20 years, it has not lived up to this potential, particularly with regard to the assessment of human cerebral pathology. Early results with spectral analysis of human EEG were disappointing. Human spectra appeared to have a considerable degree of variability (both intersubject and intrasubject over time), and this significantly diminished the sensitivity of statistical comparisons. We believe, however, that the variability observed in these early studies was more apparent than real. The problem is not that the EEG is inherently insensitive to pathophysiology; rather, the problem lies in the high degree of complexity of the signal. The EEG signal is complex in the frequency domain and varies both as a function of brain region and behavioral state. Therefore, to obtain a reliable measure of the EEG, one must not only analyze the frequency content of the signal, but one must also analyze its spatial distribution while controlling for the behavioral state of the subject.

To accomplish these tasks, we have developed a technique of electrophysiological analysis known as "Brain Electrical Activity Mapping" or BEAM.[2] This technique records EEG under rigidly controlled behavioral conditions from 20 scalp electrodes and performs spectral analysis for each individual electrode derivation. Then, from these multielectrode spectral data, topographic maps are constructed which define the spatial distribution of voltage over the scalp for any given frequency or band of frequencies in the EEG. These distributions of quantitative EEG frequency data are characteristic and reproducible.

We applied the BEAM technique of quantitative EEG analysis to the question of low-level lead toxicity by analyzing EEGs from two populations of children differing with respect to their exposure to lead. Neither group had overt signs of lead poisoning. All subjects were drawn from a large population of normal school children. These were the same children whose psychologic and classroom performance were assessed by Needleman et al.[3] The present study extends these behavioral evaluations by comparing high- and low-lead children on a number of quantitative EEG measures. In addition, we examined the power of selected EEG and psychologic measures, considered both separately and in combination, to discriminate between high- and low-lead exposure by multivariate statistical analyses and automated classification techniques.

II. METHODS

A. SUBJECTS

Subjects were drawn from a population of 3329 first and second grade school children in Chelsea and Sommerville, Massachusetts. Deciduous teeth were collected from these children and analyzed for their dentine lead content. Those children whose dentine lead concentration fell within the uppermost tenth percentile of the population (greater than 24 ppm) were designated as "high lead", and those in the lowest tenth percentile (less than 6 ppm) were designated as "low lead". The high- and low-lead children examined in the present study were a subset of the population evaluated in the study of Needleman et al.,[3] and their report should be consulted for details.

The present sample population consisted of 41 children: 22 low lead and 19 high lead.

B. EEG EVALUATION

All subjects were studied in a standard EEG laboratory. Twenty electrode positions were measured on the scalp according to the International 10-20 System,[4] and a standard gold cup electrode was applied to each location with collodion. EEG was recorded separately from each electrode in a monopolar montage using a common reference consisting of linked electrodes on the two earlobes. Signals were amplified via a Grass Model 78 polygraph and recorded on a 28-channel FM tape recorder (Honeywell Model 5600E). Simultaneously, a standard EEG tracing was produced on the oscillograph of the polygraph.

Epochs of spontaneous EEG were recorded from each subject under four controlled behavioral states: (1) relaxed, alert with eyes open (EO); (2) same, but with eyes closed (EC); (3) during 3 min of hyperventilation (HV); and (4) during the first 4 min immediately posthyperventilation (PHV).

Quantitative EEG analysis was performed using the BEAM topographic mapping technique.[2] First, spectral analysis was performed for each of the 20 channels of EEG in each of the four behavioral states. Analog EEG data were played back from FM tape, analog bandpass filtered between 0.5 and 35 Hz (24 dB/octave), and analog-to-digital converted at 256 Hz. The digital EEG data were divided into consecutive 2-s segments and each individual segment was inspected for artifact. Any segment containing artifact was eliminated from further analysis. A fast Fourier transform was computed for each 2-s EEG segment, and an ensemble of 50 segments was averaged to yield a final frequency spectrum representing a total of 1 min 50 s of EEG. This final spectrum covered the frequency range 0 to 64 Hz, but only the 0.5 to 32 Hz range was subsequently analyzed. Each average spectrum was divided into the four traditional EEG frequency bands: delta (0.5 to 3.5 Hz), theta (4.0 to 7.5 Hz), alpha (8 to 11.5 Hz) and beta (12 to 32 Hz). Spectral energy was summed within each of these bands and expressed as a percentage of the total spectral energy over the full frequency range 0.5 to 32 Hz.

Next, spectra were averaged for the 22 low-lead children and the 19 high-lead children, respectively. Averages were computed for each electrode in each of the four behavioral states. Finally, from these average spectra, topographic maps were constructed which defined the spatial distribution of voltage within each EEG frequency band for each behavioral state.

In summary, we defined the EEG from the low-lead children and the high-lead children by a series of quantitative topographic maps. Each map was the average representation for the high-lead group or the low-lead group of the spatial distribution of EEG voltage within a given frequency band for a given behavioral state during recording. Thus, the EEG for each group was characterized by a set of 16 average, quantitative, topographic maps: the four behavioral states of EO, EC, HV, and PHV, each with four maps representing the respective EEG frequency bands of delta, theta, alpha, and beta.

Statistical comparison between topographic maps was done by Student's t-test and by the nonparametric Wilcoxin-Mann-Whitney two-sample rank test.[5] Homologous pairs of quantitative maps from the high- and low-lead groups were compared electrode-by-electrode. This comparison yielded a statistical value of group separation at each electrode location. These statistical values were then used to construct a new topographic map depicting the magnitude and spatial distribution of difference between the high and low lead groups. This topographic technique of depicting statistical differences between groups is known as significance probability mapping (SPM).[6] Thus, each of the 16 homologous pairs of maps produced a corresponding statistical map defining the separation between the high and low lead groups.

C. PSYCHOLOGIC DATA

For use in multivariate statistical analysis, subjects' scores on psychologic performance tests were obtained from the data of Needleman et al.[3] We chose nine measures as having potentially high discriminating value based on their significant differences between the high- and low-lead populations. These were the best discriminators within the total population of 160 subjects studied by Needleman et al.; their discriminating power for the subset of 41 subjects used in the present study had not previously been determined. The nine measures were as follows: (1) full-scale IQ; (2) verbal IQ; (3) performance IQ; (4) Seashore rhythm test; (5) sentence-repetition test; (6) reaction time under varying intervals of delay, block 2 (tested at 12 s); (7) reaction time, block 3 (12 s); (8) reaction time, block 4 (3 s); and (9) token test. In addition, parent's IQ score on the Peabody picture vocabulary test was evaluated as a possible confounding variable.

D. MULTIVARIATE STATISTICAL ANALYSIS

The major goal of multivariate data analysis was to determine the power of EEG and psychologic variables to discriminate between the high- and low-lead children. To this end, multivariate parametric analyses and nonparametric pattern recognition procedures were performed on a PDP-11/45 computer using the taxonomic intracellular analytic system (TICAS) of statistical programs.[7-9] This comprehensive system was originally developed for the automatic recognition of cancer cells based on morphologic features of their microscopically displayed images. However, its statistical evaluation techniques are not data specific; they can be applied to any set of descriptive features, in this case, EEG and psychologic measures.

Multivariate statistical analyses proceeded in a sequential manner, which is described in the following sections.

1. Reduction of the Dimensions of the Data Set

Statistical theory specifies a limit to the number of features per subject that can be used to demonstrate group differences or to develop classification rules to define a subject as belonging to a given diagnostic group.[10] If this limit is exceeded, the analysis becomes excessively individualized to the immediate data set and, consequently, does not perform well on prospective tests of repeated measure on other data sets. As a rough approximation, this limiting number is derived by dividing the number of subjects in the smallest group by three. In the present case this equals 6.33 (19/3). Therefore, the initial task was to reduce the number of features to six or less.

Data reduction was performed in two steps. First, an initial screening of features was done by univariate analysis. Those variables that did not reach a high level of significance by the Student's t-test or the Wilcoxin-Mann-Whitney two-sample rank test ($p < 0.05$) were excluded from further analysis.

Second, the surviving variables were subjected to merit value analysis. The merit value as formed by TICAS is a relative measure of a feature's ability to distinguish between two

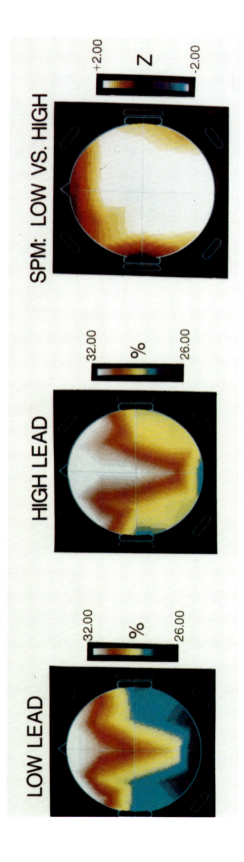

PLATE 1. Comparison of average EEG delta activity between high and low lead children. EEG data were recorded while subjects were in a relaxed, alert state with their eyes closed. Energy in the delta frequency band (0.5 to 3.5 Hz) is expressed as a percentage of total energy in the frequency range 0.5 to 32.0 Hz. Topographic maps of percentage delta activity are constructed by linear interpolation based on measured values at 20 electrode locations (see Plate 2 and text). High lead children show a generalized increase of delta activity in comparison to low lead children. This is statistically significant over a majority of the scalp surface as shown in the final panel (SPM: LOW VS. HIGH). The SPM plot[6] is a topographic map of Student's t values between the high and low lead groups. Values of t above 2.0 are shown in white.

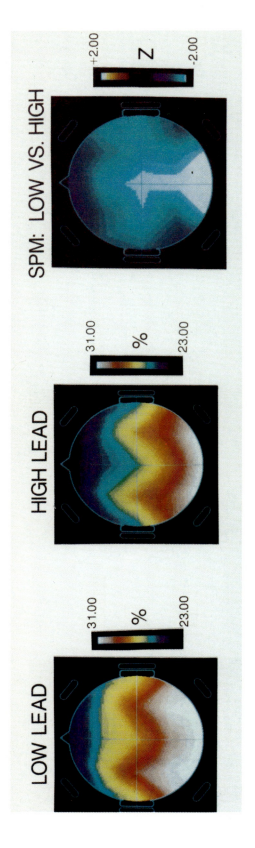

PLATE 2. Comparison of average EEG alpha activity (8.0 to 11.5 Hz) between high and low lead children. Same recording conditions and topographic map construction are shown as in Plate 1. High lead children show a significant generalized decrease of EEG energy in the alpha band. The white area in the SPM plot indicates the region in which t values between the high and low lead groups exceed 2.0. Together, Plates 1 and 2 indicate that in comparison to their low lead classmates, the high lead children have a higher percentage of delta slowing and a lower percentage of alpha rhythm in their EEG spectra.

populations. It is derived as the average of three independent factors. The first two are measures of the degree to which a given variable separates two populations. One is "d-prime", the measure of detectability calculated from the receiver operating characteristic (ROC) curve.[11,12] The value of this statistic ranges from 0 (no detectability of group separation) to 0.5 (complete detectability of group separation). The other factor is the "ambiguity" value based on the ambiguity function of Genchi and Mori.[13,14] A feature which shows no overlap between two populations has an ambiguity value of zero. Conversely, exact overlap of the populations yields the maximum ambiguity value of one. The final factor contributing to the merit value is a measure of the correlation of a given feature with the other features under consideration. This was calculated as the average of the correlation coefficients between a given feature and each of the other features.[15] This average correlation statistic was used to detect the presence of redundancy among features.

The merit value was calculated as follows:

$$\text{merit value} = \frac{2(\text{d-prime}) + (1 - \text{ambiguity}) + (1 - \text{average correlation coefficient})}{3}$$

Merit value, therefore, ranges from 0 to 1. The higher a variable's merit value, the better its potential usefulness for demonstrating group separation or for developing rules to classify subjects into one of two populations. Features with high merit value possess a combination of high detectability, low ambiguity, and low correlation with other variables.

2. Tests of Separation of High- and Low-Lead Populations and Retrospective Classification of Subjects

The features with the best merit value scores were submitted to direct and stepwise discriminant analysis[15,16] to assess the collective ability of these features to separate the high- and low-lead groups. The significance of group separation was tested by Wilks' lambda[17] and the Mahalanobis distance between group centroids.[18] The former statistic may be considered as a multivariate analog of the variance ratio (F) test; therefore, an approximation of the F-statistic was calculated according to the technique of Rao.[17] For stepwise discriminant analysis, retrospective classification of subjects was carried out by the Bayes classification procedure using the linear discriminant function.[16]

3. Prospective Classification of Subjects

Prospective classification was performed by the nonparametric, supervised learning programs of TICAS.[7,19,20] These algorithms develop classification rules for assigning subjects to one of two populations by the sequential determination of decision boundaries along the feature axes within multivariate feature space. The technique is nonparametric because subjects are arrayed along each axis by rank, rather than by absolute value of a given variable. Thus, each subject is represented by a vector whose elements are the relative rank the subject occupies within the sample population for each variable.

The classification rules are developed on a training set of subjects whose membership in one of the two populations is known. The procedure involves examining subjects in rank order along a given feature axis. Sampling continues as long as all subjects belong to the same population; however, as soon as the first subject of the other population is encountered, sampling stops and a decision boundary is established. The sampled subjects of the first population are then removed, and sampling proceeds in the same way with the remaining subjects on another feature axis. This process continues until all subjects of one population have been removed or until the end ranks of each axis contain subjects of the other population.

Successive runs of this procedure are carried out, and the order of feature selection is varied based on the classification success and efficiency of previous runs. This procedure is repeated until the optimum set of classification rules is developed. It differs from the parametric discriminant techniques in that it selects features according to their ability to classify subjects correctly from the remaining subset of, as yet, unclassified subjects. Thus, it may select features that are not weighted heavily by parametric selection methods, which always consider a feature's usefulness in classifying subjects from the full sample.

Where a significant separation exists between the centroids of the two populations in multivariate space, a set of classification rules that will classify subjects of the training set with nearly 100% accuracy can be developed. Therefore, the true test of the diagnostic power of the classification rules is to prospectively evaluate them on a test set of subjects who were not part of the original training set. Ideally, for this purpose one would like to divide the study population into separate training and test sets. In the present case, however, the population was not large enough to allow this. Instead, we established test sets by the "jackknifing" or "leaving-one-out" procedure.[21] Diagnostic classification rules were developed on a training set of 37 of the 41 subjects and tested on a test set of the remaining 4 subjects, who had not been used for development of the classification rules. This procedure was repeated 11 times with a different set of four left out each time. The final time only one subject was left out. To be considered successful, such rules should correctly classify substantially more than 50% of the test-set subjects. The jackknifing technique is generally taken as a method to appraise the prospective performance of diagnostic classification rules when the rules are developed on small data sets. In its original form, jackknifing referred to the leaving out of a single subject for testing. Our procedure of leaving out a larger test set of four subjects is a more conservative approach.

III. RESULTS

A. EEG EVALUATION

The spontaneous EEG of the 19 children who were in the highest tenth percentile of dentine lead concentration differed significantly from that of their 22 classmates who were in the lowest tenth percentile. This difference occurred most dramatically in the behavioral state of EC. The average EEG spectra of the high lead children had a significantly greater proportion of slow wave activity in the delta frequency range. Spatially, this increase in delta slowing occurred in a generalized manner. This finding is clearly illustrated in the topographic maps of Plate 1.* The typical spatial distribution of delta is maximal in the midline frontal-central region and falls off smoothly and symmetrically into the parasagittal, temporal, and occipital regions. The high percentage of delta activity in the frontal polar leads is an artifact due to eye movements. Essentially the same spatial distribution was exhibited by both the high- and low-lead children, but the high-lead group had a significantly greater percentage of delta activity within this distribution. All electrode locations within the high lead group showed more delta, and the magnitude of this increase was graded across the distribution: those regions which normally have more delta showed greater increases. The differences between the high- and low-lead children attained statistical significance in the maximal regions of the distribution in the midline and posterior parasagittal areas. This can be seen clearly in the topographic map of t-scores (labeled SPM in Plate 1).

The increase in the percentage of delta slowing in the high lead children occurred mainly at the expense of alpha frequency activity. The high lead group showed a statistically significant decrease in the magnitude of the percentage alpha distribution (Plate 2*). As with the delta increase, this alpha decrease occurred in a generalized manner. The spatial distribution of alpha was not different between the two groups; the high lead children simply had

---

* Plates 1 and 2 follow page 212.

a lesser percentage of alpha in their average EEG spectra at all electrode locations. As can be seen in the topographic map of t-scores (SPM in Plate 2), the maximal differences between the high- and low-lead groups involved the maximal regions of the alpha distribution in the occipital and midline parietal-central areas.

In the EO state qualitatively similar differences were seen between the average EEG spectra of the high- and low-lead children. High-lead children had a greater percentage of delta slowing and a lesser percentage of alpha activity. However, these differences were quantitatively smaller than those in the EC state. Statistically significant increases of delta occurred only at the most maximal electrodes of the distribution in the midline central and parietal regions. In a slight departure from the pattern, significant alpha decreases did not occur in the maximal posterior electrodes of the distribution; rather, there was a significant shrinkage of spatial extent of the distribution in the midline frontal area.

In either the EO or EC state, the average EEG spectra of the high- and low-lead children did not differ significantly in the theta- or beta-frequency bands. Finally, there were no significant differences in any frequency band between the average EEG spectra of the high- and low-lead groups in response to HV or during the PHV recovery phase.

In summary, quantitative EEG analysis showed that within a population of asymptomatic first- and second-grade school children, the resting background EEG of those children with dentine lead concentrations in the highest tenth percentile had significantly greater percentages of slow frequency delta activity and significantly lesser percentages of alpha activity in comparison to their classmates whose dentine lead concentrations were in the lowest tenth percentile. These differences were most pronounced in a completely inactivated state with the eyes closed. Relatively mild activation of EEG by eye opening reduced these differences, and the differences disappeared entirely during a greater degree of EEG activation induced by HV and PHV.

B. DISCRIMINATION BETWEEN HIGH- AND LOW-LEAD CHILDREN BASED ON EEG AND PSYCHOLOGIC VARIABLES

Merit value analysis culled from the available features those with the greatest potential for discriminating between the high- and low-lead groups. EEG and psychologic features were considered both alone and in combination. Table 1 lists the four best features in each category with the highest merit value.

The power of these features to discriminate between high- and low-lead exposure was tested with direct and stepwise discriminant analyses. For both types of analysis, three separate calculations were performed, each with a different set of input variables. First, the four best EEG features were used as input variables; next, calculations were repeated using the best psychologic features; and finally, the discriminating power of the four best combined EEG and psychologic features was determined.

The most striking result of these discriminant analyses was the marked difference in discriminating power using a combination of EEG and psychologic variables as opposed to using either type of variable alone. The results of linear discriminant analysis are presented in Table 2. This shows that the separation of the high- and low-lead groups using combined variables was highly significant ($p = 0.005$). The group separation using psychologic variables alone was also statistically significant but with a confidence level an order of magnitude smaller ($p = 0.041$). The p-value for EEG features alone was of the same order as that for psychologic features alone but was of equivocal statistical significance ($p = 0.079$).

Significant separation between high- and low-lead children was also demonstrated by stepwise discriminant analysis (Table 3). Again, the greatest magnitude of group separation was obtained with the combined EEG and psychologic variables. This reaches a highly significant value by either Wilks' lambda ($p = 0.0002$) or the Mahalanobis distance between

TABLE 1
Merit Value Analysis of EEG and Psychologic Features[a]

Feature	Mean high lead[b]	Mean low lead[b]	Merit value[c]
EEG[d] alone			
Delta PZ EC	31.14	29.53	0.374
Alpha FZ EO	24.27	25.39	0.368
Delta OZ EC	27.71	26.11	0.358
Alpha OZ EC	31.61	33.86	0.327
Psychologic alone			
Seashore rhythm test	19.68	22.77	0.493
Sentence-repetition test	10.90	13.81	0.410
Performance IQ	103.63	111.50	0.410
Token test	22.90	25.23	0.406
Combined EEG and psychologic			
Delta PZ EC	31.14	29.53	0.493
Alpha PZ EC	28.20	30.00	0.485
Seashore rhythm test	19.68	22.77	0.484
Verbal IQ	97.26	106.55	0.441

[a] Listed are the four best features selected from EEG variables alone, psychologic variables alone, and combined EEG and psychologic variables.
[b] Values for EEG features are percentage of total spectral voltage. Values for psychologic features are test results (see Reference 3 for details).
[c] Merit value is a relative measure of a feature's ability to discriminate between the high-lead and low-lead populations. It is calculated from three independent factors: d-prime, ambiguity value, and average correlation coefficient (see Methods for more detail). Merit value ranges from zero (no ability to discriminate) to one (complete ability to discriminate).
[d] Designation of EEG features is by frequency band, electrode location, and behavioral state during recording. Thus, Delta PZ EC indicates percentage of total spectral energy in the delta frequency band (0.5—3.5 Hz) for EEG recorded from the PZ electrode during the state of relaxed wakefulness with eyes closed.

group centroids ($p = 0.001$). By contrast, the significance of group separation was smaller (but still substantial) using EEG features alone (Mahalanobis distance, $p = 0.01$) or psychologic features alone (Wilks' lambda, $p = 0.015$; Mahalanobis distance, $p = 0.025$). Note that for EEG features alone, the total discriminatory power resided in a single variable; thus, Wilks' lambda is without meaning.

Discriminant analysis defines a boundary in multivariate space which optimally separates the two populations. Multivariate space, therefore, is divided into two population domains, and any subject whose feature vector falls into one of these domains would be classified as a member of that population. Obviously, the more significant the group separation, the fewer the errors that will be made in such classification. Complete separation would yield 100% accuracy of classification; whereas, two identical populations would result in classification at a chance level. Hence, the success rate of retrospective classification based on the discriminant function is another test of the significance of group separation for a set of features.

Retrospective classification of high- and low-lead children was significantly better than chance (Table 4). This was true for all categories of features. Based on EEG features alone, 70.7% of subjects were classified correctly. For psychologic features alone, accuracy was 68.3%. Combined psychologic and EEG features yielded the most successful classification, with 75.6% correct.

TABLE 2
Separation of High- and Low-Lead Groups by Linear Discriminant Analysis[a]

	Wilks' lambda	Significance of group separation by Rao's approximation	
		F(4,36)	p-Value
EEG alone	0.80	2.27	0.079
Psychologic alone	0.76	2.79	0.041
Combined EEG and psychologic	0.66	4.61	0.005

[a] Calculations were done separately for EEG variables alone, psychologic variables alone, and combined EEG and psychologic variables. In each category, input variables were the four best features determined by merit value analysis in Table 1.

TABLE 3
Separation of High- and Low-Lead Groups by Stepwise Discriminant Analysis[a]

Feature entered (F to enter)	Wilks' lambda (p-Value)	Significance of mahalanobis distance	
		F(df)	p-Value
EEG Alone			
Step 1 Delta PZ EC (8.33)	NA	8.33 (1,39)	0.010
Psychologic alone	0.76 (p = 0.015)	3.97 (3,37)	0.025
Step 1 Full-scale IQ (7.13)			
Step 2 Sentence-repetition test (2.85)			
Step 3 Seashore rhythm test (1.47)			
Combined EEG and psychologic	0.56 (p = 0.0002)	7.15 (4,36)	0.001
Step 1 Delta PZ EC (8.33)			
Step 2 Full-scale IQ (14.70)			
Step 3 Verbal IQ (1.27)			
Step 4 Performance IQ (1.12)			

[a] Calculations were done separately for EEG variables alone, psychologic variables alone, and combined EEG and psychologic variables. In each category, input variables were the four best features determined by merit value analysis in Table 1.

Note: df = degrees of freedom. NA = not applicable. Since total discriminatory power resides in a single feature, Wilks' lambda has no meaning.

C. PROSPECTIVE CLASSIFICATION OF SUBJECTS INTO HIGH- OR LOW-LEAD GROUP BASED ON EEG AND PSYCHOLOGIC VARIABLES

A better test of the discriminating power of a set of features is prospective classification. In retrospective classification one defines and tests a discriminant function using the same subjects. In prospective classification, on the other hand, rules are developed on one set of subjects (the training set) and tested on another set (the test set). Clearly, for prospective classification to succeed at a level greater than chance, the subjects must fall into two distinct clusters in multivariate space.

On the basis of the best combined EEG and psychologic features, we were able to classify test sets of children as high or low lead with an overall accuracy of 65.9%, well above the chance level (Table 5). Moreover, the additive discriminating power of EEG and

TABLE 4
Retrospective Classification of High- and Low-Lead Children Based on Stepwise Discriminant Analysis

Features	Percentage of children correctly classified		
	Low lead	High lead	All
EEG alone	68.2	73.7	70.7
Psychologic alone	72.7	63.2	68.3
Combined EEG and psychologic	77.3	73.7	75.6

TABLE 5
Prospective Classification of High- and Low-Lead Children Using Nonparametric Supervised Learning Programs and the Jackknifing Procedure[a]

Features	Percentage of children correctly classified		
	Low lead	High lead	All
EEG alone	36.4	63.2	48.8
Psychologic alone	50.0	31.5	41.5
Combined EEG and psychologic	59.1	73.7	65.9

[a] Input variables were the four best features determined by merit value analysis in Table 1.

psychologic features was clearly demonstrated by this procedure; using EEG features alone or psychologic features alone yielded only chance levels of prospective classification (48.8% and 41.5%, respectively). This indicates that the separation of high- and low-lead children into two respective clusters is much more distinct in a multivariate space consisting of both EEG and psychologic features than it is in one consisting of either type of feature alone.

IV. DISCUSSION

Conventional wisdom has been that lead exposure does not affect the EEG except at very high levels during acute intoxication or as an accompaniment of a chronic encephalopathy in which there is symptomatic evidence of neurological damage.[22-24] The present data, however, clearly demonstrate that EEG is altered as a function of low levels of lead exposure in children who are overtly asymptomatic. Using sensitive techniques of quantitative spectral analysis and topographic mapping we found significant EEG differences between normal school children with high and low dentine lead concentrations. The spontaneous, resting EEG of high lead children had higher percentages of low frequency delta (0.5 to 3.5 Hz) activity and reduced percentages of alpha (8.0 to 12.0 Hz) activity. These alterations occurred in a generalized spatial distribution over the scalp without evident focal features or significant interhemispheral asymmetry. Qualitatively, these EEG changes are similar to those seen with acute lead intoxication, and in general, such findings of diffusely increased slow frequency activity and reduced alpha are common findings in a variety of toxic encephalopathies.[25] These qualitative similarities suggest that lead may induce neuropathological effects along a continuum of exposure. There may be no absolute threshold for neuropathology: any exposure of the CNS to lead may produce neuronal alterations, and the magnitude of these alterations may accumulate quantitatively with increasing levels of exposure.

These EEG alterations we have demonstrated have not been reported previously in asymptomatic children. We believe that the major reason for this has been a lack of adequate sensitivity of previously employed EEG measures. Most early studies relied entirely on visual inspection of polygraphic EEG tracings. Such analysis is adequate for detection of gross alterations of EEG background resulting from severe encephalopathies or structural lesions or for the detection of paroxysmal events associated with epilepsy. However, visual inspection is very insensitive to more subtle shifts of frequency distribution within the EEG background. Indeed, two experienced electroencephalographers blindly reviewed the EEG tracings recorded from the children in the present study and were unable to classify them into high- and low-lead groups above a chance level. The vast majority of the EEGs were read as completely normal with only a few considered to have borderline or mildly abnormal findings of a nonspecific nature. These nonnormal findings were equally distributed between the high- and low-lead groups.

Quantitative spectral analysis significantly increases the sensitivity of EEG measures. However, in order to take full advantage of this sensitivity, such quantitative techniques must be applied carefully to EEG data. The frequency content of the human EEG exhibits significant spatial variation over the scalp; therefore, each electrode location will have a different, relatively unique, frequency spectrum. Consequently, an adequate definition of the EEG frequency spectrum requires that spectral analysis be done on samples of EEG from several electrode locations covering the scalp. The necessity for such spatial sampling can be seen by looking at the topographic maps of delta and alpha activity shown in Plates 1 and 2. It is evident from even a casual examination of these maps that no single electrode's data (or even those from a few electrodes) could yield an adequate representation of "the EEG spectrum". For example, the proportion of alpha in the EEG ranges from less than 25% in the anterior temporal areas to over 34% in the occipital area. Likewise, the percentage of delta in different leads ranges from 27 to 32%. Furthermore, the leads in which alpha is maximal are different from those in which delta is maximal. Most importantly, it is evident from the topographic maps of the t-statistic differences between the high- and low-lead groups (SPM maps in Plates 1 and 2) that alterations within the EEG spectrum did not occur uniformly at all electrode locations.

In general, different brain regions underlying the scalp electrodes vary significantly in the degree to which they exhibit alterations of their EEG background in response to either normal or pathological conditions. Furthermore, these regional variations are different for different portions of the EEG spectrum. Thus, adequate spatial sampling becomes mandatory if one hopes to measure EEG changes with high sensitivity. For example, in the present study had we sampled EEG only from frontal and temporal leads, we would not have detected any spectral differences between high- and low-lead children. Likewise, had we sampled only from parasagital central and parietal leads, we would have detected a small difference in the delta frequency band, but no difference in the alpha band.

We believe that this lack of adequate spatial sampling of EEG has been the major reason why other studies employing quantitative analytic techniques have failed to detect EEG alterations associated with low-level lead exposure. A notable case in point is the study of Benignus and Otto and their co-workers.[26] These authors examined EEG spectra from asymptomatic children with blood lead levels ranging from 7 to 59 μg/dl at the time of testing. They analyzed data from only two electrodes (parasagital parietal in each hemisphere) and found no correlation between spectral content and blood lead concentration. In our study we found that these two electrodes had only borderline discriminating power in the delta frequency band and showed no significant differences in the alpha band.

In contrast to the negative results with EEG spectra, Otto, et al.[27] were able to demonstrate a significant effect of lead in the same group of asymptomatic children using a classical conditioning paradigm. In order to test the very young children (13 to 75 months) in their study, they devised a passive analog of the foreperiod warning task commonly used to elicit a conditioned negative variation (CNV). This passive paradigm elicited a slow wave conditioned response which resembled the CNV, and the voltage of this slow wave varied as a linear function of blood lead concentration. Significantly, these children showed the same interaction of slow wave voltage and blood lead concentration in a 2-year follow-up study,[28] indicating a consistent and persistent effect of lead body burden on this electrophysiological parameter.

The limited spatial sampling of EEG, which most likely prevented the Otto and Benignus group from detecting an effect of lead body burden on EEG spectra, does not appear to have been a factor during the classical conditioning paradigm. The CNV elicited by this paradigm (and, presumably, the analogous conditioned slow wave elicited in their modified passive protocol) has a relatively restricted spatial distribution over the scalp with a distinct maximum occurring at the vertex. Fortunately, their data sampling during the conditioning task included midline frontal and central electrodes in addition to the parietal electrodes used for EEG spectral analysis. Thus, with a limited number of electrodes, they were able to adequately detect and characterize the conditioned slow wave.

The data of the Otto and Benignus group complement our results and provide independent evidence of significant electrophysiologic alterations resulting from low-level lead exposure. It is particularly notable that they were able to find the same lead-related effects at a 2-year follow-up. Recently, these investigators conducted another follow-up of the study group at 5 years.[29] These results contrasted with the earlier findings: they failed to detect a significant variation of slow wave voltage with blood lead levels. However, this lack of replication appears to reflect primarily an age-dependent decrease in the sensitivity of the paradigm. In the earlier studies, the slope of the linear relationship between slow wave voltage and blood lead concentration was found to vary systematically with age. In older children the slope progressively decreased, suggesting diminishing sensitivity with increasing age. Therefore, a likely explanation for the negative result at the 5-year follow-up is that the study group was outside the age range for which the passive conditioning paradigm is a useful discriminator. This interpretation is reinforced by the finding that the slow-wave voltage lost sensitivity to all factors: not only did it fail to vary with blood lead, but it also failed to show any continued variation with age.

The failure of the Otto and Benignus group to replicate their findings with the conditioned slow-wave paradigm in the 5-year follow-up study exemplifies the general problem of choosing sensitive measures for neurotoxicologic assessment. In the present study, we have increased our potential sensitivity to neuropathology by combining quantitative electrophysiologic measures with psychologic measures. Our EEG data complement the psychologic findings of Needleman et al.[3] In the same population of children, they found significant deficits in verbal behavior, attention, and classroom performance among children with high dentine lead levels. These psychologic data provide independent evidence of significant neurologic involvement at low levels of lead exposure. The independent and additive nature of EEG and psychologic features for defining the effects of low-level lead exposure was demonstrated clearly by the results of our multivariate and automated classification analyses. Greater significance of group separation and fewer errors of classification resulted from the use of a combination of EEG and psychologic features than from the use of either type of variable alone. These results indicate that each type of variable provided unique and independent information for the discrimination of high- and low-lead children.

In summary, our data, together with those of Needleman et al.,[3] lead to the following conclusions: (1) lead, at levels of absorption resulting from everyday environmental exposure, can alter the function of the developing brain; (2) in studies looking for effects of low levels of environmental contaminants in ostensibly asymptomatic subjects, it is essential to use highly sensitive measurement and analytical tools. For EEG measures, this includes adequate definition of the topographic spatial distribution of spectral frequency content. (3) Finally, in assessing CNS pathology, the sensitivity with which effects can be detected is likely to be enhanced significantly by using measures of both behavior and electrophysiology.

REFERENCES

1. **Niedermeyer, E. and Lopes da Silva, F.**, *Electroencephalography,* 2nd ed., Urban and Schwarzenberg, Baltimore, 1987, 579.
2. **Duffy, F. H., Burchfiel, J. L., and Lombroso, C. T.**, Brain electrical activity mapping (BEAM): a method for extending the clinical utility of EEG and evoked potential data, *Ann. Neurol.,* 5, 309, 1979.
3. **Needleman, H. L., Gunnoe, C., Leviton, A., Reed, R., Peresie, H., Maher, C., and Barrett, P.**, Deficits in psychologic and classroom performance of children with elevated dentine lead levels, *N. Engl. J. Med.,* 300, 689, 1979.
4. **Jasper, H. H.**, Report of committee on methods of clinical exam in EEG, *Electroenceph. Clin. Neurophysiol.,* 10, 370, 1958.
5. **Lindren, B. W.**, *Statistical Theory,* 2nd ed., MacMillan, New York, 1968, 436.
6. **Duffy, F. H., Burchfiel, J. L., and Bartels, P. H.**, Significance probability mapping: an aid in the topographic analysis of brain electrical activity, *Electroenceph. Clin. Neurophysiol.,* 51, 455, 1981.
7. **Bartels, P. H. and Wied, G. L.**, Extraction and evaluation of information from digitized cell images, in *Mammalian Cells: Probes and Problems,* Richmond, C. R., Peterson, D. F., Mullaney, P. F., and Anderson, E. C., Eds., ERDA, Washington, D.C., 1975, 15.
8. **Bartels, P. H. and Wied, G. L.**, Computer analysis and interpretation of microscopic images: current problems and future directions, *Proc. IEEE,* 65, 252, 1977.
9. **Wied, G. L., Bartels, P. H., Bahr, G. F., and Oldfield, D. C.**, Taxonomic intracellular analytic system (TICAS) for cell identification, *Acta Cytol.,* 12, 180, 1968.
10. **Foley, D. H.**, Consideration of sample and feature size, *IEEE Trans. Inf. Theory,* IT-18, 618, 1972.
11. **Selin, I.**, *Detection Theory,* Princeton University Press, Princeton, N.J., 1965.
12. **Sherwood, E. M., Bartels, P. G., and Wied, G. L.**, Feature selection in cell image analysis: use of the ROC curve, *Acta Cytol.,* 20, 254, 1976.
13. **Genchi, H. and Mori, K.**, Evaluation and feature extraction on automated pattern recognition system (in Japanese), *Denki-Tsuchin Gakkai,* Part 1, 1965.
14. **Tanaka, N., Ikeda, H., Ueno, T., Watanabe, S., Imasato, Y., and Kashida, R. K.** Fundamental study of automated cytoscreening system I., in *The Automation of Uterine Cancer Cytology,* Wied, G. L., Bahr, G. F., and Bartels, P. H., Eds., Tutorials of Cytology, Chicago, 1976, 223.
15. **Afifi, A. and Azen, S. P.**, *Statistical Analysis: A Computer Oriented Approach,* Academic Press, New York, 1972.
16. **Cooley, W. W. and Lohnes, P. R.**, *Multivariate Statistical Analysis,* Wiley, New York, 1971.
17. **Rao, C. R.**, *Advanced Statistical Methods in Biometric Research,* 2nd ed., Hafner Press, New York.
18. **Duda, R. D. and Hart, P. E.**, *Pattern Classification and Scene Analysis,* Wiley, New York, 339.
19. **Bartels, P. H. and Bellamy, J. C.**, Self-optimizing, self-learning system in pictorial pattern recognition, *Appl. Opt.,* 9, 2453, 1970.
20. **Bartels, P. H., Bahr, G. F., Bellamy, J. C., Bibbo, M., Richards, D. C., and Wied, G. L.**, A self-learning computer program for cell recognition, *Acta Cytol.,* 14, 486, 1970.
21. **Lachenbach, P. A.**, *Discriminant Analysis,* Hafner Press, New York, 1975, 36.
22. **Perstein, M. A. and Attalla, R.**, Neurologic sequelae of plumbism in children, *Pediatrics,* 5, 292, 1966.
23. **Seshia, S. S., Rajani, K. R., Boeckx, R. L., and Chow, P. N.**, The neurological manifestations of chronic inhalation of leaded gasoline, *Dev. Med. Chil. Neurol.,* 20, 323, 1978.
24. **Smith, H. D.**, Pediatric lead poisoning, *Arch. Environ. Health,* 8, 256, 1961.
25. **Markand, O. N.**, Electroencephalography in diffuse encephalopathies, *J. Clin. Neurophysiol.,* 1, 357, 1984.
26. **Benignus, V. A., Otto, D. A., Muller, K. E., and Seiple, K. J.**, Effects of age and body lead burden on CNS function in young children. II. EEG spectra, *Electroenceph. Clin. Neurophysiol.,* 52, 240, 1981.
27. **Otto, D. A., Benignus, V. A., Muller, K. E., and Barton, C. N.**, Effects of age and body lead burden on CNS function in young children. I. Slow cortical potentials, *Electroenceph. Clin. Neurophysiol.,* 52, 229, 1981.

28. **Otto, D., Benignus, V., Muller, K., Barton, C., Seiple, K., Prah, J., and Schroeder, S.,** Effects of low to moderate lead exposure on slow cortical potentials in young children: two year follow-up study, *Neurobehav. Toxicol. Teratol.*, 4, 7333, 1982.
29. **Otto, D., Robinson, G., Bauman, S., Schroeder, S., Mushak, P., Kleinbaum, D., and Boone, L.,** Five-year follow-up study of children with low-to-moderate lead absorption: electrophysiological evaluation, *Environ. Res.*, 28, 168, 1985.

Chapter 13

LEAD, BLOOD PRESSURE, AND CARDIOVASCULAR DISEASE

Joel Schwartz

TABLE OF CONTENTS

I. Introduction 224

II. Animal Data 224
    A. *In Vivo* Results 224
    B. *In Vitro* Results 224

III. Human Studies 226
    A. Recent Epidemiologic Studies of Lead and Blood Pressure 226

References 229

I. INTRODUCTION

Historically, the focus of studies of lead toxicity has been on neurotoxic effects in children. Recent research has documented health effects of lead at much lower levels than previously expected, and the pervasive nature of low-level lead toxicity in children, coupled with a growing understanding that the physiologic basis for those effects likely involves interference with basic metabolic mechanisms, has produced renewed interest in the health effects of low-level lead exposure in adults. High-level lead exposure has long been implicated in renal hypertension, but an accumulation of animal and human data has indicated that low-level exposure, probably through different mechanisms, produces small, but important increases in blood pressure. These studies have been extensively reviewed by the U.S. Environmental Protection Agency (EPA), and were the subject of an international conference jointly sponsored by the EPA, the Lead Industries Association, and the American Heart Association in 1987.[1]

II. ANIMAL DATA

Recent experimental evidence, both *in vivo* and *in vitro*, have demonstrated that moderate levels of lead cause elevations in blood pressure in animals. These findings will be discussed. Only lower-level exposure studies are discussed, as the higher exposures, often accompanied by lead poisoning or kidney damage, do not appear relevant to the possible effects of lead exposure in the general population.

A. *IN VIVO* RESULTS

Victery et al.[2] have demonstrated large, statistically significant elevations of systolic blood pressure in rats whose blood lead levels were 40 µg/dl, in comparison to controls. This effect seemed saturable, with rats at 70 µg/dl showing no effect. Iannaccone and co-workers[3] demonstrated the same effect at 38 µg/dl. Carmignani and colleagues[4] repeated these findings, and also reported elevations of diastolic blood pressure. Similar elevations were also reported by Webb and co-workers.[5] Iannaccone et al. and Carmignani et al. have both also reported increased response of their lead exposed rats to *in vivo* α-adrenergic stimulation. At significantly lower dietary lead levels, but unmeasured blood lead levels, both Perry and Erlanger[6] and Kopp[7] have reported blood pressure elevations in rats due to lead. Revis[8] has also reported elevations of blood pressure in pigeons fed lead, compared to controls. Calcium supplementation reduced the lead effect. This interaction with calcium is supported by the large effect at low dose found by Perry and Erlanger using a lower calcium diet than, for instance, Victery.

Boscolo and Carmignani[9] recently reported results for chronically exposed rats at lower blood lead levels. Four exposure regimes for 10 rats each produced mean blood lead levels 3.9, 7.4, 11.5, and 16.7 µg/dl. A dose dependent increase in blood pressure was noted, as shown in Figure 1. The hypotensive response to clonidine also showed a dose dependent increase, suggesting that lead modifies the central control of sympathetic tone.

Skoczynska et al.[10] reported that lead-treated rats had augmented and prolonged pressor responses to epinephrine and norepinephrine, and less pronounced depression of arterial pressure in response to isoproterenol.

Lead exposure has been reported to accelerate the development of hypertension in the spontaneously hypertensive rat,[11] and has also produced elevated blood pressure in dogs at a blood lead of 36 µg/dl.

B. *IN VITRO* RESULTS

Webb et al.[5] reported *in vitro* analysis of helical strips of smooth muscle lining the arteries of lead-exposed and control rats. They found that the lead-exposed tissue manifested

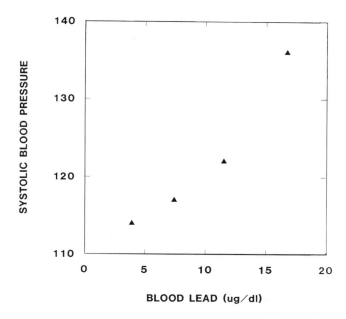

FIGURE 1. Each point represents the mean systolic blood pressure and mean blood lead level for an exposure group of 10 rats, using data from Boscolo et al.[1]

increased contractile response to α-adrenergic stimulation, similar to the *in vivo* findings of Iannaccone[3] and Carmignani.[4] Further, the time required to reach half maximal relaxation after stimulation was also significantly increased. Skoczynska et al.[10] also reported that in perfusion studies of isolated rat arteries, lead caused more pronounced vasoconstriction in response to norepinephrine. Piccinini et al.[12] reported that the infusion of lead into the media flowing through isolated tail arteries from unexposed rats produced a dose dependent contraction of the arterial muscle, restricting the flow of fluid through the artery. Subsequent analysis of the smooth muscle cells showed increased intracellular calcium stores, and the pressor effect was absent if extracellular calcium was not present, implicating a disturbance of calcium homeostasis in the cell as an essential feature of the lead effect. Since calcium is the second messenger for smooth muscle contraction, such a mode of action is also consistent with the increased sensitivity to α-adrenergic stimulation reported above.

Lead-induced increases in intracellular calcium concentrations have been noted in other studies. Goldstein[13] showed that 100 μM lead tripled calcium accumulation in the brain capillaries of rats. Kim[14] found evidence for increased calcium accumulation in brain cells, and Pounds[15,16] found clear evidence of increased calcium accumulation in all three compartments of hepatocytes. Interestingly, Pounds reported that at low doses, the lead-induced increases in intracellular calcium appeared uniform across all three compartments, while at high doses the smaller compartments became saturated, and the increase was predominantly in the deep compartment, thought to be the mitochondria.

Markovac and Goldstein[17] have demonstrated that lead activates protein kinase C at picomolar concentrations and lower. This is five orders of magnitude greater activation potency than calcium. The effect of lead was saturable, with a higher slope at lower concentrations. Chai and Webb[18] have noted that protein kinase C plays a role in smooth muscle contraction, and investigated the effect of lead on this system. They showed that isolated strips of rabbit arteries contracted in response to lead in the muscle bath, in concordance with previous findings in the rat. This contractile response was increased by pretreatment with a protein kinase C activator, and reduced when pretreated by a selective inhibitor of protein kinase C. Lead has also been shown to be more potent than calcium in activating calmodulin,[19] which through phosphodiesterase, modulates the concentration of cyclic AMP (cAMP) in the cell. cAMP can affect muscle contraction by phosphorylating myosin light chain kinase.

Taken together, these experimental studies suggest that lead leads to a disturbance of intracellular calcium metabolism in the vascular smooth muscle. This disturbance appears to involve both direct increases in intracellular calcium stores and the activation by lead of calcium dependent proteins. Since calcium is the messenger for contraction in smooth muscle, increases in intracellular calcium may result in increased resting tone, hyper reactivity, or both. Reduced ability to exclude calcium, possibly due to increased mitochondria stores, or slowed uptake[20] could explain the slowed relaxation time after stimulation.

III. HUMAN STUDIES

High levels of lead exposure have been reported to be associated with hypertension since Lorimer's paper in 1886.[21] These high exposures were almost certainly accompanied by renal damage, which likely accounted for the effect on blood pressure. At more recent occupational exposure levels, studies have yielded mixed results. Small sample sizes and inadequate control of covariates confound many of these studies. Seniority rules often lead to the assignment of older workers to lower exposure jobs, creating a confounding with age, for example.

Recently, epidemiologic studies in populations with low to moderate exposure to lead have more consistently identified a positive association between blood lead levels and blood pressure. These lead levels, while possibly sufficient to modify the renin-angiotensin system, are not currently associated with significant renal impairment, and the mechanisms for the effects of these exposures on blood pressure are likely to be quite different than the high exposure effects. Possible mechanisms were previously discussed.

A. RECENT EPIDEMIOLOGIC STUDIES OF LEAD AND BLOOD PRESSURE

Khera et al.[22] compared 38 patients being treated in a Birmingham, U.K. hospital for cardiovascular disease to 48 hospital controls with no known cardiovascular disease. Both blood lead levels and 24 h urinary lead levels were substantially higher in the cardiovascular patients.

Kirkby and Gyntelberg[23] compared 96 smelter workers to controls matched on age, sex, weight, height, socio-economic status, alcohol consumption, and tobacco use. Mean diastolic blood pressure was 5 mmHg higher in exposed subjects, who had mean blood lead levels of 51 µg/dl compared to controls, with mean blood lead levels of 11 µg/dl. Of the smelter workers, 20% had abnormal resting electrocardiograms (ischemic changes under the Minnesota codes) versus only 6% in the controls.

Moreau et al.[24] and Orssaud et al.[25] reported results of a study of 431 civil servants in Paris, France. Significant associations between both systolic and diastolic blood pressure and blood lead were found, after controlling for age, alcohol, tobacco, and weight. The mean blood lead level in this population was about 18 µg/dl, and the relationship was strongest at the lower blood lead levels.

Kromhout and co-workers[26,27] reported a significant association between blood lead and blood pressure in a cohort of 152 elderly men in the Netherlands, again with a mean blood lead of about 18 µg/dl.

Pocock et al.[28,29] in reports from the British Regional Heart Study, found significant associations between blood lead levels and both systolic and diastolic blood pressure in 7371 men aged 40 to 59 years. Mean blood lead levels were about 15 µg/dl in this study.

Harlan et al.[30] and Pirkle et al.[31] reported significant associations between blood lead and blood pressure in males in the Second National Health and Nutrition Examination Survey (NHANES II). These analyses held across the age range 12 to 74[30] and for the subgroup 40 to 59 years old.[31] Later analyses[32,33] reported significant associations in males and females aged 20 to 74 years.

The Pirkle et al.[31] and Schwartz[32] papers considered multiple functional forms, and found the best fit was with the natural logarithm of blood lead, implying a relationship with a higher slope at lower blood lead levels. This is consistent with the findings of Orssaud[25] and Pocock.[29] NHANES II was noteworthy for the large number of potential covariates available, and Pirkle et al.[31] and Schwartz[32] both report a very stable blood lead coefficient even when insignificant covariates are included in the model.

Weiss and co-workers[34,35] in a longitudinal study of policemen exposed to higher levels of lead reported that blood lead levels in excess of 30 μg/dl were associated with subsequent increase in blood pressure.

In 1987, Parkinson et al.[36] reported a positive, but significant association between blood lead and blood pressure in an occupational case control study. In contrast, DeKort and Zwinnis[37] found a significant association in another occupational study.

Elwood and colleagues[38,39] reported results from two general population surveys in Wales. Lead was not a significant predictor of blood pressure in either survey. The signs of the correlations were positive however.

Neri et al.[40] reported results from two studies. The first was the Canadian Health Survey, where a marginally significant to marginally insignificant association between diastolic blood pressure and blood lead was reported, depending on the model used. This study is complicated by the large number of blood lead measurements reported as "zero", however. Neri also reported on a longitudinal study of foundry workers, where changes in blood lead levels were significantly associated with changes in blood pressure.

Sharp and co-workers[41] reported an analysis of medical data from San Francisco bus drivers. The significance of the association varied depending on the model. However, a bootstrap analysis, which is more robust than normal regression, found a significant association with diastolic blood pressure, controlling for covariates.

Rabinowitz and co-workers[42] reported an association between umbilical cord lead levels and both blood pressure during delivery and the presence of pregnancy hypertension in a large sample of women with mean blood lead levels below 10 μg/dl.

Finally, Grandjean and colleagues[43] reported an initially significant association between lead and blood pressure which became insignificant after controlling for cofactors.

When drawing conclusions from multiple studies, simple "vote counting" of significance levels is less robust statistically, and less convincing epidemiologically, than examination of the concordance of the effect sizes. To address this, the estimated change in blood pressure for a reduction in mean blood lead levels from 10 μg/dl to 5 μg/dl was computed from those studies with sufficient information. This reduction represents a plausible estimate of the achievable reduction in adult blood lead levels from those prevailing in the early 1980s, due to the elimination of lead from gasoline, reduction of lead in food, and control of lead in drinking water. Figure 2 shows the results for diastolic blood pressure, and Figure 3 for systolic blood pressure in males. Fewer studies are available for women, and they show somewhat smaller effect sizes. These figures show reasonably consistent results across the 10 studies, suggesting that the 5 μg/dl reduction in blood lead levels would reduce diastolic blood pressure by 1 mmHg in males, with a slightly larger effect for systolic. The average effect size in females is about 0.6 mmHg.

Obviously lead is not the principle determinant of blood pressure in the U.S. However, cardiovascular and cerebrovascular disease, which are principally determined by blood pressure, is responsible for approximately 50% of all U.S. mortality. Hence small changes in blood pressure can produce large attributable risks, even when the relative risk is small. For

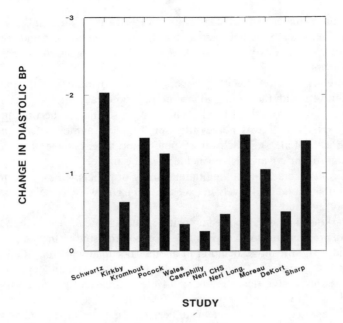

FIGURE 2. Estimated reduction in diastolic blood pressure for a 5 µg/dl reduction in blood lead level, by study.

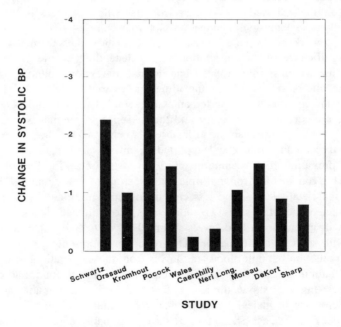

FIGURE 3. Estimated reduction in systolic blood pressure for a 230 g/dl reduction in blood lead level, by study.

example, the Hypertension Detection and Followup Study found a 5 mmHg reduction in diastolic blood pressure was associated with a 20% reduction in follow-up mortality. Moreover Framingham and other cardiovascular studies have shown that the risk of cardiovascular disease varies continuously with blood pressure, and does not begin only above 90 mmHg. Viewed in this light, a 1 mmHg reduction in the mean blood pressure of the entire adult population may produce substantial cardiovascular benefits.

The inference that increase in blood pressure due to lead will increase cardiovascular risk makes physiologic sense, and is supported indirectly by the prospective cardiovascular studies such as Framingham. A few papers address this more directly. Voors et al.[44] in a study of a North Carolina autopsy series, reported a highly significant association between tissue lead levels and whether the death was from cardiovascular disease. Kirkby and Gyntelberg,[23] as noted above, reported a significant association between electrocardiogram abnormalities and lead. Schwartz[33] found a significant association between blood lead and left ventricular hypertrophy in men and women in NHANES II, as diagnosed from electrocardiograms. Finally, Pocock[29] reported that the risk of Ischemic heart disease was about 10% higher in subjects in the British Regional Heart Study who had blood lead levels above 12 μg/dl, compared to those with blood leads below that level. These differences were not statistically significant. However, the number of events in the study was quite small, and from the blood pressure differences one would not expect a larger relative risk.

Overall, the animal and human evidence is consistent with a small, but important effect of lead on blood pressure, and with that effect having the expected cardiovascular consequences. In 1987 a conference jointly sponsored by the American Heart Association, the U.S. EPA, and the Lead Industries Association reviewed the evidence and reached a similar conclusion.[1]

REFERENCES

1. **Victery, W., Ed.,** Symposium on Lead-Blood Pressure Relationships, *Environ. Health Perspect.*, 78, 1988.
2. **Victery, W., Vander, A. J., Markel, H., Katzman, L., Shulak, J. M., and Germain, C.,** Lead exposure begun *in utero* decreases renin and angiotensin II in adult rats, *Proc. Soc. Exp. Biol. Med.*, 170, 62—67, 1982.
3. **Iannaccone, A., Carmignani, M., and Boscolo, P.,** Cardiovascular reactivity in the rat following chronic exposure to cadmium and lead, *Ann. 1st Super Sanita*, 17, 655—660, 1981.
4. **Carmignani, M., Boscolo, P., Ripanti, G., and Finalli, V. N.,** Effects of chronic exposure to lead and/or cadmium on some neurohumoral mechanisms regulating cardiovascular function in the rat, *Proc. 4th International Conf. on Heavy Metals in the Environment*, CEP Consultants, Edinburgh, 1983.
5. **Webb, R. C., Winquist, R. J., Victery, W., and Vander, A. J.,** *In vivo* and *in vitro* effects of lead on vascular reactivity in rats, *Am. J. Physiol.*, 214, H211—216, 1981.
6. **Perry, H. M., Jr. and Erlanger, M. W.,** Pressor effects of chronically feeding cadmium and lead together, in *Trace Substances in Environmental Health -XII*, Proceedings of the University of Missouri's 12th Annual Conference on Trace Substances in Environmental Health, Hemphill, D. D., Ed., University of Missouri, Columbia, MO, 1978, 268—275.
7. **Kopp, S. J., Barany, M., Erlanger, M., Perry, E. F., and Perry, H. M., Jr.,** The influence of low level cadmium and/or lead feeding on myocardial contractility related to phosphorylation of cardiac myofibrillar proteins, *Toxicol. Appl. Pharmacol.*, 54, 48—56, 1980.
8. **Revis, N. W., Zinsmeister, A. R., and Bull, R.,** Atherosclerosis and hypertension induction by lead and cadmium ions: an effect prevented by calcium ion, *Proc. Nat. Acad. Sci. (U.S.A.)*, 78, 6494—98, 1981.
9. **Boscolo, P. and Carmignani, M.,** Neurohumoral blood pressure regulation in lead exposure, *Environ. Health Perspect.*, 78, 101—106, 1988.
10. **Skoczynska, A., Juzwa, W., Smolik, R., Szechinski, J., and Behal, F. J.,** Response of the cardiovascular system to catecholamines in rats given small doses of lead, *Toxicology*, 39, 275—289, 1986.
11. **Evis, M. J., Dhaliwal, K., Kane, K. A., Moore, M. R., and Parratt, J. R.,** The effects of chronic lead treatment and hypertension on the severity of cardiac arrhythmias induced by coronary artery occlusion or by noradrenalin in anesthetized rats, *Arch. Toxicol.*, 59, 336—340, 1987.
12. **Piccinini, F., Favalli, L., and Chiari, M. C.,** Experimental investigations on the contraction induced by lead in arterial smooth muscle, *Toxicology*, 8, 43—51, 1977.
13. **Goldstein, G. W.,** Lead encephalopathy: the significance of lead inhibition of calcium uptake by brain mitochondria, *Brain Res.*, 136, 185—88, 1977.
14. **Kim, C. S., O'Tuama, L. A., Cookson, S. L., and Mann, J. D.,** The effects of lead poisoning on calcium transport by the brain in 30-day old albino rabbits, *Toxicol. Appl. Pharmacol.*, 52, 491—96, 1980.

15. **Pounds, J. G., Wright, R., and Kodell, R. L.**, Effect of lead on calcium homeostasis in the isolated rat hepatocyte, *Toxicol. Appl. Pharmacol.*, 63, 389—401, 1982.
16. **Pounds, J. G.**, Effect of lead intoxication on calcium homeostasis and calcium mediated cell function: a review, *Neurotoxicology*, 5, 295—332, 1984.
17. **Markovac, J. and Goldstein, G. W.**, Picomolar concentrations of lead stimulate brain protein kinase C, *Nature*, 334, 71—73, 1988.
18. **Chai, S. and Webb, R. C.**, Effects of lead on Vascular Reactivity, *Environ. Health Perspect.*, 78, 85—89, 1988.
19. **Goldstein, G. W. and Ar, D.**, Lead activates calmodulin sensitive processes, *Life Sci.*, 33, 1001—1006, 1983.
20. **Kapoor, S. C. and van Rossum, G. D. V.**, Effects of Ph^{++} added *in vitro* on Ca movements in isolated mitochondria and slices of rat kidney cortex, *Biochem. Pharmacol.*, 1984.
21. **Lorimer, G.**, Saturnine gout, and its distinguishing marks, *Br. Med. J.*, 2(1334), 163, 1886.
22. **Khera, A. K., Wibberley, D. G., Edwards, K. W., and Waldron, H. A.**, Cadmium and lead levels in blood and urine in a series of cardiovascular and normotensive patients, *Int. J. Environ. Stud.*, 14, 309—312, 1980.
23. **Kirkby, H. and Gyntelberg, F.**, Blood pressure and other cardiovascular risk factors of long term exposure to lead, *Scand. J. Work Environ. Health*, 11, 15—19, 1985.
24. **Moreau, T., Orssaud, G., Juget, B., and Busquet, G.**, Blood lead levels and arterial pressure: initial results of a cross sectional study of 431 male subjects, *Rev. Epidemiol. Sante Publique*, 30, 395—397, 1982.
25. **Orssaud, G., Claude, J. R., Moreasu, T., Tellouch, J., Juget, B., and Festy, B.**, Blood lead concentrations and blood pressure, *Br. Med. J.*, 290, 244, 1985.
26. **Kromhout, D., Wibowo, A. A. E., Herber, R. F. M., Dalderup, L. M., Heerduink, H., de Lezenne Coulander, C., and Zielhuis, R. L.**, Trace metals and coronary heart disease risk indicators in 152 elderly men (the Zutphen study), *Am. J. Epidemiol.*, 122, 378—385, 1985.
27. **Kromhout, D.**, Blood lead and coronary heart disease risk among elderly men in Zutphen, the Netherlands, *Environ. Health Perspect.*, 78, 43—46, 1988.
28. **Pocock, S. J., Shaper, A. G., Ashby, D., and Delves, T.**, Blood lead and blood pressure in middle-aged men, in *Int. Conf. Heavy Metals in the Environment*, CEP consultants, Edinburgh, U.K., 1985, 303—305.
29. **Pocock, S. J., Shaper, A. G., Ashby, D., Delves, H. T., and Clayton, B. E.**, The relationship between blood lead, blood pressure, stroke, and heart attacks in middle aged British men, *Environ. Health Perspect.*, 78, 23—30, 1988.
30. **Harlan, W. R., Landis, J. R., Schmouder, R. L., Goldstein, N. G., and Harlan, L. C.**, Blood lead and blood pressure: relationship in the adolescent and adult U.S. population, *J. Am. Med. Assoc.*, 253, 530—534, 1985.
31. **Pirkle, J. L., Schwartz, J., Landis, Jr., and Harlan, W. R.**, The relationship between blood lead levels and blood pressure and its cardiovascular risk implications, *Am. J. Epidemiol.*, 121, 246—258, 1985.
32. **Schwartz, J.**, The relationship between blood lead and blood pressure in the NHANES II survey, *Environ. Health Perspect.*, 78, 15—22, 1988.
33. **Schwartz, J.**, Blood lead, blood pressure, and cardiovascular disease in men and women, *Environ. Health Perspect.*, 1990, in press.
34. **Weiss, S. T., Munoz, A., Stein, A., Sparrow, D., and Speizer, F. E.**, The relationship of blood lead to blood pressure in a longitudinal study of working men, *Am. J. Epidemiol.*, 123, 800—808, 1986.
35. **Weiss, S. T., Munoz, A., Stein, A., Sparrow, D., and Speizer, F. E.**, The relationship of blood lead to systolic blood pressure in a longitudinal study of policemen, *Environ. Health Perspect.*, 78, 53—56, 1988.
36. **Parkinson, D. K., Hodgson, M. J., Bromet, E. J., Dew, M. A., Connell, M. M.**, Occupational lead exposure and blood pressure, *Br. J. Ind. Med.*, 44, 744—748, 1987.
37. **de Kort, W. L. A. M. and Zwennis, W. C. M.**, Blood lead and blood pressure: some implications for the situation in the Netherlands, *Environ. Health Perspect.*, 78, 67—70, 1988.
38. **Elwood, P. C., Davey-Smith, G., Oldham, P. D., and Toothhill, C.**, Two Welsh surveys of blood lead and blood pressure, *Environ. Health Perspect.*, 78, 119—121, 1988.
39. **Elwood, P. C., Yarnell, J. W., Oldham, P. D., Catford, J. C., Nutbeam, D., Davey-Smith, G., and Toothhill, C.**, Blood pressure and blood lead in surveys in Wales, *Am. J. Epidemiol.*, 127, 942—945, 1988.
40. **Neri, L. C., Hewitt, D., and Orser, B.**, Blood lead and blood pressure: analysis of cross sectional and longitudinal data from Canada, *Environ. Health Perspect.*, 78, 123—126, 1988.
41. **Sharp, D. S., Osterloh, C. E., Becker, B., Bernard, B., Smith, A. H., Fisher, J. M., Syme, S. L., Holman, B. L., and Johnston, T.**, Blood pressure and blood lead concentration in bus drivers, *Environ. Health Perspect.*, 78, 131—138, 1988.

42. **Rabinowitz, M., Bellinger, D., Leviton, A., Needleman, H., and Schoenbaum, S.,** Pregnancy hypertension, blood pressure during labor, and blood lead levels, *Hypertension,* 10, 447—451, 1986.
43. **Grandjean, P., Hollnagel, H., Hedegaard, L., Christensen, J. M., and Larsen, S.,** Blood lead-blood pressure relations: alcohol intake and hemoglobin as confounders, *Am. J. Epidemiol.,* 129, 732—4, 1989.
44. **Voors, A. W., Johnson, W. D., and Shuman, M. S.,** Additive statistical effects of cadmium and lead on heart-related disease in a North Carolina autopsy series, *Arch. Environ. Health,* 37, 98—102, 1982.

Chapter 14

LOW LEVEL HEALTH EFFECTS OF LEAD: GROWTH, DEVELOPMENTAL, AND NEUROLOGICAL DISTURBANCES

Joel Schwartz

TABLE OF CONTENTS

I. Effects of Lead on the Growth and Development of Children 234
    A. Epidemiological Studies 234
        1. Correlation Between Lead and Stature 234
        2. Correlation Between Lead, Birth Weight, and Gestational Age 234
        3. Correlation Between Lead and Growth Rates 236
        4. Correlation Between Lead and Gestational Age 237
        5. Lead and Congenital Anomalies 237
        6. Fetal Death 237
    B. Metabolic Studies 238
        1. Correlation Between Lead and Vitamin D Metabolism 238
        2. Correlation Between Lead and Pituitary-Thyroid Function 238
        3. Interaction of Lead with Iron Deficiency 238

II. Other Neurological Effects of Lead 239
    A. Correlation Between Lead and Hearing 239
    B. Lead Induced Peripheral Neuropathy 239

III. Conclusions 239

References 240

I. EFFECTS OF LEAD ON THE GROWTH AND DEVELOPMENT OF CHILDREN

While cognitive effects have long been central to the field of lead toxicity, recent studies have examined the effects of low-level lead exposure on the growth and development of children. Simultaneously, increased attention has been given to the fetus as a target of low-level lead exposure. The effects studied include stature, growth rates, hormonal metabolism, and heme synthesis in children, and birthweight, gestational age, and congenital anomalies in the fetus. The overall pattern of these studies suggests that any threshold for the effects of lead on the fetus or young children is so low as to be inconsequential. While these effects are far less severe than the encephalopathy present in acute lead poisoning, the blood lead levels at which they occur indicate that large segments of the population are affected. Programs aimed solely at preventing high level exposure run the risk of allowing important, although more moderate, health impairment to occur in orders of magnitude more children than those suffering the profound disturbances of high level lead exposure. These new findings are discussed in more detail below.

A. EPIDEMIOLOGICAL STUDIES
1. Correlation Between Lead and Stature

Short stature has been associated with lead poisoning since the 1920s in Australia,[1] and more recently in asymptomatic cases in the U.S.[2,3] However, these studies focused primarily upon relatively high levels of lead exposure.

In early 1986, Schwartz et al.[4] published cross-sectional analyses of data from the second National Health and Nutrition Examination Survey (NHANES II), showing a relationship between children's blood lead levels and their stature. This survey was a representative sample of the U.S. population, and the study covered 2695 children aged 6 months to 7 years. Nutritional intake was obtained from a diet recall and a nutritional data bank. This allowed control for 15 nutritional factors, as well as hematocrit, transferrin saturation, and socio-economic factors. This relationship is illustrated in Figures 1 and 2, after controlling for all the other significant variables.

No threshold for the relationship was found down to the lowest observed levels of blood lead (4 μg/dl). At the mean age of the children studied (59 months), the mean blood lead level of the children was associated with a reduction of about 1.5% below the height that would be expected if their blood leads had been zero. At 25 μg/dl, a reduction in height of about 3% appeared to have occurred. By itself, a cross-sectional epidemiological study cannot definitively establish a causal link. However, these findings have stimulated longitudinal studies as well as animal experiments, that will be discussed, that together form a relatively consistent picture.

2. Correlation Between Lead, Birth Weight, and Gestational Age

The association between lead and stature may well begin with fetal exposure. A number of recent studies have addressed this issue. Dietrich et al.,[5] reporting on the Cincinnati prospective study of lead exposure, found an association between maternal-blood lead levels during pregnancy and reduced birth weight. They also found an association between maternal-blood lead levels and reduced gestational age. By the use of structural equation modeling, they were able to demonstrate that the effect of lead on birth weight was both indirect, through the reduction in gestational age, and also had a direct component, after controlling for gestational age. Recently, Bornschein et al.[6] have reported further analysis of this data. The new results confirm the earlier findings, and add the interesting additional observation that the negative effect of maternal alcohol use on birthweight is not additive with lead. The total effect exceeds the effect of either exposure, but by less than an additive amount. They

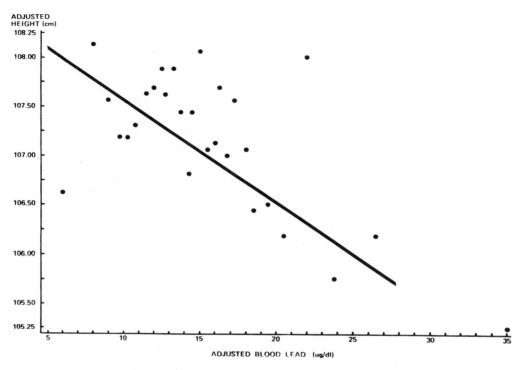

FIGURE 1. This figure plots the mean height and mean blood lead level for children aged 7 years or less in the National Health and Nutrition Examination Survey, after adjusting for effects of age, race, sex, and nutritional intake. Each point represents the mean of approximately 100 children, ranked in increasing order of blood lead.

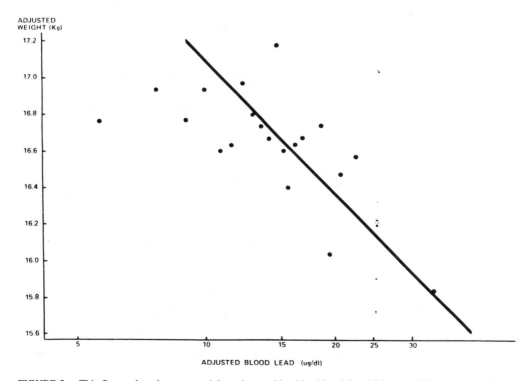

FIGURE 2. This figure plots the mean weight and mean blood lead level for children aged 7 years or less in the National Health and Nutrition Examination Survey, after adjusting for effects of age, race, sex, and nutritional intake. Each point represents the mean of approximately 100 children, ranked in increasing order of blood lead.

have hypothesized that in nutritionally sufficient pregnancies, there is a natural limit to the negative impact of environmental factors, producing the nonlinearity.

In their study, Dietrich et al.[5] reported that a change of 1 in blood lead on the log scale (e.g., from 3 to 8 µg/dl) was associated with a 0.6-week decrease in gestational age. That decrease produced an estimated 46 g decrease in birth weight, and in addition, a 180 gm reduction in birth weight was associated with that lead exposure through the direct route. The mean maternal-blood lead level in this cohort was 8.3 µg/dl, with a standard deviation of 3.8 µg/dl, so these effects are occurring at extremely low blood lead levels.

Bryce-Smith[7] has recently reported an association between both birth weight and head circumference and placental lead levels in a cohort of 100 normal infants. Ward et al.[8] have recently reported more detail from this study. Placental lead levels in children with birth weights under 2500 g were almost triple those in children above 4000 g. Placental lead correlated better than maternal blood lead levels with these outcome.

Fahim et al.[9] also reported higher blood lead levels in women with premature delivery than in women with full-term pregnancies, although covariate control was minimal.

McMichael et al.[10] have reported the results of a prospective study in Australia, in which they found a higher incidence of birth weights under 2500 g in their high-lead area than in their low-lead area. However, blood lead levels at delivery were actually lower, although insignificantly so, in the mothers of the low birth weight children. One possible explanation of this phenomenon is the role of the fetus as a sink for maternal lead. Ong et al.[11] have shown that in 25% of their cases, cord-blood lead levels exceeded maternal-blood lead at delivery, suggesting that such gradients can exist. In those cases, low maternal-blood lead levels may be indicative of increased fetal exposure.

Moore et al.[12] also reported a significant association between blood lead levels and reduced gestational age in a study in Glasgow, Scotland.

Ernhart and co-workers[13] did not find any association between lead and birthweight in a study of 185 children, half of whom had alcoholic and/or drug abusing mothers. It is often difficult to assess the impact of lesser factors in the presence of such major perturbers of health, however.

Bellinger et al.[14] also reported an exposure related trend in small for gestational age infants, but no significant association with birth weight was found in their study. Again, almost all of the subjects had blood lead levels below 20 µg/dl. Taken together, these studies present a qualitatively coherent picture of an association between blood lead levels and reduced birth weight even at blood lead levels well below the current CDC guidelines, and indeed usually at half that level.

3. Correlation Between Lead and Growth Rates

Two recent prospective studies on children with low blood lead levels have examined the effects of lead on postnatal growth. Shukla et al.[15,16] followed a cohort of 260 infants from the Cincinnati study to examine growth between ages 3 and 15 months. He found an interactive effect between postnatal and prenatal exposure. High prenatal exposure suppressed growth, but this effect was reversible if high postnatal exposure was avoided. High pre- and postnatal exposure had the strongest effect. Unfortunately, the term "high" in this context refers to exposure above the median prenatal exposure of 7.7 µg/dl, a quite low level. In children whose prenatal exposure was above the median, postnatal increases in blood lead levels were strongly associated with slow growth. There was a 2 cm difference in stature at age 15 months between those infants who averaged 3 µg/dl during the previous year and those who averaged 15 µg/dl during that period.

Similarly, Lyngbye et al.[17] reported on a population based study of school children in Aarhus, Denmark. Children were classified into high and low groups based on lead concentrations in circumpulpal dentine. The children were measured three or four times between ages 6 and 10 years. Dietary factors were evaluated and data on maternal smoking during pregnancy obtained, as were socio-economic factors and medical histories. After considering all of these factors, lead remained associated with lower growth in this population, whose mean blood lead level was below 6 μg/dl.

Amgle and Kuntzelman[18] have reported that a group of children with blood lead levels above 30 μg/dl showed growth retardation compared to a control group from ages 24 months to 42 months.

While many animal studies have shown reduced weight or length gain in young animals following high-level lead exposure, the lead levels were generally high enough to inhibit dietary intakes. Grant et al.[19] in a study that avoided that problem, did find growth retardation in young rats following lead exposure. Recently Burger and Gochfeld[20] reported that single doses of 0.1 mg Pb/g (as lead nitrate) produced significant reductions in growth rates for herring gulls. These differences persisted until the end of development. The authors reported a similar finding for terns.[21] High doses of lead have also been shown to retard growth in chicks.[22] Hammond and co-workers[23] in a paired feeding study, have shown that lead reduces growth in rats, with part of the reduction coming from appetite suppression, and part existing even at the same caloric input.

4. Correlation Between Lead and Gestational Age

As noted above, the Cincinnati study found a significant association between maternal lead exposure and gestational age. In addition, the McMichael et al.[10] study found a significant association between maternal lead levels and pre-term delivery. In the study, mothers with blood lead levels above 14 μg/dl had more than a fourfold relative risk of premature delivery compared to mothers with blood lead levels below 8 μg/dl. A recent study by Savitz and co-workers[24] using the National Natality and Fetal Mortality survey, a probability sample of live births in the U.S., reported an odds ratio of 2.3 for preterm birth for women with occupational lead exposure.

5. Lead and Congenital Anomalies

The assessment of a relationship between lead and congenital anomalies is difficult, because the prevalence of malformations is low. Therefore, large sample sizes are required to detect effects. Nonetheless, three studies have looked for an association between lead levels and malformations. Needleman et al.[25] reported that lead was associated with increased relative risk of minor malformations of all kinds, with a relative risk at cord blood lead levels of 15 μg/dl more than twice that for 0.7 μg/dl. This study, with a sample size of over 4000, had a relatively high power, particularly because all malformations were grouped together. It does lose specificity because of that grouping, making interpretation more difficult. On the other hand, McMichael et al.,[10] in their study of 774 subjects, did not find any association with congenital anomalies. Ernhart et al.,[13] studying 185 subjects, also failed to find an effect. It is not clear whether the difference between the studies reflects primarily sample size, or a contrary finding, since the insignificant studies have not published regression coefficients to compare to those of Needleman et al. No definitive conclusion can be drawn under these circumstances.

6. Fetal Death

When one identifies a toxin associated with reduced fetal growth and/or gestational age, and which may be associated with congenital anomalies, one must consider the possibility of an association with fetal loss. High doses of lead were used as an abortificant during the beginning of this century. Less data is available at the doses currently of concern. Because of the low rate of identified fetal loss, the prospective lead studies have not had sufficient sample size to assess the issue, and case-control studies are required. Wibberley et al.[26]

found that lead levels in the placenta from still births or neonatal deaths were higher than those from the placenta of normal births. Bryce-Smith and colleagues[7] reported bone lead concentrations in still births were higher than in bone from infants who died from accidents and other causes. Savitz et al.,[24] reporting from the National Natality and Fetal Mortality Survey, found maternal occupational exposure was associated with an odds ratio of 1.6 for stillbirth. While the lack of precision in exposure assessment suggests caution in interpreting these results, they gain considerable support from the evidence of fetal growth disturbance cited above. Overall, the evidence for a causal link between lead exposure and stillbirths is highly suggestive, but not yet fully convincing.

B. METABOLIC STUDIES
1. Correlation Between Lead and Vitamin D Metabolism

Rosen[27] and Mahaffey[28] have reported strong correlations between blood lead levels and circulating levels of 1,25-$(OH)_2$ vitamin D in children. These correlations are seen across the whole range of blood lead values from 12 to 120 $\mu g/dl$. They appear to result from lead inhibiting the hydroxylation of the 25 to the active 1,25 form of the vitamin. At 35 $\mu g/dl$, the suppression of vitamin D activity reaches levels comparable to those seen in severe kidney dysfunction and several genetic disorders.[29] The consequent disturbance in calcium metabolism that is produced by lead may relate to the growth and developmental effects of lead. Because calcium serves as the second messenger for numerous cellular metabolic processes, these results, as well as the findings of Habermann and co-workers[30] that lead can replace calcium in activating calmodulin, suggest a more widespread disturbance in metabolism in the exposed child.

2. Correlation Between Lead and Pituitary-Thyroid Function

Several studies have suggested that lead impairs the pituitary-thyroid endocrine system in a manner that may be related to its effect on growth. At high lead levels Sandstead[31] has shown a lead induced impairment of the iodine-concentrating mechanism by lead in rats and in men. The effect was reversible upon injection of thyroid stimulating hormone (TSH), a pituitary hormone, suggesting the effect was mediated by suppression of TSH. More recently, Huseman et al.[32] reported that two lead intoxicated children had decreased TSH release in response to TRH. They incubated rat pituitary cells, and found that those exposed to lead again showed decreased TSH release in response to TRH. This interference in pituitary-thyroid functioning may relate to the correlation between lead exposure and growth. Tuppurainen and colleagues,[33] in a regression analysis of lead workers, reported significant negative associations between duration of lead exposure and both total and free thyroxine in serum.

3. Interaction of Lead with Iron Deficiency

Two recent analyses of the NHANES II data indicate that low-level lead exposure can interact with iron deficiency, yielding enhanced effects and effects at lower levels in iron deficient children. Recently, Mahaffey and Annest[34] analyzed the relationship between blood lead levels and free erythrocyte protoporphyrin levels (FEP). They reported greater increases in the proportion of children with elevated FEP levels as blood lead increased, when the children were iron deficient (transferrin saturation below 16%).

Marcus and Schwartz[35] have also analyzed this data, fitting a toxicokinetic model to the relationship between blood lead and FEP, stratifying on iron status. Their model allows the prediction of the concentration of lead in plasma and body fluid, as opposed to simply in the blood, where over 95% of the lead is bound to the erythrocyte and less toxicologically available. They also report an interactive effect with iron status. In children with transferrin saturation below 14%, the threshold for lead elevating FEP levels is at 12 $\mu g/dl$; for high iron children, it did not begin until 23 μg/dl. These results should be compared to Piomelli et al.[36] threshold of 17 μg/dl for all children.

These effects are unlikely to be restricted merely to FEP elevation. The relationship between plasma lead and blood lead has been shown to be nonlinear. The parameters estimated in Marcus and Schwartz[35] indicate that the plasma levels obtained at a blood lead of 25 μg/dl in children with average iron levels occur at a blood lead of 20 μg/dl in the iron deficient children, and not until 30 μg/dl in children whose transferrin saturation exceeds 31%. Given the large difference in the toxicological availability of erythrocyte and plasma lead, it is likely that heme synthesis is not the only effect of lead that will occur at lower blood lead levels in iron deficient children. Because both lead exposure and iron deficiency are associated with poverty, these children may be doubly at risk. While FEP elevation begins at lead levels below 20 μg/dl in blood, anemia is usually not noted in children until blood lead levels exceed 40 μg/dl. However, a recent report by Schwartz and co-workers[36] found evidence that lead induced anemia can occur at blood lead levels between 20 and 39 μg/dl in children 1 year old and younger. This may reflect in part the interaction with iron deficiency, which is greater at that age.

II. OTHER NEUROLOGICAL EFFECTS OF LEAD

A. CORRELATION BETWEEN LEAD AND HEARING

In 1985, Robinson et al.[37] reported a linear increase in the 2 KHz pure tone hearing threshold as blood lead levels increased from 6 to 47 μg/dl. There was no sign of a threshold. To confirm those results, and to examine hearing thresholds at other frequencies, Schwartz and Otto[38] examined the hearing data from the NHANES II study. Lead was positively associated with hearing loss at 500, 1000, 2000, and 4000 Hz. Children with blood lead levels of 25 μg/dl had an average 3 dB hearing loss compared to children with 5 μg/dl of blood lead. A recent analysis of data from the Hispanic HANES survey has confirmed that finding.[39]

Hearing loss at higher exposure levels has been reported in occupationally exposed workers. However Repko and Corum[40] have criticized those studies for inadequate matching by age.

B. LEAD INDUCED PERIPHERAL NEUROPATHY

While the peripheral nervous system is generally less sensitive to toxicants than the central nervous system, a recent study in adults showed decreased peripheral nerve conduction velocity at blood lead levels as low as 40 μg/dl.[41] This prompted Schwartz et al.[42] to re-examine an earlier study in children to determine the blood lead level where peripheral neuropathy began. Two forms of analysis were used. In the first they postulated a uniform threshold for all children, and fit two lines to the data on nerve conduction velocity above and below the threshold lead value. The threshold value was varied to find the level with greatest explanatory power. In the second model, they postulated that threshold levels vary among children, following a normal distribution. This produces a logistic shaped dose response curve for the whole population. This curve was then fit to the data.

In both cases they found that blood lead levels of 30 μg/dl and higher were associated with decreased nerve conduction velocity in children. The logistic model indicated that in the more sensitive children, the threshold may be in 20 to 30 μg/dl range. Again, these levels are lower than those previously thought to cause demyelinization.

III. CONCLUSIONS

The accumulated evidence points to a role of lead in disturbing physical development in the fetus and the child at levels well below those once thought to be safe. The evidence from studies of birth weight, gestational age, malformations, postnatal growth, and postnatal attained stature, when taken together, show a remarkable consistency.

Recent toxicological studies suggest that disturbances in heme biosynthesis and heme related enzymes, in hormonal production, and in the function of calcium in mediating metabolic processes all occur at blood lead levels below 25 µg/dl. Again, the likely causal role of these mechanisms in the effects cited above strengthens the conclusions of each separate set of studies.

Given the extensive role of calcium in regulating metabolic processes, and the equally ubiquitous role of heme related enzymes, it is possible that further investigations will identify other subtle, but potentially important disturbances caused by low-level lead exposure. The finding of peripheral nerve disturbances at lower than expected levels, and of hearing loss as an additional possible effect of lead only adds to this developing concern.

While medical intervention for children below 25 µg/dl is not an option, this is a reflection of the inability of medicine to provide therapy safe enough to make its administration prudent at those levels, not a reflection of lack of toxicity. While this situation is likely to continue for the foreseeable future, environmental intervention to prevent these exposures is clearly needed to protect children and developing fetuses. Lead poisoning and prevention programs need to incorporate these interventions in the future, and national action to reduce common exposure sources is also needed.

REFERENCES

1. **Nye, L. J. J.,** An investigation of the extraordinary incidence of chronic nephritis in young people in Queensland, *Med. J. Aust.,* 2, 145—159, 1929.
2. **Johnson, N. E. and Tenuta, K.,** Diets and blood lead levels of children who practice pica, *Environ. Res.,* 18, 369—376, 1979.
3. **Mooty, J., Ferand, D. E., Jr., and Harris, P.,** Relationship of diet to lead poisoning in children, *Pediatrics,* 55, 636—639, 1975.
4. **Schwartz, J., Angle, C., and Pitcher, H.,** Relationship between childhood blood lead levels and stature, *Pediatrics,* 77, 281—288, 1986.
5. **Dietrich, K. N., Krafft, K. M., Shukla, R., Bornschein, R. L., and Succop, P. A.,** The neurobehavioral effects of prenatal and early postnatal lead exposure, in *Toxic Substances and Mental Retardation: Neurobehavioral Toxicology and Teratology,* Schroeder, S. R., Ed., Washington, D.C., AAMD monograph series, 1986.
6. **Bornschein, R. L., Succop, P. A., Dietrich, K. N., Krafft, K. M., et al.,** Prenatal lead exposure and pregnancy outcomes in the Cincinnati lead study, in 6th Int. Conf. Heavy Metals in the Environment, Lindberg, S. E. and Hutchinson, T. C., Eds., New Orleans, 1987.
7. **Bryce-Smith, D., Deshpunde, R. R., Hughes, J., and Waldron, H. A.,** Lead and cadmium levels in stillbirths, *Lancet,* 1, 1169, 1977.
8. **Ward, N. I., Watson, R., and Bryce-Smith, D.,** Placental element levels in relation to fetal development for obstetrically normal births: a study of 37 elements. Evidence for effects of cadmium, lead, and zinc on fetal growth, and for smoking as a source of cadmium, *Int. J. Biol. Res.,* 9(1), 63—81, 1987.
9. **Fahim, M. S., Fahim, Z., and Hall, D. G.,** Effects of subtoxic lead levels on pregnant women in the state of Missouri, *Res. Commun. Chem. Pathol. Pharmacol.,* 13, 309, 1976.
10. **McMichael, A. J., Vimpani, G. V., Robertson, E. F., Baghurst, P. A., and Clark, P. D.,** The Port Pirie cohort study: maternal blood lead and pregnancy outcome, *J. Epidemiol. Commun. Health,* 40, 18—25, 1986.
11. **Ong, C. N., Phoon, W. O., Law, H. Y., Tye, C. Y., and Lim, H. H.,** Concentrations of lead in maternal blood, cord blood, and breast milk, *Arch. Dis. Child.,* 60, 756—759, 1985.
12. **Moore, M. R., Goldberg, A., Pocock, S. J., Meredith, A., Stewart, I. M., Macanespie, H., Lees, R., and Low, A.,** Some studies of maternal and infant lead exposure in Glasgow, *Scott. Med. J.,* 27, 113—22, 1982.
13. **Ernhart, C. B., Wolf, A. W., Kennard, M. J., et al.,** Intrauterine exposure to low levels of lead: the status of the neonate, *Arch. Env. Health,* 1986.

14. **Bellinger, D., Needleman, H. L., Leviton, A., Waternaux, C., Rabinowitz, M. B., and Nichols, M. L.**, Early sensory-motor development and prenatal exposure to lead, *Neurobehav. Toxicol. Teratol.*, 6, 387—402, 1984.
15. **Shukla, R., Bornschein, R. L., Dietrich, K. N., et al.**, Effects of fetal and early postnatal lead exposure on child's growth in stature — the Cincinnati lead study, in Proc. 6th Int. Conf. Heavy Metals in the Environment, Lindberg, S. E. and Hutchinson, T. C., Eds., New Orleans, 1987.
16. **Shukla, R., Bornschein, R. L., Dietrich, K. N., Buncher, C. R., Berger, O., Hammond, P. B., Succop, P. A.**, Effects of fetal and infant lead exposure on growth in stature, *Pediatric*, 1989, in press.
17. **Lyngbye, T., Hansen, O. N., and Grandjean, P.**, The influence of environmental factors on physical growth in school age: a study of low level lead exposure, in Proc. 6th Int. Conf. Heavy Metals in the Environment, Lindberg, S. E. and Hutchinson, T. C., Eds., New Orleans, 1987.
18. **Angle, C. R. and Kuntzelman, D. R.**, Increased erythrocyte protoporphyrins and blood lead — a pilot study of childhood growth patterns, *J. Toxicol. Environ. Health*, 26, 149—56, 1989.
19. **Grant, L. D., Kimmel, C. A., West, G. L., Martinez-Vargas, C. M., and Howard, J. L.**, Chronic low level toxicity in the rat. II. Effects on postnatal physical and behavioral development, *Toxicol. Appl. Pharmacol.*, 56, 42—58, 1980.
20. **Burger, J. and Gochfeld, M.**, Effects of lead on growth in young herring gulls *(larus argentatus)*, *J. Toxicol. Environ. Health* 25(2), 227—36, 1988a.
21. **Burger, J. and Gochfeld, M.**, Effects of lead on growth and feeding behavior of young common terns *(sterna hirundo)*, *Arch. Environ. Contam. Toxicol.* 17(4), 513—17, 1988b.
22. **Cupa, M. A. and Donaldson, W. E.**, Effect of lead and niacin on growth and serotonin metabolism in chicks, *J. Nutr.*, 118, 107—19, 1988.
23. **Hammond, P. B., Chernausek, S. D., Succop, P. A., Shukla, R., and Bornschein, R. L.**, Mechanisms by which lead depresses linear and ponderal growth in weaning rats, *Toxicol. Appl. Pharmacol.*, 99, 474—476, 1989.
24. **Savitz, D. A., Whelan, E. A., and Kleckner, R. C.**, Effects of parents' occupation exposures on risk of stillbirth preterm delivery, and small-for-gestational-age infants, *Am. J. Epidemiol.*, 129, 1201—18, 1989.
25. **Needleman, H. L., Rabinowitz, M., Leviton, A., Linn, S., and Schoenbaum, S.**, The relationship between prenatal exposure to lead and congenital anomalies, *JAMA*, 151, 2956—2959, 1984.
26. **Wibberly, D. G., Khera, A. K., Edwards, J. H., and Rushton, D. I.**, Lead levels in human placentae from normal and malformed births, *J. Med. Genetics*, 14, 339, 1977.
27. **Rosen, J. F., Chesney, R. W., Hamstra, A. J., DeLuca, H. F., and Mahaffey, K. R.**, Reduction in 1,25 dihydroxy vitamin D in children with increased lead absorption, *N. Engl. J. Med.*, 302, 1128—1131, 1980.
28. **Mahaffey, K. R., Rosen, J. F., Chesney, R. W., et al.**, Association between age, blood lead concentration, and serum 1,25 dihydroxycholecalciferol levels in children, *Am. J. Clin. Nutr.*, 35, 1327—1331, 1982.
29. **Rosen, J. F. and Chesney, R. W.**, Circulation calcitriol concentrations in health and disease, *J. Pediatr.*, 103, 1—7, 1983.
30. **Habermann, E., Cropwell, K., and Janicki, P.**, Lead and other metals can substitute for Ca^{++} in calmodulin, *Arch. Toxicol.*, 54, 61—70, 1983.
31. **Sandstead, H. H., Orth, D. N., Abe, K., and Steol, J.**, Lead intoxication: its effect on pituitary and adrenal function in man, *Clin. Res.*, 18, 76, 1970.
32. **Huseman, C. A., Moriarty, C. M., and Angle, C. R.**, Childhood lead toxicity and impaired release of thyrotropin stimulating hormone, *Environ. Res.*, 42, 524—33, 1987.
33. **Tuppurainen, M., Wagar, G., Kurppa, K., Sakari, W., Wambugu, A., Froseth, B., Alho, J., and Nykyri, E.**, Thyroid function as assessed by routine laboratory tests of workers with long-term lead exposure, *Scand. J. Work Environ. Health*, 14, 175—80, 1988.
34. **Mahaffey, K. R. and Annest, J. L.**, Association of erythrocyte protoporphyrin level and iron status in the Second National Health and Nutrition Examination Survey, 1976—1980, *Environ. Res.*, 41, 327—338, 1986.
35. **Marcus, A. and Schwartz, J.**, Dose-response curves for erythrocyte protoporphyrin vs. blood lead: effect of iron status, *Environ. Res.*, 44, 221—227, 1987.
36. **Schwartz, J., Landrigan, P. J., Baker, E. L., Orenstein, W. A., and van Lindern, I. H.**, Lead induced anemia: dose-response relationships and evidence for a threshold, *Am. J. Public Health.*, in press, 1989.
37. **Robinson, G., Baumann, S., Kleinbaum, D., et al.**, Effects of low to moderate lead exposure on brainstem auditory evoked potentials in children, *Environ. Health Doc.*, Vol. 3, World Health Organization, Copenhagen, 1985, 177—182, 1985.
38. **Schwartz, J. and Otto, D.**, Blood lead, hearing thresholds, and neurobehavioral development in children and youths, *Arch. Env. Health*, 42(3), 153—160, 1987.
39. **Schwartz, J. and Otto, D.**, Lead and minor hearing impairment, *Arch. Environ. Health*, 1991, in press.
40. **Repko, J. and Corum, C.**, Critical review and evaluation of the neurological and behavioral sequelae of inorganic lead absorption, *CRC Crit. Rev. Toxicol.*, 6, 135, 1979.

41. **Seppalainen, A. M., Hernberg, S., Vesanto, R., and Kock, B.,** Early neurotoxic effects of occupational lead: a prospective study, *Neurotoxicology,* 4, 181—192, 1979.
42. **Schwartz, J., Landrigan, P. J., Feldman, R. G., Baker, E., Silbergeld, E., and van Lindern, I.,** Does lead-induced peripheral neuropathy exhibit a threshold?, *J. Pediatr.* 112, 12—17, 1988.

Chapter 15

RECENT EPIDEMIOLOGIC STUDIES OF LOW-LEVEL LEAD EXPOSURE AND THE IQ OF CHILDREN: A META-ANALYTIC REVIEW

Constantine A. Gatsonis and Herbert L. Needleman

TABLE OF CONTENTS

I. Introduction 244

II. Methodological Issues 244
    A. Selecting Adequate Markers of Exposure or Internal Dose 245
    B. Measuring Outcome With Instruments of Adequate Sensitivity 245
    C. Identifying, Measuring, and Controlling for Factors Which Might Confound the Lead Effect 245
    D. Recruiting and Testing a Sample Large Enough to Provide Adequate Statistical Power to Detect a Small Effect 246
    E. Designing a Study Which Avoids Biases in Sample Selection 247
    F. Assessing the Effects of Measurement Error 247

III. Meta-Analysis 247
    A. Methods 247
        1. Data Collection 247
        2. Data Analysis 249
    B. Results 249
        1. Study Cohort Characteristics 249
        2. Lead Effect 249
        3. Combining p-Values 250
        4. Estimation of Effect Size 250
        5. Selection Bias and the "File Drawer" Problem 252

IV. Discussion 253

References 254

I. INTRODUCTION

The neurotoxic effects of lead at high doses have long been recognized[1] but at low dose the presence and magnitude of effects have been the subject of considerable debate. Since the early 1970s, epidemiologic studies have been conducted on children in several sites around the world, with results that often conflicted. The early studies involved relatively simple comparisons of intelligence test performance data between a control and an exposed group. Little attention was paid to what were later recognized to be important covariates, e.g., socioeconomic background, parental education, and intelligence. As experience accumulated, the experimental design and the data analytic strategies became more sophisticated. Beyond measuring lead burden, researchers collected additional data on a variety of socioeconomic, familial, and environmental factors, and assessed the influence of these covariates using multiple regression techniques. Despite progress in design and analytic technique the results have not been completely concordant. When lead exposure and neuropsychological deficit were the only variables in the model, most studies tended to find a significant statistical association. However, in several studies the statistical significance of the association disappeared when further covariates were introduced into the model.

The effects of conflicting evidence are reflected in the conclusions of the reviews published to date.[2-5] These reviews, which focused on psychometric intelligence as the major outcome of interest, were mostly narrative summaries. They paid little attention to statistical and methodological problems (an exception is the work by Pocock and Ashby[5]) and made only rudimentary attempts to provide a numerical synthesis of the results. The techniques of quantitative review of meta-analysis offer improved methods for synthesizing studies.[6-8]

The early efforts to carry out a meta-analytic review of the evidence supported the presence of a link between lead exposure and IQ deficit.[9,10] Schwartz et al.[9] included six studies in their meta-analysis, three using tooth lead and three using blood lead to measure exposure. The analytic strategy in each of the six studies was an analysis of covariance in which control and exposed groups of children were identified and their IQ scores compared while controlling for several covariates. Using Fisher's aggregation technique[8] the authors found a joint p-value of .004 for the effect of lead on IQ in the six studies. The same aggregation technique was used by Needleman and Bellinger[10] on a larger set of studies and resulted in an extremely small joint p-value for the lead effect.

An extensive meta-analysis of all modern studies was recently published by us in the *Journal of the American Medical Association*.[6] The review process began with 24 studies, half of which finally met our specific entry criteria for the meta-analysis. Studies were grouped together according to the type of tissue analyzed for lead exposure (tooth or blood) and results were combined within each group. The meta-analysis also included a discussion of the "file-drawer" problem and an exploration of the sensitivity of the findings to the influence of any one of the individual studies.

The present chapter is largely devoted to describing the results from that paper.[6] In the second section we discuss statistical and methodological problems that have confronted researchers in this area. The third section describes the methods used in our meta-analysis, the results of which comprise section four. Our conclusions and suggestions for further research are presented in the last section.

II. METHODOLOGICAL ISSUES

Researchers conducting retrospective epidemiologic studies of the neurotoxic sequelae of lead exposure have encountered a host of methodological difficulties. Accounts of these difficulties were provided in several of the published papers and reviews.

A categorization of the methodological problems includes virtually all major aspects of the design and analysis of epidemiological studies.

A. SELECTING ADEQUATE MARKERS OF EXPOSURE OR INTERNAL DOSE

Lead exposure has been measured by analyzing either blood samples or teeth. The concentration of lead in blood tends to be transient, while tooth lead content is cumulative and reflects cumulative exposure more accurately. The correspondence between the two types of measurement is not well understood. Moreover, there were further variations within each tissue group. For example, in some studies blood samples were drawn by venipuncture, while in others blood was obtained by fingerprick. Among tooth lead studies, some used the entire tooth while others used only part of it. The resulting lack of comparability among measures of lead exposure is a series impediment for any synthesis of these studies.

B. MEASURING OUTCOME WITH INSTRUMENTS OF ADEQUATE SENSITIVITY

A clear problem in this category was presented by some of the early studies which employed group tests to detect differences between exposed and unexposed subjects.[13,16] Other studies did not use definitive tests but relied on screening tests to measure neuropsychological deficit.[13,20]

C. IDENTIFYING, MEASURING, AND CONTROLLING FOR FACTORS WHICH MIGHT CONFOUND THE LEAD EFFECT

This is probably the most difficult and controversial set of issues confronting researchers in this area. The fundamental difficulty is that lead exposure could be a proxy for a host of other variables representing socioeconomic background, parental characteristics, and environmental factors. Any apparent correlation between lead burden and neuropsychological deficit may be due in part to these other variables. Conversely, the effect of some of these variables in a regression model may act to mask the significance of lead burden, especially if the study population was not chosen judiciously.

A number of the studies we reviewed appeared to have collected inadequate data on the possible confounders of lead. Most of the early studies did not address the issue at all and were excluded from the main body of the meta-analysis. Among those which were included, two[23,34] controlled only for social class. On the other end of the spectrum, there were also instances of overcontrol. A clear case of this phenomenon is the inclusion of "presence of peeling paint" and "pica" as covariates in a regression model along with the variable measuring lead burden directly.[28] It would be no surprise if the effect of lead did not appear significant in such a regression model.

Even assuming that appropriate data had been available on all covariates of interest, the task of assessing the significance of lead burden would be by no means simple. The initial number of possible covariates was often very large compared to the overall sample size of the study, pointing to the need for choosing a parsimonious model. In regression theory, the problem has been approached by first choosing an optimality criterion (e.g., minimize the residual mean square) and then applying this criterion to the problem at hand.[11] A thorough treatment of the problem would require applying the criterion to all possible subsets of the covariates and then selecting the one which is optimal. In practice this requires extensive computing, although algorithms have been developed which are efficient at least with problems of moderate size. Stepwise regression methods have been developed in response to these computational difficulties. These methods, while they are computationally efficient and easy to use, have some serious drawbacks. On the one hand, they do not necessarily lead to models that are optimal by any formal criterion and, on the other, they foster a tendency on the part of the user to overinterpret the results from each step.

Most of the studies under review have relied on some stepwise regression computer program to select a final model, but have paid little attention to the intricacies of the process. An alternative approach was implemented by Pocock and Ashby[5] using Mallows' "Cp" as

TABLE 1
Power Calculations for "Small" Effects of Lead[a]

Author	Sample size	Covariates (final)	Power
Blood Lead Studies			
Fulton	501	14	0.87
Hatzakis	509	8	0.88
Hawk	75	5	0.21
Schroeder	104	7	0.28
Yule	129	2	0.35
Lansdown	86	2	0.25
Ernhart	80	4	0.23
Tooth Lead Studies			
Needleman	218	5	0.53
Fergusson	724	8	0.96
Smith	388	10	0.78
Winneke	115	4	0.31
Hansen	156	7	0.40

[a] $\alpha = 0.05$; $r = 0.14$.

the criterion for selecting the best model among a group of 11 covariates. The Cp criterion is applicable to situations in which it is desired to select a model that minimizes the prediction mean squared error. Calculations are practically feasible for 11 covariates and, in fact, using the algorithm implemented in BMDP subroutine 9R, the computing required for 20 or less covariates is of the same order of magnitude as that needed for a stepwise procedure. Assuming that Mallows' criterion, minimizing the prediction error, is a desirable property for the final model, the model selection process implemented by Pocock and Ashby[5] offers a theoretically satisfying alternative to the use of stepwise regression. After the final ("optimal") model is selected, importance of lead as a predictor of IQ deficit can be judged on the basis of the corresponding t-statistic.

D. RECRUITING AND TESTING A SAMPLE LARGE ENOUGH TO PROVIDE ADEQUATE STATISTICAL POWER TO DETECT A SMALL EFFECT

A variety of practical considerations have limited the sample size obtained in many of the studies under review. This is especially true in the earlier studies and has resulted in substantially low power to detect what would be a "small" (but not biologically unimportant) effect. We have carried out calculations of the *a priori* power to detect a partial r of .14 for lead exposure, which corresponds to lead accounting for 2% of the remaining variability of IQ, after all other covariates have been included in the model. The magnitude of this effect is termed "small" by Cohen[43] but, as he points out, "[it] all too frequently in practice represents the true order of magnitude of the effects being tested."

The power figures for all the studies included in this meta-analysis are given in Table 1. They are optimistic for two reasons. First, they were calculated on the basis of the number of covariates included in the final model, which was usually a fraction of the full range of covariates considered in each study. Second, they utilized the initial sample size reported in each study, without taking into consideration the missing data. Unfortunately, almost none of the published reports contained any information about the magnitude of the missing value problem and none gave any details about how it was handled. Of the 12 studies included in Table 1, only 4 had power above 60% to detect a partial r of .14.

E. DESIGNING A STUDY WHICH AVOIDS BIASES IN SAMPLE SELECTION

Several types of bias may enter into the process of selecting the study sample. It is possible, for example, that children of parents who did not consent to participation in the study may differ in relevant attributes from those which were enrolled. Parents who are concerned about their child's development may avoid, or conversely seek, participation in a study of IQ.

F. ASSESSING THE EFFECTS OF MEASUREMENT ERROR

The potential effects of measurement error on the robustness of the conclusions from each study are important. This issue is often overlooked. The question of errors-in-variables is especially relevant when measuring exposure at low levels. Other covariates that purport to represent arbitrary constructs (e.g., marital relationships, parental interest, parental involvement in school, etc.) are also particularly vulnerable to errors-in-variables problems. A discussion of the issues would take us too far afield. However, readers will find a very useful expose of both the technical and the subject matter considerations in the provocative work by Klepper et al.[12] These authors apply a set of errors-in-variables diagnostics to one of the studies[22] included in this review and derive bounds on the coefficients of lead under a set of assumptions about the error structure in the covariates. The bounds include the value zero in this case. The results from such an analysis are heavily dependent on the appropriateness of the assumptions made and, in this sense, one may disagree with the final conclusions. However, this does not diminish the need for a careful consideration of the problem.

III. META-ANALYSIS

A. METHODS
1. Data Collection

Using a computerized MEDLINE subject search, as well as searching dissertation abstracts and programs of meetings on metals, neurotoxicology, lead, pediatrics, and public health, we identified all studies on lead exposure and children's neurobehavioral development which were published since 1972. A list of all studies identified by our search is presented in Table 2.

The 24 studies resulting from our original search were then narrowed down to 12 by applying the following exclusion criteria:

1. Inadequate control of covariates reflecting socioeconomic, and familial factors — Most of the early studies[13,17-21,25,29] were excluded on the basis of this criterion.
2. Inclusion of covariates which reflect lead exposure (overcontrol) — One study[28] was excluded since it controlled for pica and peeling paint.
3. Inclusion of subjects with defined clinical lead poisoning (i.e., blood lead levels >3.9 μmol (80 μg/dl) — This criterion excluded studies 13, 15, 20.
4. Reported data either did not permit any further quantification, as in study 16 or did not enable us to calculate the coefficient of lead in a multiple regression model. The latter was the case in References 13, 15, 17, 18, 24, 25.

As it can be seen in Table 2, several of the studies were excluded on the basis of more than one criteria. In keeping with our intention to combine results from studies which treated exposure as a continuous variable in their data analysis, we chose to use a published regression-based re-analysis for two of the studies[22,26] which were originally analyzed by dichotomizing lead exposure. Moreover, we had to ask for supplementary information about the regression analysis from the authors of two studies.[27,39]

TABLE 2
Candidate Studies for Meta-Analysis

Author	Year	No. subjects	Tissue	Lead range	Data analysis	Included/reason for exclusion	Comments	Lead effect $P < .05$
Kotok[13]	1972	C = 25; E = 24	BLOOD	C = 30; E = 81	t-Test	No/G,D	A,D	No
Perino and Ernhart[14]	1974	C = 50; E = 30	BLOOD	C < 30; E = 40—70	Mult regress	No/H		Yes
Rummo et al.[15]	1974	C = 45; E = 45	BLOOD	C = 23; E = 61—98	Anova	No/G,D	A	Yes
de la Burde and Choate[16]	1975	C = 67; E = 70	BLOOD	RANGE: 30—100	Chi square	No/D,E,B	C	Yes
Landrigan et al.[17]	1975	C = 78; E = 46	BLOOD	C < 40; E = 40—68	t-Test	No/D,E		Yes
McNeil et al.[18]	1975	C = 37; E = 101	BLOOD	C = 29; E = 59	T-test	No/D	E,B,H	No
Yamins[19]	1976	80	BLOOD	PbBx = 33.2 = 9.1	Mult regress	No/E	A,D	Yes
Kotok et al.[20]	1977	C = 36; E = 24	BLOOD	C = 38; E = 81	t-Test	No/G,D	A,D	No
Ratcliffe[21]	1977	C = 23; E = 24	BLOOD	C = 28; E = 44	t-Test	No/E	A,B	No
Needleman et al.[22]	1979	C = 100; E = 58	TOOTH	PbC < 24; PbE = 36 PbTC < 10; PbTE > 20	Anacova	No/H		Yes
Yule et al.[23]	1981	166	BLOOD	C < 13; E = 13—32	Mult regress	Yes		Yes
Winneke et al.[24]	1982	C = 26; E = 26	TOOTH	PbTCx = 2.4; PbTEx = 9	t-Test	No/B	A	No
McBride et al.[25]	1982	108	BLOOD	C = 2—9; E = 19—29	Anova	No/D,E		No
Smith et al.[26]	1983	402	TOOTH	PbTx = 5.1 = 2.8	Anacova	No/H		No
Winneke et al.[27]	1983	115	TOOTH	PbTx = 6.2; PbBx = 14	Mult regress	Yes		No
Harvey et al.[28]	1984	48	BLOOD	RANGE = 6.2—26.8	Mult regress	No/F	A	No
Shapiro and Maracek[29]	1984	193	TOOTH	RANGE = 30—150ppa	Mult regress	No/E		Yes
Needleman et al.[30]	1985	218	TOOTH	PbTx = 12.7	Mult regress	Yes		Yes
Ernhart et al.[31]	1985	80	BLOOD	C = 30; E = 40—70	Mult regress	Yes	A	No
Schroeder et al.[32]	1985	104	BLOOD	MEDIAN = 30	Mult regress	Yes		Yes
Hawk et al.[33]	1986	75	BLOOD	PbBx = 21; R:6—47	Mult regress	Yes	A	Yes
Lansdown et al.[34]	1986	C = 80; E = 80	BLOOD	C = 7—12; E = 13—24	Mult regress	Yes		No
Hatzakis et al.[35]	1987	509	BLOOD	PbBx = 23; R:7—63	Mult regress	Yes		Yes
Pocock et al.[36]	1987	402	TOOTH	PbTx = 5.1 = 2.8	Mult regress	Yes		Yes
Fergusson et al.[37]	1987	724	TOOTH	PbTx = 6.2 = 3.8	Mult regress	Yes		No
Fulton et al.[38]	1987	501	BLOOD	GM = 11.5; R = 3—34	Mult regress	Yes		Yes
Hansen et al.[39]	1987	156	TOOTH	PBTx = 10.7; PbBx = 5	Mult regress	Yes		Yes

Note: Comments on design/reasons for exclusion: (A) small sample; (B) weak outcome measures; (C) poor exposure measures; (D) inadequate data analysis or reporting; (E) inadequate or no covariate control; (F) overcontrol; (G) clinical levels of lead exposure (PbB>80 μg/dl); (H) later reanalysis substituted: Needleman et al. 1984 for Needleman et al. 1979; Pocock et al. 1987 for Smith et al. 1983; Ernhart et al. 1985 for Perino and Ernhart 1974.
Code: (PbTx) = mean tooth lead value; (PbBx) = mean blood lead value; (R) = range; (PbTC) = values for control group; (PbTE) = values for high lead group; (GM) = geometric mean.

2. Data Analysis

To ensure an acceptable level of homogeneity, the studies were divided into two groups, according to the type of tissue analyzed for lead (blood or tooth). Before combining the p-values within each group, a test was carried out for homogeneity.[8] Joint p-values for lead were calculated for each of the two groups, using Fisher's procedure as well as the procedure of Mosteller and Bush.[8] Whereas the former combines unweighted p-values, the latter procedure effectively weights each study by the number of subjects involved. Such weighting is particularly desirable in this analysis because of the wide range of sample sizes (N = 75 to 724).

The partial correlation coefficient of lead (derived from the corresponding t-value) was used as a measure of effect size. Their z-scores were derived using Fisher's transformation[8] and were then compared via a chi-squared statistic.[7] When the hypothesis of homogeneity was not rejected, this was regarded as evidence that the values of partial r from each study can be treated as independent estimates of a common (group) partial correlation. The group correlation was then estimated by a weighted z-score average, which was further used to construct a 95% confidence interval.[7]

The "file-drawer" problem was addressed using the original approach suggested in Reference 8. A more elaborate treatment of the same issue was also given based on the recent work by Iyengar and Greenhouse.[44] The sensitivity of the conclusions to any one study was investigated by sequentially deleting the data from each individual study and then computing the aggregate indices. Finally, in order to assess whether the exclusion of 12 of the original 24 studies was partly responsible for our findings, we used Fisher's aggregation technique to derive joint p-values for all 24 studies.

B. RESULTS

1. Study Cohort Characteristics

Twelve studies were included in the final meta-analysis. These were divided into two groups: the blood lead group (seven studies), and the tooth lead group (five studies). Methods for measuring IQ differed across studies. The WISC-R scales were used in eight studies and the Stanford Binet scales in two. The remaining two studies used the British Ability Scales and the McCarthy Scales. Methods for measuring lead level also differed, particularly in the tooth lead group. This made it difficult to compare the distribution of lead exposure across studies. This difficulty was compounded by the differences in summary descriptions of lead exposure provided in each study. As best we could determine, the lead exposure in Hawk et al.[34] (mean 62 μmol/l, 12.75 μg/dl) was among the lowest in the blood lead group, while the exposure in Schroeder et al.[32] (median = 1.46 μmol/l, 30 μg/l) was among the highest in that group. In the tooth lead group, the lead exposure in the cohort of Smith et al.[26] was among the lowest (248 of 402 children with tooth lead concentration <5.5 ppm), while the exposure in Needleman et al.[22] (mean = 12.7 ppm) was among the highest.

2. Lead Effect

The overlap among the sets of covariates examined in each study appeared to be substantial. However, important differences remained and were further accentuated by the fact that methods for measuring similar constructs differed across studies. A summary of the information on covariates used in each study is presented in Table 3, in which we classify the various covariates into seven groups: socio-economic status (SES); parental factors (e.g., parent health score); parent IQ; parental rearing measures; perinatal factors (e.g., birthweight, length of hospital stay after birth); physical factors (e.g., age, weight, medical history) and gender. The unadjusted coefficient of lead is also included in this table (where available), and is contrasted to the coefficient of lead resulting from the final model. In several studies the logarithm of the lead measurement was used in the regression equations.

TABLE 3
Lead Coefficients for Full Scale IQ Scores

Author	Coefficient	S.E.	t-Value	p(one-sided)	Sample size	Partial r	Total R2
Blood Lead Studies[a]							
Hatzakis	−0.27	0.07[c]	−3.86[c]	0.0001	509	−.17	0.25
Hawk	−0.25	0.15	−1.67	0.05	75	−.20	0.21
Schroeder	−0.2	0.07[c]	−2.78	0.003	104	−.27	NA
Fulton[b]	−3.7	1.37	−2.77	0.003	501	−.12	0.46
Yule[b]	−8.08	4.63	−1.75	0.04	129	−.16	NA
Lansdown[b]	2.15	4.48[c]	0.48	0.68	86	.05	NA
Ernhart	NA	NA	−1.8[c]	0.04	80	−.20	NA
Tooth Lead Studies[d]							
Needleman	−0.21	0.07	−3	0.001	218	−.20	0.35
Hansen	−4.27	1.91	−2.23[e]	0.01	156	−.18	0.2
Winneke	−0.13	4.66	−0.03[e]	0.49	115	−.003	0.13
Pocock[b]	−0.77	0.63	−1.22	0.11	388	−.06	NA
Fergusson[b]	−1.46	1.25	−1.17	0.12	724	−.04	NA

Note: NA = Not available.

[a] Average weighted partial, r = −0.152; 95% CL = −0.2, −0.1.
[b] log Transforms.
[c] Estimated from data in paper.
[d] Average weighted partial, r = −0.08; 95% CL = −0.13, −0.03.
[e] Obtained from the author.

3. Combining p-Values

The t-value of the regression coefficient for lead was negative in all but one study, ranging from −3.86 to .48 in the blood lead group and from −3.0 to −.03 in the tooth lead group. The corresponding p-values for the common directional hypothesis that lead exposure is negatively correlated with IQ are shown in Table 4. The chi-squared statistics for testing homogeneity of the p-values were 11.02 (df = 6, p = .09) and 5.13 (df = 4, p = .26) for the blood lead and tooth lead group respectively. Thus, the hypothesis of homogeneity was not rejected for either group and joint p-values were subsequently derived. Both the Fisher and the Mosteller and Bush method for combining probabilities resulted in a combined p-value of less than .0001 for the blood lead group. The combined p-values for the tooth lead group were <.0005 and .004, respectively.

In order to investigate the sensitivity of these findings to the inclusion of any single study, we removed the studies one by one from the analysis and recalculated combined p-values. For the tooth lead group the highest combined p-value was .025 and the lowest was .0001. The corresponding figures for the blood lead group were below .0001. On the basis of these results, no single study appeared to be responsible for the significance of the final finding.

4. Estimation of Effect Size

The usual meta-analytic measure of effect size in multiple regression/correlation studies, is the partial correlation coefficient (partial r).[7,8,43] Admittedly, it would be more useful to derive an estimate of the "raw effect" of lead (i.e., average change in IQ units per unit change in lead exposure). However, we could not make such an estimate on the basis of the data available. As noted earlier, these data contain substantial differences in model

TABLE 4
Covariates Entered Into the Final Multiple Regression Model

	SES	Parental factors	Perinatal factors	Physical factors	Gender	Parent IQ	Parental rearing	Lead coefficients unadjusted	Final model
Yule(2)[a]	#			(Age)				NA	−8.08(4.63)
Lansdown(2)	#			(Age)				NA	2.15(4.48)
Winneke(52)	#		#	#	#			NA	−.125(466)
Needleman(5)	#	#	#			#		NA	−.21(.07)
Ernhart(3)		#	#	(Age)		#		NA	NA
Schroeder(7)	#	#			#	#		NA	−.199(.07)
Hawk(1)	#	#			#	#	#	−.456[c]	−.255(.15)
Fergusson(7)	#	#	#		#	#		NA	−1.46(1.25)
Fulton(21)	#	#		#	#	#	#	−5.45(1.5)	−3.70(1.31)
Hatzakis(10)	#	#	#			#	#	−.376	−.266(.07)
Smith(18)[b]	#	#	#		#		#	−2.66(.86)	−.77(.63)
Hansen(6)	#	#	#		#			NA	−4.27(1.21)

[a] The number of coefficients entered into the initial model are indicated in parentheses after the author's name.
[b] The standard error of the coefficients in the Smith study was estimated from the data.
[c] Where available, coefficients for lead are given for the unadjusted bivariate model and the final multivariate model.

TABLE 5
Results of Synthesis of 12 Studies

	Weighted t values		Fisher's technique	
	Z-Score	p-Value (one-sided)	×2	p-Value
Blood Lead Studies				
All studies	−5.46	<.0001	61.29	<.0001
Eliminating one study at a time				
Study eliminated:				
Hatzakis	−3.88	<.0001	42.87	<.0001
Hawk	−5.34	<.0001	55.3	<.0001
Schroeder	−5.15	<.0001	49.68	<.0001
Fulton	−4.87	<.0001	49.68	<.0001
Yule	−5.25	<.0001	54.86	<.0001
Lansdown	−5.56	<.0001	60.52	<.0001
Ernhart	−5.31	<.0001	54.86	<.0001
Combining studies using log transformed values (Fulton, Yule and Lansdown)			18.83	0.005
Tooth Lead Studies				
All studies	−2.65	0.004	33.11	<.0005
Eliminating one study at a time				
Study eliminated:				
Needleman	−1.97	0.024	19.29	<.025
Hansen	−2.3	0.011	23.9	<.005
Winneke	−2.67	0.004	31.68	<.0005
Smith	−2.36	0.009	28.69	<.0005
Fergusson	−3.04	0.001	28.88	<.0005
Combining studies using log transformed values (Smith and Fergusson)	−1.61	0.001	8.66	<.0005

specification among the studies, as well as in units and methods of measuring both lead exposure and outcome.

The partial r's for lead, derived from the published data are presented in Table 5. They range from −.27 to .05 and from −.2 to −.003, respectively for the two groups. A z-score was derived for each partial r using Fisher's z-transform, and the hypothesis of homogeneity was tested. The corresponding chi-squared statistics were 5.78 (df = 6, p >.4) for the blood lead and 6.44 (df = 4, p >.1) for the tooth lead group. Thus the hypothesis of homogeneity was not rejected for either group and combined effect sizes were derived. The weighted z-score averages were −.152 (se = .027) and −.08 (se = .025), respectively for the two groups of studies.

The resulting (approximate) 95% confidence intervals for the group partial r were −.15 ± .05 for the blood lead group and −.08 ± .05 for the tooth lead group. Both confidence intervals lie entirely to the left of zero, indicating that the analysis in terms of the partial r's supports the results obtained from the analysis of the p-values.

5. Selection Bias and the "File Drawer" Problem

The investigation of possible biases has received considerable attention in meta-analytic theory and practice. In this meta-analysis, we investigated the possiblity of bias in both phases of the study selection process: (1) study retrieval, and (2) formulation and application of exclusion criteria to the retrieved studies.

The effect of bias resulting from step (1) is known in meta-analytic literature as the "file drawer problem".[8] Such bias may result from at least two sources beyond faults in the retrieval process: the failure of some investigators to report their results and/or the failure of journals to publish some results submitted. It is a well-known truism that studies which show a statistically significant result tend to be published more frequently.

An estimate of the magnitude of the "file drawer" problem can be obtained by calculating the number of unpublished nonsignificant studies that would be necessary to raise the overall p-value to $>.05$. Calculations based on Rosenthal's procedure[8] showed that an additional 26 null result studies would be needed to negate the finding for the tooth lead group and 67 for the blood lead group. Rosenthal's procedure assumes that the mean z-score of the unseen studies is zero. If one takes a more conservative approach and only assumes that all unseen studies are not significant at the .05 level, then according to the method of Iyengar and Greenhouse,[44] 16 such studies would be needed to overturn the finding for the tooth lead group and 35 for the blood lead group. It seems unlikely that such large numbers of negative studies have escaped our notice, especially if one considers the expense of conducting human studies of lead exposure and the amount of attention directed to this question.

In order to address the possibility of bias resulting from our own exclusion criteria (step (2) as previously described), we calculated joint p-values for the entire set of the orginal 24 retrieved studies. In the few cases where a p-value was not available, we conservatively assumed that it was .5. The combined p-values, based on Fisher's procedure, were less than .0001 for the blood lead group, less than .001 for the tooth lead group and less than .0001 for all 24 studies together. We concluded from these calculations that the application of the exclusion criteria did not appear to create significant bias in this meta-analysis.

IV. DISCUSSION

Two main themes emerge from this review of studies on the link between low-dose lead exposure and intellectual deficit in children. First, the combined evidence from these studies argues in favor of the existence of such an association. Second, the methodological difficulties in these studies are substantial and pose limits on the credibilty of the evidence they can provide and, consequently, on the utility of more studies of this type for the future. To put it another way, the ensemble of retrospective studies we reviewed has established the epidemiological basis for the argument that an association exists between low-lead exposure and intellectual deficit. This has generated knowledge for use both in setting current policy and future research agendas. However, the review also focused on a series of important methodological problems which have constrained these studies and, in some limiting cases,[28] may have totally invalidated their main results. Thus, the evidence from the published studies (and, of course, from this meta-analytic review of them) has to be evaluated in light of their limitations.

Even under ideal conditions, the studies we reviewed could, by themselves, offer only partial evidence for the existence of a link. As we discussed in a previous article,[6] the requirements for showing that the link is a casual one go considerably further than what could be obtained from a meta-analytic review. However, it should be pointed out that the results from this review are concordant with findings from animal studies,[45,46] and from recent prospective studies in this area.[40-42] In our estimation, the combined body of the available evidence argues convincingly that low-level lead exposure is neurotoxic.

REFERENCES

1. **Byers, R. K. and Lord, E.**, Late effects of lead poisoning on mental development, *Am. J. Dis. Child.*, 66, 471—94, 1943.
2. **Rutter, M.**, Raised lead levels and impaired cognitive/behavioral functioning: a review of the evidence, *Dev. Med. Child. Neurol.*, 22(Suppl. 42) 1—26, 1980
3. **Bornschein, R., Peterson, D., and Reiter, L.**, *Behavioral effects of moderate lead exposure in children and animal models,* CRC Critical Reviews in Toxicology, 1983.
4. **Needleman, H. L. and Bellinger, D.**, The developmental consequences of childhood exposure to lead: recent studies and methodological issues, *Advances in Clinical Child Psychology,* Lahey, B. B. and Kazdin, A. E., Eds., Plenum, Press, New York, 1984, 195—220.
5. **Pocock, S. J. and Ashby, D.** Environmental lead and children's intelligence: a review of recent epidemiological studies, *Statistician,* 35, 31—44, 1985.
6. **Needleman, H. L. and Gatsonis, C. A.**, Low-level lead exposure and the IQ of children: a meta-analysis of modern studies, *JAMA,* 263, 673—678, 1990.
7. **Hedges, L. V., and Olkin, I.**, *Statistical Methods for Meta-analysis,* Academic Press, Orlando, FL, 1985.
8. **Rosenthal, R.**, *Meta-Analytical Procedures for Social Research,* Sage Publications, Beverly Hills, CA., 1984.
9. **Schwartz, J., Pitcher, H., Levin, R., Ostro, B., and Nicholas, A. L.,** Cost and Benefits of Reducing Lead in Gasoline: Final Regulatory Impact Analysis, Office of Policy Analysis, U.S. Environmental Protection Agency, Washington DC, 1985.
10. **Needleman, H.L. and Bellinger, D.**, Type II errors in studies of low level lead exposure, a critical and quantitative review, in *Lead Exposure and Child Development: An International Assessment,* Smith, M., Grant, L., and Sors, A., Eds., Kluwer Academic Publishers, Boston, 1988, 293—304.
11. **Weisberg, S.** *Applied Linear Regression.* John Wiley & Sons, New York, 1980.
12. **Klepper, S., Kamlet, M. S., and Frank, R. G.**, Regressor diagnostics for the error-in-variables model: an application to the health effects of pollution, Carnegie-Mellon University, Department of Economics, Technological Report, May 1987.
13. **Kotok, D.**, Development of children with elevated blood levels, a controlled study, *J. Pediatr.*, 80, 57—61, 1972.
14. **Perino, J., and Ernhart, C.**, The relation of subclinical lead level to cognitive and sensory impairment in black preschoolers, *J. Learn. Disab.,* 7, 26—30, 1974.
15. **Rummo, J. H., Routh, D. R., Rummo, N. J., and Brown, J. F.**, Behavioral and neurological symptomatic and asymptomatic lead exposure in children, *Arch. Env. Health,* 34, 120—124, 1979.
16. **de la Burde, B. and Choate, M. S.**, Early asymptomatic lead exposures and development at school age, *J. Pediatr.*, 87, 638—64, 1975.
17. **Landrigan, P., Baloh, R., Whitworth, R.**, Staeling, N., and Rosenblum, B., Neuropsychological dysfunction in children with chronic low level lead absorption, *Lancet,* 1, 708, 1975.
18. **McNeil, J., Ptasnik, J., and Croft, D.**, Evaluation of long-term effects of elevated blood lead concentrations in asymptomatic children, *Arch. Ind. Hygiene Toxicol.*, 26, 97—118, 1975.
19. **Yamins, J. G.**, The relationship of subclinical lead intoxication to cognitive and language function in preschool children, Unpublished doctoral thesis, 1976.
20. **Kotok, D., Kotok, R., and Heriot, T.**, Cognitive evaluation of children with elevated blood levels, *Am. J. Dis. Child.,* 131, 791—793, 1977.
21. **Ratcliffe, J. M.**, Developmental and behavioral functions in young children with elevated blood lead levels, *Br. J. Prev. Soc. Med.* 31, 258—64, 1977.
22. **Needleman, H., Gunnoe, C., Leviton, A., Reed, R. R., Peresie, H., Maher, C., and Barrett, P.,** Deficits in psychologic and classroom performance in children with elevated dentine lead levels, *N. Engl. J. Med.,* 300, 584—695, 1979.
23. **Yule, W., Lansdown, R., Millar, I., and Urbanowicz, M.**, The relationship between blood lead concentration, intelligence, and attainment in a school population: a pilot study, *Dev. Med. Child. Neurol.,* 23, 567, 1981.
24. **Winneke, G., Kramer, G., and Brockhaus, A.** Neuropsychological studies in children with elevated tooth lead concentration. I. Pilot study, *Int. Arch. Occup. Environ. Health,* 51, 169—183, 1982.
25. **McBride, W. G., Black, B. P., and English, B. J.,** Blood lead levels and behaviour of 400 preschool children, *Med. J. Aust.,* 2, 26—29, 1982.
26. **Smith, M., Delves, T., Lansdown, R., Clayton, B., and Graham, P.,** The effects of lead exposure on urban children: The Institute of Child Health/Southampton study, *Dev. Med. Child. Neurol.* 25 (Supplement 47),1, 1983.
27. **Winneke, G., Kramer, G., Brockhaus, A., Ewers, U., Kujanaek, G., Lechner, H., and Janke, W.**, Neuropsychological studies in children with elevated tooth lead concentration, *Int. Arch. Occup. Environ. Health,* 51, 231—252, 1983.

28. **Harvey, P. G., Hamlin, M. W., Kumar, R., and Delves, H. T.**, Blood lead, behaviour and intelligence test performance in preschool children, *Sci. Total Environ.*, 40, 45—60, 1984.
29. **Shapiro, I. M. and Marecek, J.**, Dentine lead concentration as a predictor of neuropsychological functioning in inner-city children, *Biol. Trace Elem. Res.*, 6, 69—78, 1984.
30. **Needleman, H. L., Geiger, S. K., and Frank, R.**, Lead and IQ scores: a reanalysis, *Science*, 227, 701—704, 1985.
31. **Ernhart, C. B., Landa, B., and Wolf, A. W.**, Subclinical lead level and developmental deficit: reanalyses of data, *J. Learn. Disab.* 18, 475-479, 1985.
32. **Schroeder, S. R., Hawk, B., Otto, D. A., Mushak, P., and Hicks, R. E.**, Separating the effects of lead and social factors on IQ, *Environ. Res.*, 91, 178—183, 1985.
33. **Hawk, B. A., Schroeder, S. R., Robinson, G., Otto, D., Mushak, P., Kleinbaum, D., and Dawson, G.**, Relation of lead and social factors to IQ of low SES children: a partial replication, *Am. J. Ment. Def.*, 91, 178-183, 1986.
34. **Lansdown, R., Yule, W., Urbanowicz, M. A., and Hunter, J.**, The relationship between blood-lead concentrations, intelligence, attainment and behavior in a school population: the second London study, *Int. Arch. Occup. Environ. Health*, 57, 225—235, 1986.
35. **Hatzakis, A., Kokevi, A., Katsouyanni, K. et al.**, Psychometric intelligence and attentional performance deficits in lead exposed children, *Int. Conf. Heavy Metals in the Environment*, CEP Consultants, Edinburgh, 1987, 204.
36. **Pocock, S. J., Ashby, D., and Smith, V.**, Lead exposure and children's intelligence, *Int. J. Epidemiol.*, 16, 57—67, 1987.
37. **Fergusson, D. M., Fergusson, J. E., Horwood, L. J., and Kinzett, N. G.**, A longitudinal study of dentine lead levels, intelligence, school performance and behavior. III. Dentine lead levels and cognitive ability, *J. Child. Psychol. Psychiatry*, 29, 793—809, 1988.
38. **Fulton, M., Raab, G., Thomson, G., Laxen, D., Hunter, R., and Hepburnm, W.**, Influence of blood lead on the ability and attainment of children in Edinburgh, *Lancet*, 1, 1221—1226, 1987.
39. **Hansen O. N., Trillingsgard, A., Beese, I., Lyngbye, T., and Grandjean, P.**, *Int. Conf. Heavy Metals in the Environment*, CEP Consultants, Edinburgh, 1987, p.54.
40. **Bellinger, D., Leviton, A., Waternaux, C., Needleman, H., and Rabinowitz, M.**, Longitudinal analyses of prenatal leads exposure and early cognitive development, *N. Engl. J. Med.*, 316, 1037—1043, 1987.
41. **Deitrich, K., Krafft, K., Beir, M., Bornschein, R., et. al.**, Early effects of fetal lead exposusre: developmental findings at 6 months, *Int. Conf. Heavy Metals Environ.* Cet Consultants, Edinburgh, Scotland, 1987, 63.
42. **McMichael, A. J., Baghurst, P. A., Wigg, N. R., Vimpani, G., V., et al.**, Port Pirie cohort study: environmental exposure to lead and children's abilities at four years, *N. Engl. J. Med.*, 319, 468—75, 1988.
43. **Cohen, J.**, *Statistical Power Analysis for the Behavioral Sciences*, Academic Press, New York, 1977.
44. **Iyengar, S. and Greenhouse, J.**, Selection models and the file drawer problem, *Stat. Sci.*, 3, 109—135, 1988.
45. **Rice, D. C., and Willes, R. F.**, Neonatal low level exposure in monkeys (macac fasciculosis): effect on two-choice non-spatial form discrimination, *J. Environ. Pathol. Toxical.*, 2, 1195—1203, 1979.
46. **Cory-Schlecta, D. A., Bissen, S. T., Young, A. M., and Thompson, T.**, Chronic post-weaning lead exposure and response during performance, *Toxicol. Appl. Pharmacol.*, 60, 78—84, 1981.

Section IV: Sociological and Legal Issues

Chapter 16

ECONOMIC ASPECTS OF THE LITIGATION FOR HARM DUE TO LEAD POISONING

Richard G. Frank

TABLE OF CONTENTS

I.	Introduction	260
II.	Lead-Paint Damage and Legal Remedies	260
III.	Application of the Law to the Case of Lead	260
	A. The Case of Samuel Brown	262
IV.	Conclusions	265
Acknowledgment		265
References		266

I. INTRODUCTION

Lead paint continues to be the single most important source of undue lead absorption in children in the U.S.[1] The social cost to society of excess lead exposure in children was estimated to be roughly $1 billion annually in 1978.[2] Taking into account inflation and population changes implies an annual cost of approximately $2.3 billion in 1989. Lead removal programs and other direct regulatory approaches have been difficult to implement for a variety of practical and political reasons. Use of the legal system has become an increasingly popular method for making claims with regard to individual harm caused by exposure to lead paint in dwellings.* This paper is concerned with the degree to which litigation for harm caused by lead paint may create incentives for efficiently lowering exposure to lead paint in dwellings in the U.S.

A large portion of the exposure to lead paint in the U.S. occurs in rental housing. The presence of lead paint is generally unknown to the occupant. Thus, the economic transaction between landlord and tenant usually involves the exchange of living space and fixtures for cash payments. Property owners that rent dwellings do so in the pursuit of income. This means that a property owner makes decisions regarding the amount of investment and the level of upkeep expenses. Such decisions are made relative to the market rental rates available. Elimination of the risk of lead exposure in many properties is costly and is hereby likely to reduce the income earning potential of a property. Nevertheless, the presence of lead paint, particularly when children are present, can result in a variety of negative health effects which are costly to the affected individuals and to society at large.[3] If landlords do not take account of the health costs of renting dwellings with lead paint, there is an economic incentive to maintain levels of lead exposure that are too high from the standpoint of societal efficiency.[4] The primary concern of this paper is with the degree to which use of tort law forces landlords to take account of the social costs of lead paint.

The remainder of the paper is organized into three major sections. The first section discusses the economic concept of an external (or side) effect and applies it to the case of torts related to lead-based paint exposure. The second section of the paper reviews economic issues related to using tort law to advance damage claims for injury suffered from lead exposure. A final section presents some conclusions and policy issues.

II. LEAD-PAINT DAMAGE AND LEGAL REMEDIES

Exposure to lead based paint can create permanent losses in the health and abilities of individuals (frequently children). Health effects of high levels of lead exposure in children can lead to permanent reductions in intellectual ability (as measured by IQ), hyperactivity, and motor deficiencies. Low-level lead exposure has also been linked to intellectual deficits and behavioral problems.[5]

Exposure to lead in a dwelling constitutes an externality because a landlord makes a decision regarding the condition of the dwelling that effects the tenant's well being and is not incorporated into the transaction between the two parties. That is, the tenant involuntarily receives more exposure to lead than he would have knowingly chosen. The tenant's rights to a safe and healthy dwelling is thereby violated. The landlord in this case has generated a "negative externality" for the tenant. Property law aims to protect owners of entitlements (the right to a safe and healthy dwelling) against adverse externalities. The destruction of the initial entitlement (the harm to health) is protected by liability rules or tort law. Under a liability rule an objective standard of value is applied relative to the transfer of entitlements. In the case of lead paint, a landlord may fail to take adequate precautions to protect the

---

* While no reliable data on the growth in lead related lawsuits have been compiled, it seems clear from casual observation of television advertising by lawyers that lead poisoning cases are an important source of activity for personal injury attorneys.

health of his tenants. This causes entitlements to be taken from the tenant (health); liability rules go into effect and guide payment of damages.

The area of the law that applies to damaged health due to lead paint is property law, more specifically habitability laws. Many large American cities have specified responsibilities for the landlord regarding repair and maintenance of leased properties in habitable condition. These laws replaced the historical doctrine of *caveat emptor* (buyer beware). The landlord is placed at risk under habitability laws. The landlord is made responsible for dangers that may not be readily apparent to the tenant. In the case of lead based paint, obtaining information on the presence of the paint and the conditions under which lead paint is most dangerous to children is not the tenant's responsibility. It is the landlord's responsibility. Thus, habitability laws establish rights and responsibilities with respect to the condition of dwellings. Damages are assessed under tort law.

A tort is a civil wrong. This type of "wrong" occurs when one party damages another party's initial entitlement by imposing a negative externality on them. There are two goals of tort law: (1) to act as a deterrent to certain conduct, and (2) to compensate victims. To prove a tort case, the lawyer must show: (1) there was a duty, (2) the duty was breached, (3) the victim was injured, and (4) the breach of duty was the proximate cause of the injury. The role of the law is to assess the damage and determine compensation to the party that has been harmed. As mentioned in the introductory discussion one can view tort law as enhancing imperfect market price systems. For example, if a landlord unintentionally fails to adequately protect his tenants from undue exposure to lead, the injured tenant may claim damages. The purpose of the claim is to compensate the injured party at the expense of the landlord who has failed to meet his responsibility for providing a safe and healthy dwelling. This creates an economic incentive for landlords to take responsibility for providing habitable dwellings by making them bear at least part of the social costs of the externality they generate.

The degree to which the application of liability rules to violations of habitability laws will decrease the risk of lead-paint poisoning depends on how the expected financial losses to landlords from not taking measures to protect tenants compare to the costs of taking the appropriate measures.* We will return to this point in considerable detail in the next section.

Liability rules determine whether a defendant that has harmed another party's initial entitlement, is liable and therefore will be required to compensate the harmed party. There are a number of different liability rules that have been identified in the literature.[6] We briefly review the eight types of liability rules reviewed by Brown.[6]

- *No liability* and *strict liability* represent the extreme cases where the victim is liable in all cases and where the injurer is liable in all cases respectively.
- Under *negligence rules* the victim is liable unless the injurer has been negligent.
- *Strict liability* with *contributory negligence* implies that the injurer is liable unless the victim has been negligent.
- *Strict liability with dual contributory negligence* makes the victim liable if he was negligent and the injurer was not, otherwise the injurer is liable.
- Under *relative negligence* the injurer is liable if an additional dollar devoted to avoidance by him has a larger effect on the likelihood of damage than would an additional dollar of effort by the victim.
- The most common rule is that of *comparative negligence*. Under this rule liability is divided between the parties according to their relative responsibility for the harm. The vast majority of states have adopted comparative liability rules.

---

* Expected financial losses in this case are equal to the probability that a suit will be brought times the payout from a suit that is settled (either by a jury or via negotiation).

The decision rule in a case considered under a comparative negligence rule reduces the amount of damages a victim may recover according to the degree which the victim's negligence contributed to the harm. There are variations in the manner in which comparative negligence rules are applied across states in the U.S. One common modification to the "liability in proportion to fault" principle is to apply the principle up to the point where the majority of fault lies with the victim. After that point, no recovery is received.[7]

The areas of property and tort law are those most relevant to the majority of cases involving children suffering adverse health effects because of undue exposure to lead. These legal mechanisms provide a means through litigation to redress losses stemming from inadequate attention to the health hazards of rental properties. In addition, it signals landlords that inattention may be costly. The possibility of being compensated for losses also provides victims with an incentive to bring legal action against landlord, again creating a signal. An effective signal from the legal system will stimulate landlords with properties containing lead paint to take actions to reduce lead exposure. Alternatively, the signal may create an inducement to seek another use for the property. Recent work by Ford and Gilligan[8] suggests that costs associated with lead abatement in cities such as Baltimore, would have to be extremely high to result in significant levels of property abandonment. In either case the lead exposure of children is likely to fall, however, the implications for the size of the stock of rental housing may be less positive. We now turn to the determination of damages which defines the strength of the signal sent out by the legal system to landlords.

III. APPLICATION OF THE LAW TO THE CASE OF LEAD

Damages that stem from the imposition of externalities require that the injurer who is found guilty must compensate the victim. The role of the judicial system is to provide the victim appropriate compensation from the injurer or his insurer. The basic rule in establishing levels of compensation is that the injurer must compensate the victim for the dollar value of damages incurred. Calculating the compensatory damages that just return the victim to his prior condition revolves around the evaluation of the various dimensions of injury. Some dimensions of damage are relatively straightforward to measure, such as direct medical costs incurred. Others are considerably more difficult. For example, for a child who has had a significant reduction in intellectual ability how does one value the diminished competence in noneconomic activities and the consequent reduction in quality of life. Similarly, the valuation of pain and suffering poses important difficulties.

Economic theory offers some limited guidance on this subject. Economists refer to the compensation principle. This principle states that when an activity is taken on where some parties gain while others lose, an efficient solution requires the winners to offer sufficient compensation to the losers to make them equally well off under either the status quo or the new activity. In the case of damage estimation this involves considering the amount of wealth it would require for an individual to accept a specified period of time during which he will endure a particular level of pain. Beyond this notion establishing an empirical basis for such calculations remains rather arbitrary.

A. THE CASE OF SAMUEL BROWN

We illustrate the estimation of damages by relying on an example from a case of lead poisoning that was recently resolved.* Samuel Brown was hospitalized at age 7 for severe lead intoxication (blood lead of over 70 μg/dl). Samuel was hospitalized in a tertiary care hospital, where chelation therapy was administered. The duration of the hospital stay was 4 weeks. A psychological evaluation was performed subsequent to hospital discharge and

---
* The identity of the individual involved in this case is disguised. The facts of the case presented here are those that were the basis for the complaint.

it concluded that Samuel had suffered an 18 point IQ loss (from 109 Weschler full scale IQ to 91). This represented the loss in intellectual ability attributable to the lead intoxication. Monetary damages were viewed as comprising of three main dimensions: (1) medical costs, (2) reduced lifetime earnings, and (3) intangible losses in quality of life.

Medical costs of roughly $40,000 were incurred by the Brown family and their insurance plan. These included the costs of hospitalization for lead intoxication, ancillary medical services, physician services, and follow-up visits to the treating physician. These costs are relatively easy to ascertain since audited financial statements for hospitals are readily available as are the billings, by the hospital, for Samuel's treatment.

The estimation of reduced lifetime earnings attributable to decreased intellectual ability stemming from lead intoxication is somewhat more complicated than the estimation of medical expenses. The complications arise because it is necessary to estimate the impact of IQ on earnings and then to project the lifetime earnings of a child, such as Samuel. Since the loss in IQ occurred at an early age there are two major components to the impact of reduced IQ on earnings. There is a direct effect of IQ on earnings which may be viewed as the impact of intellectual ability on productivity, holding other factors constant. There is also an indirect effect that occurs through the impact of IQ on educational attainment.

There has been a great deal of research on the direct effect of IQ on earnings.[9] Only a subset of this work has examined both the direct and indirect impacts. The study by Behrman, Hrubec, Taubman and Wales[10] examined these linkages in a very complete manner. The estimated linkages between IQ, educational attainment, and earnings for males reported in the study by Behrman and colleagues were used to calculate the percent reduction in earnings due to a reduction in full scale IQ from 109 to 91.* For the 18 point reduction to full scale IQ experienced by Samuel Brown, the estimates from the study by Behrman and colleagues implies a 19.5% reduction in lifetime earnings. It is important to note that the direct impact of IQ on earnings dominates the indirect effect through education attainment.

In order to calculate the value of the reduced IQ for an individual such as Samuel Brown we obtained data from a sample survey of men from his city of residence. A regression model of annual earnings was estimated for a sample of men between the ages of 18 and 65. The explanatory variables included each respondents age, race, marital status, and educational attainment. The sample was based on a household survey and included males who were employed, unemployed and not in the labor force. This regression model was used to project the lifetime earnings for an African-American male (Samuel Brown's race) with the average characteristics of his community absent any reduction in IQ (i.e., assuming an IQ of 109). This was done by: (1) projecting educational attainment for Samuel Brown using the Behrman study (12 years), (2) adjusting projected IQ for the fact that an IQ of 109 is 9 points above average, (3) using the age variation in the sample to project the age-earnings profile for an individual similar to Samuel, and (4) adjusting for the mortality probability prior to age 65.

Since a dollar of future income is worth less than a dollar of current income, the projected lifetime stream of earnings needs to be converted into present dollars (1989 in this case). This involves choosing an appropriate discount factor. Choosing the appropriate discount rate is a controversial area in the economics literature. It has been the subject of debate for at least 25 years.[11] The approach taken in the Brown case was to use the average interest rate for short term U.S. government treasury notes for the past 15 years (after adjusting for inflation). This represents the value of a dollar received today at a future point in time if it

---

* It should be pointed out that the Behrman et al.[10] study relied on the GCT measure of intellectual ability. This scale has a mean of 50 and a standard deviation of 10.3. This was converted into IQ equivalents which has a mean of 100 and a standard deviation of 15. The impact estimates were adjusted to account for the differences in sample variances.

is invested in a relatively risk free asset. The projected age-earnings profile for Samuel Brown was converted into 1989 dollars using a 3% discount factor (e.g., $1 today is worth $1.03 in one year after adjusting for inflation).

The 1989 discounted value of Samuel Brown's lifetime earnings absent any loss in IQ was estimated to be approximately $842,000. This figure represents only the wage component of total compensation, it does not include fringe benefits. Average fringe benefits in the U.S. amount to about 15% of wage income. Thus the projected total compensation for Samuel Brown was estimated at approximately $968,000 in 1989. Applying the estimated impact of the reduction in IQ attributable to lead exposure (19.5%) to the total compensation figure allows one to estimate the lost lifetime earnings component of the monetary damages. The estimated losses in lifetime earnings are estimated to be $189,000.

The third component involves placing a value on a variety of intangible factors related to the quality of life reductions attributable to the undue exposure to lead. Measuring the value of the intangible impacts of undue lead exposure is consistent with the compensation principle described above. One of three approaches have generally been taken to measuring the cost of intangibles. The *surrogate market* approach relies on finding situations where individuals have a choice between making expenditures or suffering the negative effect in question. Housing markets have sometimes been used to estimate the value of a variety of "amenities". For example, housing prices for identical houses differ depending on traffic conditions on the street where the house is located. Obtaining such estimates for lead paint is likely to be quite difficult. A second approach has been to estimate the full amount of money that it would take to reverse the ill effect. In the case of IQ loss due to lead exposure the process is probably not reversible making the technique of limited use. A final approach is to rely on surveys for eliciting the value individuals place on various types of damage. This approach is also wrought with difficulties. However, some successful work has been done in this area.[12] The approach used in the case of Samuel Brown relies loosely on the survey method in that it uses values of health found from similar hazards and makes the lead effects roughly comparable to "similar" risks. The estimate in this case was approximately $70,000.

The result of this series of estimates was a damage estimate of $309,000. This estimate is close to the final settlement amount in the case. In order to evaluate the degree to which such litigation sends an efficient signal to potential injurers we must examine the factors that determine expected costs to landlords of having lead paint in their rental units. The ideal economic signal from the legal system to landlords would be achieved only in a special set of circumstances. That is, when every significant case of damages due to lead poisoning results in a claim and when every valid claim leads to a full award. In this situation, the income-maximizing landlord would invest in lead abatement up to the point where the last dollar spent on abatement would equal the expected savings in damages. Further investments would increase the costs to society.

If not all cases of poisoning result in suits or not all valid suits result in awards equal to the social costs imposed, the expected costs of having lead paint in rental units will result in "too little" investment in lead paint abatement. Since it is rather clear that neither condition for the ideal signal holds, it is important to know whether expected damages to landlords approach the efficient signal. Estimates of the prevalence of undue lead exposure (>40 µg/dl) ranged in the 6% to 7% region since screening of children was initiated in the early 1970s.[13] (Recent changes in the definition of lead intoxication use 25 µg/dl). Not even 10% of such cases result in a complaint. This may be because not all of these cases result in any significant observable damage. Nevertheless, we suspect that a substantial portion of cases with legitimate damages do not result in claims. It should be noted that the increasing size of awards and the large supply of attorneys has resulted in an increase in litigation related to lead poisoning. Thus over time the expected damages to landlords may begin to approach the socially efficient level. This, of course, does not imply that the tort system is an efficient compensation mechanism, even if it has an effective deterrent impact.

IV. CONCLUSIONS

The social costs of undue lead exposure are substantial by all estimates. A substantial portion of the social costs from this environmental hazard stem from childhood ingestion of lead based paint. The application of habitability rules and tort law have the potential to cause landlords to face important economic consequences from not undertaking investments in lead abatement. We show that in cases of significant health impacts caused by lead poisoning the economic damages can be large, on the order of $300,000. However, for the legal system to send an "efficient signal" to landlords a high proportion of cases of significant injury must make claims and the settlement of those claims must reflect the social costs of the harm. Until recently, the legal system has been largely ignored as a tool of public health policy.

The author believes that encouraging legitimate cases of harm due to lead poisoning can provide a strong complementary approach to other regulatory and educational strategies that have typically been the focus of public policy. Measures that could be adopted to promote effective use of the legal system would include educating the public regarding: (1) the presence of lead paint; (2) the types of damage that could be incurred from exposure; (3) individual rights to make claims associated with violation of habitability rules; and (4) strengthening habitability laws so that rental units must be tested for lead based paint when tenants change. In this case, the legal system can be used to serve a function similar to that of effluent taxes used in other areas of environmental policy.

There are a variety of research projects that would be useful in evaluating the impacts of the legal system on lead poisoning. Most important would be understanding the response of landlords to level of expected damages. The larger the response in terms of abatement activity, the more effective the legal system is as a tool of health promotion. Other areas of relevant research would further develop survey approaches to evaluate the social costs associated with all losses due to undue lead exposure. These would include both the tangible and intangible types of harm. Finally, legal approaches should be assessed relative to regulatory strategies, taking into account relative costs, benefits, and ability to implement.

ACKNOWLEDGMENTS

The author is grateful to Linda Bailey for helpful comments on an earlier draft.

REFERENCES

1. **Lin-Fu, J.**, Lead poisoning and undue lead exposure in children: history and current status, in *Low Level Lead Exposure*, Needleman, H., Ed., Raven Press, New York, 1980, 5—16.
2. **Provenzano, G.**, The social cost of excessive lead exposure during childhood, in *Low Level Lead Exposure*, Needleman, H., Ed., Raven Press, New York, 1980, 299—315.
3. **Needleman, H. L., Schell, A., Bellinger, D., et al.**, The long-term effects of exposure to low doses of lead in childhood: an 11 year follow-up, *N. Engl. J. Med.*, 322(2), 83—88, 1990.
4. **Turvey, R.**, Side effects of resource use, in *Environmental Quality in a Growing Economy*, Jarret, H., Ed., Johns Hopkins University Press, Baltimore, 1966, 47.
5. **Needleman, H. L. and Landrigan, P. J.**, The health effects of low level exposure to lead, *Annu. Rev. Public Health*, 2, 227—298, 1981.
6. **Brown, J. P.**, Towards an economic theory of liability, *J. Leg. Stud.*, 2, 323—349, 1973.
7. **Schwartz, G.**, Contributory and comparative negligence: a reappraisal *Yale Law J.*, 87, 710—747, 1978.
8. **Ford, D. A. and Gilligan, M.**, The Effect of lead paint abatement laws on rental property values, Working Paper, University of Baltimore, undated.
9. **Jencks, C., et al.**, *Who Gets Ahead*, Basic Books, New York, 1979.
10. **Behrman, J. R., Hrubec, Z., Taubman, P., and Wales, T. J.**, *Socioeconomic Success*, Elsevier/North Holland, Amsterdam, 1980.
11. **Baumol W.J.**, On the social rate of discount, *Am. Econ. Rev.*, 58, 788—802, 1968.
12. **Randall, A., Hoehn, J. P., and Brookshire, D. S.**, Contingent valuation surveys for evaluating environmental assets, *Nat. Resour. J.*, 23, 635—647, 1983.
13. **Lin-Fu, J. S.**, The evolution of childhood lead poisoning as a public health problem, in *Lead Absorption in Children*, Chisolm, J. and O'Hara, D., Eds., Urban and Schwarzenberg, Baltimore, 1982.

Chapter 17

THE ROLE OF VALUES IN SCIENCE AND POLICY: THE CASE OF LEAD

Samuel P. Hays

TABLE OF CONTENTS

I.	Introduction	268
II.	The Context of Value Choice	268
III.	The Context of Policy Choice	269
IV.	Historical Evolution of the Debate: 1923 to 1971	271
V.	The Transformation of Lead Politics: 1971 to 1981	272
VI.	The Lead-in-Gasoline Phasedown	274
VII.	The Response of the Lead Industry	277
VIII.	Patterns of Disputes: Science and Values	278
References		280

I. INTRODUCTION

The human health effects of lead are a subject of ancient interest and of great contemporary concern. Its science and its public policy have been debated with both light and heat. This chapter will concentrate on both lead science and lead policy in their close relationship as society seeks to define an acceptable level of exposure. That acceptable level involves not just policy choices, but values expressed by scientists, citizens, and policy makers. A closer examination of the lead issue promises greater insight as to the role of values in both science and public policy.[1-4]

II. THE CONTEXT OF VALUE CHOICE

The focus of analysis is the level of lead in humans thought to be acceptable. Different terms have arisen in different fields of specialization to identify such a level — in public health, a threshold; in environmental science, a critical load; in public regulation, a standard. They refer to the goal to be achieved in regulation. Each chemical or metal in question, therefore, becomes the subject of scientific investigation and the results play a significant role in the public policy debate. What to investigate, the research design and the assessment of results are so intimately bound with policy choices that each, in turn, becomes as controversial as those choices themselves.

These debates clarify scientific research and assessment as far more than a simple process of the discovery and application of truth. Objections to the contemplated policies give rise to challenges to the underlying science, especially if policy requires changes in human behavior — either through the practices of business firms or in the daily lives of individuals. Scientific inquiry as a result becomes controversial at every step because of its continued implications for public policy.[5,6]

Disagreement and debate are also inherent in the scientific enterprise. Scientists differ in their experience and training, their commitment to profession and methodology and their personal values. As they become involved in research and its implications they choose one side or the other of scientific dispute. The interaction between specialized knowledge and personal values creates commitments which lead to persistent differences, often shaped into personal antagonisms. Scientific choices become policy choices partly because of the values which scientists bring to those choices.

Lead is a convenient case for analysis of this problem for two reasons: (1) it is far simpler than many other pollutants and (2) it has been a focus of public policy debate for so many years that its numerous episodes of dispute provide an opportunity for systematic longitudinal analysis.[7-10]

The debate over the health effects of lead over the past several decades has taken place around a progressive reduction in the acceptable level of lead in blood. Until the 1960s it was thought that the line between the acceptable and unacceptable was 80 μg/dl of blood. These standards had been formulated largely from observation of the effects of lead on able-bodied workers. With time, however, concern shifted to young children, the threshold level went down to 60, then 40, then 30 and finally 25 μg/dl. The declining concentration brought into play new problems, new measurement techniques, new methodological alternatives, new assessments of effect and new standards of acceptability.

This downward trend in the threshold level for blood lead provides a convenient format from which to analyze the scientific debate. At each point in the debate some scientists argued that the current level provided ample protection and others argued that new knowledge required that the issue be reopened. For the most part scientists who pressed one side of the argument at one blood lead level did so also at the next level. One suspects, therefore, that it is not the data itself that is the controlling factor in the debate, but the values that scientists bring to it.

Throughout these years a major shift began in environmental science and policy from concern about the acute effects of high-level exposures to pollutants to concern about the chronic effects of persistent low-level exposures. While the former are readily observable, the latter are more difficult to detect and require more sensitive measurements. Chronic low level effects involve much smaller observed differences and give rise to greater controversy. Often the methods of analysis in acute and high level exposure cases are not appropriate to the investigation of chronic and low level effects.

Environmental science and policy now face major challenges shaped by this new context of low-level exposures and subtle effects. To the scientist they present opportunities to expand the frontiers of knowledge to new realms of understanding. To the general public the opportunities lie in the expansion of benefits beyond those few acutely affected to the many who have experienced lower levels of harm. To the regulated they present the need to defend themselves against the controls established by public policies. To the policy maker they present the need to find ways to foster effective policy when difficult-to-resolve controversies can well immobilize action.

III. THE CONTEXT OF POLICY CHOICE

The public dispute over the health effects of lead was shaped by selected stages in the lead biogeochemical cycle that became focal points of concern.[11]

One of the first foci in the 20th century was lead in paint. This came about because of the discovery by Australian physicians of the connection between blood lead levels in children and paint.[12-14] For decades this information lay unused. In the 1960s community groups in poorer sections of cities demanded that action be taken to reduce the threat.[15,16] This led to legislation in the U.S. in 1971 and 1977 that lowered the level of allowable lead in paint and to proposals to remove old paint from houses.[17,18] The lead-in-paint issue long remained because of the slow pace of removal of leaded paint in housing.[19-20]

By the late 1960s the focal point of policy was shifting to lead in gasoline. Administrative responsibility for action on this issue was in the Environmental Protection Agency (EPA), created in 1970. For 4 years the issue was hotly debated and by 1974 EPA decided to take action to require unleaded gasoline in new cars and to phase-down the allowable levels of lead in gasoline over the next decade.[21] That schedule proceeded as planned until action was initiated by the Reagan administration in 1981 to reverse the phase-down schedule and to permit an increase in lead in gasoline. Faced with intense debate over the issue, the agency reversed itself and even accelerated the phase down.

Lead in the ambient air became another point of debate. After implementing regulations for six pollutants in the ambient air as required by the Clean Air Act of 1970, EPA did not take up others for action. Under court order as a result of litigation brought by the Natural Resources Defense Council, EPA developed an assessment of the health effects of lead — a criteria document — which would serve as a basis for regulation. After considerable debate the agency established an ambient air level of 1.5 $\mu g/m^3$. This survived a court challenge from the lead industry and by the early 1980s the new standard was in process of implementation.[22]

Still a fourth focal point of dispute was the "lead advisory" drawn up by the Centers for Disease Control (CDC) in the U.S. Department of Health and Human Services, intended for circulation to physicians and public health officials. Such a document had no legal standing by itself, but served as medical advice to health specialists as to the blood lead levels at which they should take action. The CDC advisory became authoritative, however, far beyond the audience of clinicians. It was taken as a highly significant scientific judgment by regulatory agencies and the courts. Hence CDC's judgment, and the way it was drawn up, become a issue in the lead debate.[23]

Closely interwoven with official policy making were debates in varied private and public settings. Two reports from the National Academy of Sciences (NAS) were among the most influential. The first, drawn up in 1971, was commissioned by the EPA and the second, issued in 1977, was underwritten by the Department of Housing and Urban Development. The first became a major center of controversy over the balance of views represented by the people selected for the review committee.[24,25]

Equally significant debate took place in scientific meetings or journals. Some of the more revealing involved questions, in one fashion or another, of bias resulting from the sources of funding, of selection of scientists to present papers and provision for rebuttal, and of the balance represented by the conference reports. Debate within scientific journals was lively, as key pieces of research became highly controversial and failure of writers to identify the sources of their funding (especially in the case of research funded by the lead industry) was challenged by opponents.[26-35]

Especially significant was the scientific assessment process carried out by EPA as the basis for establishing standards. This was done informally in preparation for the lead-in-gasoline standard, but much more formally through preparation of a criteria document for the ambient air standard. The criteria document process originated in 1963 when Congress directed the U.S. Public Health Service to draw them up at its discretion. The first attempt, with sulfur oxides prepared initially in March 1967, was criticized heavily by the coal and utility industries. Henceforth the criteria document and the personnel chosen to prepare it became the most important focal point of scientific, and hence political, choice in the regulation of air pollution.[36,37]

Litigation was another forum in which the lead issue was debated. Lawsuits received considerable media publicity in the regions in which they occurred, such as Toronto, Canada, Kellogg, ID, and El Paso, TX. The importance of these trials extended beyond the litigants; they became significant opportunities for the contending parties to marshall their scientific and technical expertise. The same expert witnesses appeared for each side in these cases, rotating from one locality to another and controversy in litigation was closely interwoven with debates in other forums.[38]

These scientific disputes all took place within the context of the appropriate environmental standard. What level of ambient concentration, or emissions, lead in gasoline or paint, should be applied in order to achieve what level of health? The issue of standards, which involved matters of costs to sources of lead in the environment and health levels to those affected, could not be dealt with independently of the personal values and views of the scientists as well as of the decision makers. Ultimately the issue came to be whether or not one placed greater value on industrial production or on improved personal and public health.

The gap between existing knowledge and a desired standard focused on debate over two questions. The first, evolving quite early and continuing through the 1970s, involved the "margin of safety". Acknowledging the fact that emerging knowledge seemed to indicate adverse health effects at lower levels of concentration, should one set the standard at a level lower than the agreed-on scientific knowledge indicated? Health protection required a preventive strategy in which one would err on the side of health benefits rather than economic protection for the emission sources. Specialists on either side of the scientific issues took predictable positions on either side of the issue of the margin of safety.[39-41]

By the late 1970s and early 1980s as newer knowledge emerged, the margin of safety issue often become absorbed into the issue of an "adverse health effect". Those seeking to restrict further regulation of lead often accepted the new knowledge and argued that the health effect it identified was not significant enough to justify regulation of lead at lower levels. Especially important was the issue of whether or not indicators such as minor malformations or changes in blood chemistry in young children had larger health meaning. What one thought about this question was as much a subject of personal values as how one thought about the margin of safety.[42]

IV. HISTORICAL EVOLUTION OF THE DEBATE: 1923 TO 1971

The debate over the environmental effects of lead went through several distinct historical stages, shaped by the state of scientific knowledge, the openness of debate, the relative influence of the lead industry and public health advocates, and the freedom and willingness of government agencies to transfer new scientific knowledge into policy. Over the years the advance of scientific knowledge, and the openness of debate enabled public health advocates to overcome the resistance of the lead industry and some public agencies and thereby permitted other public agencies to take a stronger stance.

The initial debate over lead arose in the 1920s when it appeared that workers in the new factories which produced tetraethyl lead for gasoline suffered from lead poisoning. Several cities, notably New York, passed ordinances that prohibited the use of lead in gasoline. Both the U.S. Bureau of Mines and the U.S. Public Health Service, called upon to assess the health effects of lead, decided that it presented no problem for the public. The city ordinances were declared unconstitutional and lead continued to be used in gasoline.[43,44]

This first flurry of debate outlined the basic viewpoints that were to continue to the present day: industry defended the use of lead as beneficial; the public was concerned about the health effects; industry attempted to secure verdicts from health authorities that lead was not harmful; government agencies with an interest in promoting the use of lead were on one side of the issue and those with an interest in promoting public health were potential advocates of the other.

The issue did not arise again until the 1960s when the industry sought to increase the level of lead in gasoline and approached the U.S. Public Health Service for an opinion about its safety. This led to cooperative action between industry and the service through the Lead Liaison Committee. This committee, with representatives from each side, fostered research concerning the relation between air lead and blood lead concentrations. The research was carried out under the auspices of the Public Health Service, with funding from the industry and under the direction of Lloyd Tepper who was considered friendly to the industry point of view. This "Seven-Cities Study" examined the relationship between lead in the air and blood lead levels in seven U.S. cities. It was conducted under the express provision that the results would not be made public without the permission of the industry.[37]

During these years and up until the late 1960s, research on lead was dominated by the Kettering Laboratory at the University of Cincinnati, directed by Robert Kehoe. Kehoe was also chief medical adviser to the Ethyl Corporation, the primary manufacturer of tetraethyl lead. By the 1960s Kehoe was the nation's leading expert on lead; his ideas constituted the dominant wisdom until challenged in the late 1960s.

Kehoe argued that most lead in the human body came from weathering in the soil which found its way into food and water. Since the source was widespread, human uptake of lead was commonplace, whether in industrial or nonindustrial societies, urban or nonurban settings. Lead uptake, moreover, was relatively harmless, since most lead was rapidly excreted and little remained permanently in the body. Most cases of lead poisoning, Kehoe argued, arose from improper nutrition which fostered lead uptake and hampered excretion. The appropriate approach to lead problems was to assure that those unduly exposed had proper nutrition.[45]

Occupational lead exposure was a somewhat different matter, since in this case the source did not come from ingestion but from inhalation. But, it was felt, this was only a temporary problem since it was possible to treat seriously exposed workers by chelation therapy. Once treated, it was argued, there was no permanent effect and, in fact, workers, once having undergone therapy, could return to their jobs. Chelation therapy became a standard operating procedure for workers exposed in battery plants, lead smelters, and other major industrial sources of lead.

This mode of reasoning was also applied to children. They, too, it was argued, could be treated with chelation therapy as were adults, and once the obvious clinical symptoms had disappeared then the patient was pronounced cured. Such conclusions about the long-run effects of lead came not from direct research about the health effects of lead on children, but from reasoning from research on able-bodied workers.

In 1943, Byers and Lord[46,47] argued that cases of lead poisoning, thought to be cured, later displayed significant effects on mental development. While there already had been concern about lead poisoning in children, this report emphasized its effect on the early development of the nervous system. The immediate evidence came from a follow-up study of 20 children with lead poisoning who had been discharged as "cured" yet all but one were doing poorly in school. The effect of these observations was not immediate, for it took another 30 years before controlled investigations documented the fact. But they reflected a potentially new departure in lead research that focused on children rather than adults, low-level rather than high-level exposures, and long-term rather than short-term effects.

This new departure was facilitated by research already undertaken in Australia which linked childhood lead poisoning with ingested paint. As early as 1897 Turner[13,14] of Queensland observed that many children became ill after changing residence and those who apparently recovered in hospitals suffered a recurrence of symptoms a few months after discharge. By 1904 his colleague, Gibson,[12] identified the hand-to-mouth ingestion of lead paint as the source of the problem. But these observations lay dormant and were rediscovered only in the late 1960s as the new focus on childhood chronic lead poisoning came to the fore.[12-14]

V. THE TRANSFORMATION OF LEAD POLITICS: 1971 TO 1981

During the decade after 1971 a veritable revolution took place in the politics of lead in which a range of factors came to bear on the debate that had been relatively neglected before. New researchers contributed a host of discoveries about the health effects of lead especially on children; new methods of measuring blood lead were developed which greatly reduced the cost and time of gathering data about lead levels; the debate shifted from the private realm of Industry-Public Health Service relationships to an open and public forum; the role of lead inhalation came to assume an importance far greater than ingestion; and frontier knowledge came to play a more immediate and significant effect in the assessment of lead science and hence in public policy.[48]

The role of new scientists and new science was rather crucial. Patterson[49] of the California Institute of Technology argued that the blood-lead levels of humans in urban-industrial societies were not at all natural, but were far higher than the levels in people in nonindustrial societies. Patterson showed that most lead measurements by researchers in the past had been contaminated by lead from all sorts of sources: the air in the laboratory, the hands of the experimental staff, and the materials used in research. He developed a laboratory that was far freer from lead contamination and thus was able to make much more accurate measurements.

Patterson and others[50,51] compared lead levels in humans in the U.S. with those in Nepal, away from urban-industrial air, and measured the lead in bones of "ancients" from Peru. They calculated lead levels in water in rivers upstream and in their lower estuaries, in top, middle, and lower strata of the ocean, in raw and canned tuna, in sediment cores in lakes and ice cores in Greenland. The data refuted the conventional wisdom that natural and industrial society lead levels were similar. It generated extensive descriptions of many phases of the lead biogeochemical cycle and enabled observers to place each instance of lead in a broader context of both contemporary and historical patterns of lead distribution.

Patterson's work was first presented to a public audience in 1965 before the Senate Environment Sub-Committee headed by Senator Edmund Muskie.[52] His views were sharply challenged by researchers in occupational medicine and representatives of the lead industry.[52,53] The debate fostered a considerable amount of new research, and while some continued to challenge Patterson, he convinced an increasing number of people that his data was accurate, his methological argument sound, and his results reliable.[54-55]

Other new researchers contributed to the lead debate. Lin-Fu issued her first article on lead poisoning in children in 1967 as a publication of the Department of Health, Education and Welfare and contributed a series of articles on the subject throughout the 1970s. Needleman devised a method of measuring long-term body burdens of lead, in contrast with the more short-lived blood-lead levels, by calculating the lead deposited in children's teeth, and related that level with both their IQ scores and their behavorial patterns in school. Silbergeld investigated experimentally the biochemical effects of lead on laboratory animals.[48]

Especially significant was the development by Piomelli[56] of N.Y. University Medical School of a new technique for measuring blood lead levels. Piomelli discovered a test not directly of lead in blood but of a biochemical derivative, ethyrocyte protoporphyrin, which served as a reliable marker of blood lead concentration. This greatly reduced the time and cost of monitoring and greatly expanded the range of subjects.

The initial impact of these new developments came with the first lead report by the NAS, because the report did not take them up fully. The NAS report came under considerable criticism because it was dominated by researchers knowledgeable about ingestion rather than inhalation of lead, came under the influence of Gordon Stopps of the Haskell Laboratory at Dupont,[24] and viewed the lead issue much in the same way as in the past.

The proceedings of this committee came to be widely known through articles published in *Science* by Gillette.[57] Gillette reported that concern about the committee's work led it to select outside reviewers. One of them, Harriet Hardy of the Boston General Hospital, was highly critical of the report. Especially significant was Gillette's argument that the committee was composed primarily of those with more traditional interests in lead ingestion, drew upon experts from animal nutrition and included none who were specialists in the newer work on inhalation and the health effects on children.

Even more sensational was a later report of the entire case by Boffey,[58] another reporter, in a general analysis of the work of the NAS. Boffey reported that scientists who had "sounded the alarm about alleged dangers of atmospheric lead concentrations" had been excluded from the panel: this group included John Goldsmith, head of the California Health Department's epidemiology unit; Henry A. Schroeder, head of Dartmouth College's Trace Metal Laboratory; Clair Patterson, geochemist at the California Institute of Technology; Paul P. Craig, a physicist who headed the Environmental Defense Fund's lead committee; and T. J. Chow, a geochemist at Scripps Institute of Oceanography.

Goldsmith[59] a pioneer in focusing on air-pollution science and standards, pressed for a California ambient air standard for lead of 1.5 $\mu g/m^3$. The lead industry took an active role in opposing the contemplated standard. Even though industry and the Public Health Service had previously agreed not to report the results until the work was completed, the industry now sought to inject into the proceedings selective parts of the Seven Cities Study. But due to the initiative of Goldsmith the entire study came to light.[37]

A new arena of debate arose within the EPA. The earlier Lead Liaison Committee, formed in the late 1950s, had been a major forum of internal EPA proceedings. But in the early 1970s environmental issues were shaping a more public setting for debate. In 1972 the EPA, inheriting the Lead Liaison Committee from the Public Health Service, opened its proceedings and its minutes to scrutiny from the public. From that time on, new scientific information was brought into debate more quickly than before and with greater effect.[37]

In considering the regulation of lead in gasoline, EPA had confined its science evaluations to personnel within the agency. This often led to stalemate. The logjam with EPA was broken by litigation brought by the Natural Resources Defense Council in which the court set a deadline for EPA action and knocked the issue loose from internal agency paralysis.[60]

The assessment leading up to the ambient air standard in 1978 was carried out through a more open process in which the agency's Science Advisory Committee (SAC) composed of external experts played a crucial role. Internal agency personnel, still convinced that an ambient air standard of 2 $\mu g/m^3$ (the industry argued for a 5 μg standard) was sufficient to protect health, drew up the first draft, but it was severely criticized by members of the SAC on the grounds that it did not incorporate into its findings the most recent data on lead. A second draft fared little better and only after a third draft from the staff did the SAC find it acceptable.[61]

Each draft of the document was available to the public and each session was publicized. The Natural Resources Defense Council persuaded Needleman and Piomelli[62] to write a summary document presenting the latest lead research. This strategy worked in tandem with the fact that the SAC had a balance of members that included several scientists who were vigorous in pressing the case for reduced exposures, for example, Dr. Samuel Epstein of the University of Illinois Medical School and Dr. Eula Bingham of the University of Cincinnati Kettering Laboratory. Through their efforts the Committee revised the criteria document to require an ambient standard of 1.5 $\mu g/m^3$ and EPA adopted that standard.

Closely related in time with the development of the ambient lead standard was the belief within the CDC that new scientific data required a downward revision of the acceptable blood lead threshold to 30 $\mu g/dl$. That position followed an earlier decision by the American Academy of Pediatrics to accept 30 $\mu g/dl$ as well. EPA drew on views of both the Centers and the Academy to justify its policies.[23]

Even more definitive was the decision of the federal court to accept these actions as scientifically valid. The lead industry appealed both EPA's lead-in-gasoline decision and its ambient air standard. The scientific conclusions, it argued, were highly controversial and therefore, EPA was "arbitrary and capricious" in adopting the 30 μg threshold. But, replied the court, to require "conclusive proof of harm" on which all would agree prior to a decision, was unrealistic. Rarely did such agreement exist in scientific matters; to wait until it did prevail would foster inaction. It was sufficient that a significant body of medical opinion supported the judgment and since both the American Academy of Pediatrics and the CDC did so, as well as a goodly body of scientists, then EPA's judgment should prevail. Such a judgment, even though still controversial, had a sound basis.[21,63-65]

One small but significant indicator of the new and wider context of debate and action was the second lead study completed by NAS in 1978. This time Patterson[66] was included on the review committee. He prepared a minority statement, unconventional in Academy reports, which called for even more stringent regulation of lead than that advocated by the full committee report.

VI. THE LEAD-IN-GASOLINE PHASEDOWN

The first Reagan administration constituted a new period in the politics of lead. New policy leaders identified the scheduled phasedown of lead in gasoline as one of many regulations that should be lifted from industry. But the attempt backfired and the proceedings led to the opposite result — an accelerated phasedown. This turn of events came from a combination of circumstances: the impact of new science, the desire of the administration to enhance its environmental credibility in the 1984 presidential campaign, and vigorous action by some policy-makers in EPA to take advantage of openings for action.[67,68]

Early in the first Reagan term, the EPA administrator, Anne Gorsuch, let it be known that the agency did not intend to enforce the phasedown regulations against small refiners and, in fact, would soon move to increase the allowable concentrations of lead in gasoline. Under the direction of a White House regulatory review committee headed by Vice President George Bush, the administration worked closely with industry behind closed doors to undermine the existing phasedown program.[69,70]

But while Gorsuch could convey assurances quietly and informally to the industry, new regulatory policy required a more open procedure. At the outset EPA apparently thought that the proceedings would go smoothly in its favor. However, they provided an opportunity for evidence accumulating during the 1970s to influence public debate. This evidence included data on low-level effects on the neurological development of children and on animals; statistical studies which demonstrated correlations between declining levels of lead in gasoline, in the ambient air and in blood leads during the years of the phasedown; and the second report on lead from NAS.

At the hearings these reports were featured so effectively by pediatric lead scientists and citizen groups that EPA found little support for its venture to reverse the phasedown. Initiative shifted to those who wished to take the new data seriously.[71,72] EPA abandoned its effort to increase lead levels in gasoline. A year later the agency announced that it would accelerate the phasedown and, in fact, would move to eliminate lead from gasoline entirely.

There were two stages in this process, the first during 1981 and 1982 and the second during 1983 and 1984. As a result of the first proceeding, EPA announced that it would continue the phase down and change the method of determining acceptable gasoline lead levels from an average for all gasoline, leaded and unleaded combined, to an average for leaded gasoline only. This tightened considerably the allowable level of lead.[73,74]

The second action, more extensive and dramatic, lowered the allowable level of lead in leaded gasoline from 1.1 g per leaded gallon on January 1, 1985 to 0.1 g per leaded gallon on January 1, 1986. The proposal was made public first in April 1983 and debated until the early part of 1984. It came from within the administration rather than from additional external initiatives, took participants in the lead debate by surprise, and was finally implemented with limited change.[75-77]

The controlling factor in the EPA action was the upcoming presidential campaign and the need by the administration to establish a new environmental credibility. EPA administrator Anne Gorsuch had resigned and her replacement, William Ruckelshaus now took up the task of restoring the agency's image. This led to new strategies on the part of several relevant government agencies.[78]

The most immediate took place in the EPA Criteria Assessment Office (ECAO) which, under the direction of Lester Grant, was revising the 1977 Lead Criteria Document. In doing so, two types of recent lead research became a major point of dispute. One was the work on the neurological effects of low-level lead exposures and another was the data developed at CDC indicating a correlation between declining levels of lead in gasoline and blood leads. On both these types of evidence the ECAO seemed to be mesmerized by its decision to take industry's objections seriously and to subject them to extended review.[79-87]

The work of EPA staff in drawing up the revised criteria document, however, was now influenced heavily by the Clean Air Science Advisory Committee (CASAC), a subdivision of the EPA Science Advisory Board. That Board was just emerging from a cloud of suspicion which had arisen in the early Reagan years when appointments to scientific advisory positions in the government seemed to be influenced heavily by administrative preference for those scientists who leaned toward industry views. Ruckelshaus now brought into the Board a wider range of scientific opinion that was more open to newer environmental health studies. The CASAC took up both of the subjects at issue within the ECAO and rejected industry arguments. This cleared the way for the new studies to play a major role in the new criteria document as well as regulatory policy.[88]

Even more important was an action within the CDC to revise the lead advisory last drawn up in 1978. The CDC constituted quite a different forum for assessment and action than did the EPA. Its role was advisory rather than regulatory. Its immediate audience was the pediatric clinicians across the country who observed and treated patients and its role was to help them improve the health of their patients. It had already played an important, even crucial role, in better treatment of lead cases by establishing reduced threshold levels of 35 µg/dl and then 30 µg/dl. In each case it had responded directly to the new scientific knowledge and it was prepared to do so now.

In 1983 it appeared to Dr. Vernon Houk of the CDC, that it was time once again to review the new research and consider the possibility of lowering the threshold. This standard was an even more crucial point of debate than was the criteria document.[89,90] The lead industry continued to argue that the lowest acceptable level that could be scientifically supported was 40 µg/dl. But CDC's lower lead level persuaded the court in litigation over the 1977 ambient air standards to reject the industry challenge to the ambient lead standard. The lead industry feared a lower CDC "standard" and tried to prevent CDC action but failed, and in April 1984 CDC announced a new effect level of 25 µg/dl.[91-93]

The administration's political needs and the advancing scientific knowledge as reflected in the criteria document and the CDC lead advisory provided an opportunity for EPA staff in the Office of Policy Analysis and Evaluation (OPAE) to press for a phase-down in lead in gasoline. This represented a marked change from the inability of EPA to act on lead in gasoline in 1973 and even in the ambient air quality standard in 1977. In both previous cases action was forced by external legal pressure from the Natural Resources Defense Council. But in 1983 to 1984 the OPAE and the CASAC took advantage of the opening provided by the administration's political needs to take the initiative and overcome the long-standing lethargy of the ECAO.

On February 3, 1984, 18 months after the EPA announced its policy reversal of 1982 which put lead phasedown back on track, the agency reported a draft cost/benefit study of an accelerated phasedown, and even a total ban, which reflected staff acceptance of the new lead data. In late March it released an internal OPAE cost/benefit study showing that 1988 phasedown benefits outweighed costs by some $700 million. The analysis was approved by the Office of Management and Budget which described OPAE's work as "conservative".[94] In August, Ruckelshaus approved reduction of lead in gasoline from 1.1 g/Pb gal (grams per leaded gallon) on January 1, 1985 to 0.1 g/Pb gal on January 1, 1986. Because of White House opposition the complete ban on leaded gasoline was rejected.[95,96]

The work of the OPAE was spearheaded by Joel Schwartz who became impressed with the new health effects data and worked out the cost/benefit analysis. At the same time OPAE took the initiative to undertake statistical studies of the data in the Second National Health and Nutritional Examination Survey (NHANES II) to demonstrate a correlation between blood leads and anemia in children at low levels, which highlighted and confirmed earlier studies. And in August the OPAE reported a similar relationship between blood lead and hypertension, also based upon the NHANES II data. In 1987, after the phasedown regulations were well in place staff statisticians reported a relationship between blood lead levels and hearing loss which held statistically down to a level of 5 µg/dl of blood lead.[97-98]

The health effects data brought to bear on the lead phasedown issue by the criteria document, the CDC action and the statistical analysis by the OPAE played more of a background rather than a direct role in the regulatory action. The cost/benefit analysis itself placed far more emphasis on the costs saved on automobile maintenance than on the health benefits from reduced lead levels. Moreover, only health benefits based upon protecting children with blood levels over 30 µg/dl were included and only those benefits from reduced costs of health examinations and remedial education were brought into the analysis as benefits.[99]

The supporting documents leading up to the regulatory action had provided also some useful suggestions about additional benefit implications of lower blood lead levels. One was the higher life-time earnings that would come from higher IQ levels. EPA staff analysis had referred to a well-established relationship between the two.[100] Such calculations had, in fact, been used as the basis of court awards in lead liability cases. But this benefit was not used for the analysis.

It was quite apparent that EPA was willing to draw upon the new health effects data in a background manner to provide indirect support for its action, but did not desire to work it directly into regulatory action for fear that it would draw challenges from industry and jeopardize the phasedown effort.

VII. THE RESPONSE OF THE LEAD INDUSTRY

The lead and lead-using industries, traditional opponents of lead regulation, sought to hold back this sequence of actions, but were powerless to do so. While in earlier years it had been able to work quietly within governmental and professional bodies to exercise influence, the more open political realm placed it at a major disadvantage.[101]

Fundamental to lead industry opposition was the continuing conviction that there were no adverse health effects from blood leads lower than 40 μg/dl. Lead industry representatives continually expressed their disagreement with the direction of the assessments of the health effects of lead.[102,103]

In his report on behalf of the Lead Industries Association in 1984, Jim Tozzi wrote "that cognitive and behavorial deficits occur at blood lead levels just above 30 μg/dl..." is an "erroneous assumption." In testimony before the Senate of Environment and Public Works Committee, Dr. Jerome Cole of the International Lead-Zinc Research Organization argued that the main source of lead poisoning in children was lead-based paint and that lead in the ambient air was "only a minor contributor" to the problem. And, according to Lawrence Blanchard, Vice Chairman of Ethyl Corporation, in spite of "overwhelming information to the contrary, EPA continues to allege adverse effects of low-level lead exposure on neurobehavior." Earlier he was more pointed: EPA has "followed the same process of deliberately distorting the facts as they have for the past 12 years. As usual, they have studiously overruled every piece of evidence presented by industry and have accepted every claim by the environmentalists."[104,105]

The lead industry looked to two important forums for potential influence, the CDC advisory proceedings and the EPA criteria document. The first of these provided the industry with only temporary leverage. Dr. Vernon Houk of the CDC had initially hoped to appoint an informal committee to proceed rapidly with the revision instead of the more lengthy process required by law for formal advisory committees. The industry threatened to sue on the grounds that the procedure was irregular. Houk, however, backed off, took up the more formal procedure, and despite further legal threats from the industry, revised the document.[106-108]

The EPA proceeding on the criteria document provided the industry with a more amenable and far reaching forum in which to press its case. The issues were the studies on neurological effects of lead and the relationship between the downward trends in lead in gasoline, ambient air lead and blood leads.

The EPA staff writing the criteria document took these industry contentions and subjected each one to review. This brought to the criteria document proceedings friendly-to-industry experts' arguments, some of whom had been peripheral to the lead debate. Few were pediatric clinicians; a number from England had opposed efforts to reduce lead exposure in that country. Some were biostaticians and social psychologists who while they brought their expertise to bear on the statistical analysis often did so in a manner uninformed about either the clinical problem or the meaning of the issues in dispute.[79-81]

It took the CASAC, composed of scientists rather than EPA staff, to reject these initiatives from industry and to break the logjam within the ECAO. Its meeting of April 26 to 27, 1984, crucial to the entire lead phasedown proceeding, was, in large part, a review of these arguments. One reporter described the results: the CASAC "both blessed the controversial IQ studies and overwhelmingly 'trashed' studies by Ethyl Corporation and Dupont downplaying the linkage between lead in gas and blood lead levels...." While the Office of Criteria Assessment in EPA had earlier recommended that the neurological studies be disregarded in agency action on ambient lead, now it took the cue from CASAC and incorporated them fully into its report.[109]

Having failed to dislodge the health scientists in either the CDC or the CASAC, the industry now chose to tackle the OPAE by developing a critique of its cost/benefit study. It hired as its principle analyst the former director of information and regulatory affairs in the Office of Management and Budget (OMB), Jim Tozzi. Tozzi had a reputation with environmental managers as fairminded and hence was usually taken more seriously than were direct representatives of industry. He had been involved in earlier lead analyses in OMB and after he left government in early 1984 he established his own consulting service and was hired by the Lead Industries Association to press its case.[110,111]

Tozzi proceeded to do so with a report that the EPA's cost/benefit analysis was seriously flawed and that a more proper study would indicate that "benefits represent only a fraction of the costs" — benefits of $559 million and costs of $9,922 million, much the reverse of the EPA figures. Embedded in this analysis was major objection both to the EPA conclusions about the health effects of lead and the damage from lead to automobile engines. A comment from one EPA staffer reflected the agency's reaction: "It just goes to show that for the right amount of money you can make the numbers say anything." Tozzi continued to press the industry's cost/benefit case, but it had no substantial effect on the outcome of the proposed phasedown.[112-114]

While the lead industry could not hold back the drive to phase down the level of lead in gasoline to 0.1 g/Pb gal by January 1, 1986, its strategies did have an important retarding effect on lead regulation with EPA. For while the agency did take up much of the new health effects data as background argument to its regulatory proposal, it did not bring that data directly into the regulatory rationale. It believed that such action would prompt a severe legal challenge from the industry which it did not want to risk. By its actions, therefore, the industry made the EPA quite cautious and prevented it from giving more substantial and formal support to the new health effects research and assessments. It could not prevent a certain amount of action on lead in the lead-in-gasoline phasedown, but it could and did create a climate which helped to retard future further lead action on a variety of fronts.[115]

The phase-down of lead in gasoline, completed by 1987, brought a pause to policy action on lead. The announced intention to completely eliminate lead from gasoline was not followed up,[116-119] the revision of the ambient air standard for lead proceeded slowly,[120-122] and an attempt to focus on lead in drinking water was only partially successful.[123-125] At the same time, however, knowledge about adverse effects of lead at even lower levels of exposure set in motion action to establish lower threshold levels in the range of 12 to 15 μg/dl of blood and prompted public officials to call for further action. But new policy emerged slowly.

VIII. PATTERNS OF DISPUTES: SCIENCE AND VALUES

The debate over lead involved a close connection between disputes over the meaning of science and disputes over desirable public policy. Hardly an event in the series of lead controversies failed to reflect the intense differences of opinion in the two realms. There is a sufficient number of cases of dispute and of individuals participating in the disputes to establish some historical patterns and to analyze systematically their meaning. I take three forays into this problem.

The first is to array the scientists in terms of their positions in the debates over both science and policy. This leads to the conclusion that some consistently were on one side of the issue and others consistently on the other. Some argued that newly emerging scientific data should be taken seriously and incorporated rapidly into public policy — Vernon Houk, Philip Landrigan, Herbert Needleman, and Sergio Piomelli. Others expressed skepticism about the significance of new data, stressed casual factors other than lead in human health problems and cautioned against taking new data too seriously — Robert Bornshein, Jerome Cole, Paul Hammond, Robert Kehoe, and Gordon Stopps.[126]

These scientists were on opposite sides of the issues in a variety of public proceedings: regulatory policy, litigation, federal legislation and city ordinances, scientific assessments, and public health advisories. They were arrayed similarly in scientific symposia and technical workshops. To observe these patterns in a wide range of settings over 25 years provides ample support for the notion that patterns in scientific dispute are closely linked with patterns of advocacy in public policy.

A second mode of analysis is to relate these patterns with institutional affiliations. The first group were associated more with hospitals and medical schools, obtained their research funds more from public sources such as the National Institutes of Health or the EPA, gathered around the CDC and the CASAC as their main hope for action and worked closely with public health and environmental organizations such as the American Academy of Pediatrics, the Natural Resource Defense Council or the Environmental Defense Fund.

The second group associated more closely with industry and its major funding agencies, the International Lead-Zinc Research Organization (ILZRO) or the Lead Industries Association. They tended to speak for those organizations in public proceedings, and they were active participants in symposia which those organizations sponsored and financed. In 1978, for example, the Department of Environmental Health at the University of Cincinnati co-sponsored a "Second International Symposium on Environmental Lead Research," which provided an opportunity for attenders to "hear ILZRO lead research grantees speak on all major facets of the ILZRO lead environmental health program." Among the participants were those closely associated with the lead industry's policy positions — Robert Bornshein, Paul Hammond, Donald Lynam and Henrietta Sachs.[127]

Still a third foray into the systematic analysis of the relationship between the values of scientists and policy choices involves patterns of consistency or change in view as the acceptable level of blood declined over the years and the focal point of dispute shifted. Did scientists on either side of the dispute at one level change their minds in the debate when it shifted to another level? The evidence suggests rather strongly that little such change took place. At each threshold level, one group of scientists rather consistently argued that it was not low enough and that emerging research required that it be lowered further, while another argued that the level had been reduced far enough, if not too far, and that it was quite satisfactory for health protection. At present this pattern of scientific dispute, which occurred at levels of 30 μg/dl and then 25 μg/dl is being replayed for levels below 25 μg/dl.

How does one identify the origin of these patters? It seems quite plausible to argue that institutional affiliation plays a significant role in scientific judgments on lead, as Frances Lynn demonstrated with survey data about disputes over cancer.[128] It seems also plausible to argue that this relationship has a twofold origin. On the one hand, individuals with particular values seek out institutions with which they feel compatible and institutions, in turn, seek out scientists who have made known views acceptable to them. On the other, it is also plausible to argue that personal values associated with distinctive scientific judgments change as one becomes affiliated with a given type of institution.

Still other factors are relevant. One is the role of professional specialization. It is conventional wisdom in science that because it applies objective methods it serves as a unifying force and brings together divergent views in the context of objective knowledge. Yet the general course of science is quite the opposite — increasingly specialized knowledge fractures science and fosters intense dispute. Individual scientists display considerable loyalty to particular bodies of data, methods of research and analytical emphases. The lead issue involves differences between those who specialize in nutrition or inhalation, occupational health or biogeochemical cycles, and those who are clinical pediatricians or laboratory researchers.

Another casual element is more personal and less institutional, arising from the particular psychological orientation of the scientist. Some do not feel personally secure unless they have in hand an extensive level of detailed description and a very high level of proof. Others are more willing to take a risk in their role in the scientific community by advocating a more rapid application of frontier knowledge about environmental health; they feel challenged by the possibilities that taking risks in advocating new approaches might well have great human benefit. These personal differences might well underly both the distinctions in professional commitments and institutional affiliations; at the least they reinforce each other.

In the analysis of environmental politics it has been especially difficult to systematically and dispassionately, focus on these differences in scientific views, often even to obtain acceptance of the problem as worthy of research and debate.

Yet a review of the close relationship between scientific dispute and policy dispute in the case of lead makes clear the need to conduct research on the role of values in science and their relationship to public policy. To fully bring such matters to light might lead participants in these debates to contribute to decision-making in a more informed manner.

REFERENCES

1. **Ames, M. E.**, *Science and the Political Process*, Communications Press, Washington, D.C., 1978.
2. **Primack, J. and von Hippel, F.**, *Advice and Dissent: Scientists in the Political Arena*, Basic Books, New York, 1974.
3. **Mazur, A.**, *The Dynamics of Technical Controversy*, Communications Press, Washington, D.C., 1981.
4. *Science, Technology and Human Values*, Cambridge, Mass., 1976.
5. **Brodeur, P.**, *Expendable Americans*, Viking Press, New York, 1973.
6. **Radford, E. P.**, Risks from ionizing radiation, *Technol. Rev.* 84, 66—78, 1981.
7. **Lansdowne, R. and Yule, W., Eds.**, *Lead Toxicity; History and Environmental Impact*, Johns Hopkins Press, Baltimore, 1986.
8. **Lynam, D. R., Piantanida, L. G., and Cole, J. F.**, *Environmental Lead*, Acadmic Press, New York, 1981.
9. **Needleman, H. L., Ed.**, *Low Level Lead Exposure; the Clinical Implications of Current Research*, Raven Press, New York, 1980.
10. **Rutter, M. and Jones, R. R., Eds.**, *Lead Versus Health; Sources and Effects of Low Level Lead Exposure*, John Wiley & Sons, London, 1983.
11. **Lin-Fu, J. S.**, Lead poisoning and undue lead exposure in children: history and current status, in *Low Level Lead Exposure*, Needleman, H. L., Ed., 1980, 5.
12. **Gibson, J. L., Love, W., Hardine, D., Bancroft, P., and Turner, A. J.**, Note on Lead Poisoning as Observed Among Children in Brisbane, *Transactions 3rd Intercolonial Medical Congress*, 3, 76—83, 1892.
13. **Turner, A. J.**, Lead poisoning among Queensland children, *Aust. Med. Gaz.*, 16, 475—479, 1897.
14. **Gibson, J. L.**, A plea for painted railings and painted walls of rooms as the source of lead poisoning among Queensland children, *Aust. Med. Gaz.*, 23, 149—153, 1904.
15. **Lin-Fu, J. S.**, Vulnerability of children to lead exposure and toxicity, *N. Engl. J. Med.*, 289, 1229—1233, 1289—1293, 1973.
16. **Sayre, J. W., Charney, E., and Vostal, J.**, House and hand dust as a potential source of childhood lead exposure, *Am. J. Dis. Child.*, 127, 167—170, 1974.

17. U.S. Senate Committee on Labor and Public Welfare, Subcommittee on Health, hearing, Lead-based paint poisoning, 91st Cong., 2nd Sess., Nov. 23, 1970, Washington, D.C., 1971.
18. U.S. House of Representatives, Banking and Currency Committee, Subcommittee on Housing, hearings, *To provide federal assistance for eliminating causes of lead-based paint poisoning,* 91st Cong., 2nd Sess., July 22-23, Washington, D.C., 1970.
19. *Environmental Health Letter,* Sept. 1, 1983, 8; July 1, 1984, 3—4.
20. *Environmental Reporter,* Apr. 27, 1979, 2381; Sept. 12, 1980, 683; June 29, 1984, 375; Sept. 7, 1984, 718.
21. **Davis, D. L., Anderson, F. R., Wetstone, G., and Ritts, L. S.,** Judicial review of scientific uncertainty: international harvester and ethyl cases reconsidered, draft copy, The Environmental Law Institute, Washington, D.C., 1981.
22. **Stanton Coerr,** EPA's air standard for lead, in *Low Level Lead Exposure,* Needleman, H. L., Ed., 1980, 253—258.
23. U.S. Department of Health, Education, and Welfare, Public Health Service, Center for Disease Control, Preventing lead poisoning in young children, Atlanta, GA, 1978.
24. National Research Council, *Lead: Airborne Lead in Perspective,* National Academy of Sciences, Washington, D.C., 1972.
25. National Research Council, *Lead in the Human Environment,* National Academy of Sciences, Washington, D.C., 1980.
26. **Ernhart, C. B., Land, B., and Schell, N. B.,** Subclinical levels of lead and developmental deficit — a multivariate follow-up reassessment, *Pediatrics,* 67, 911—919, 1981.
27. **Needleman, H. L., Bellinger, D., and Leviton, A.,** Does lead at low dose affect intelligence in children?, *Pediatrics,* 68, 894—896, 1981.
28. **Ernhart, C. B., Landa, B., and Schell, N. B.,** Lead levels and intelligence, letter to the editor, *Pediatrics,* 68, 903—905, 1981.
29. **Spector, S. and Brown, K. E.,** Lead study questions, letter to the editor, *Pediatrics,* 69, 134—135, 1982.
30. **Ernhart, C. B., Landa, B., and Schell, N. B.,** Letter to the editor, *Pediatrics,* 69, 135, 1982.
31. **Landrigan, P. J.,** Lead study results questioned, letter to the editor, *Pediatrics,* 69, 248, 1982.
32. **Ernhart, C. B., Landa, B., and Schell, N. B.,** Letter to the editor, *Pediatrics,* 69, 248—249, 1982.
33. **Ernhart, C. B., Landa, B., and Wolf, A. W.,** subclinical lead level and developmental deficits: reanalyses of data, *J. Learning Disabilities,* 18, 475—479, 1985.
34. **Needleman, H. L.,** letter to editor, *J. Learning Disabil.,* 19, 322—323, 1986.
35. **Ernhart, C. B.,** letter to editor, *J. Learning Disabil.,* 19, 323, 1986.
36. **Hays, S. P.,** *Beauty, Health and Permanence; Environmental Politics in the United States, 1955—1985,* Cambridge University Press, New York, 1987, 73, 74, 75, 184—185, 196—197.
37. **Wetstone, G. S. and Goldman, J.,** chronology of events surrounding the ethyl decision, in *Judicial Review of Scientific Uncertainty,* Davis, D. L., Anderson, F. R., Wetstone, G. and Ritts, L. S., Eds., Environmental Law Institute, Washington, D.C., 1981.
38. *Environment Reporter* Nov. 30, 1979, 1567—1568; Jan. 29, 1982, 1263; Feb. 26, 1982, 1376; April 15, 1983, 2321.
39. U.S. Senate, Committee on Public Works, *Legislative History of the Clean Air Act Amendments,* Vol. I, Serial 98-18.
40. *Environment Reporter,* Jan. 14, 1977, 1361—1362; Apr. 22, 1977, 1962.
41. Environmental Protection Agency, Proposed National Ambient Air Quality Standard for Lead, Federal Register, 63076, Dec. 14, 1977.
42. *Environment Reporter,* Feb. 24, 1978, 1651; June 29, 1984, 375.
43. **Rosner, D. and Markowitz, G.,** A gift of God?: the public health controversy over leaded gasoline during the 1920s, *Am. J. Public Health,* 75, 344—352, 1985.
44. *Environmental Health Letter,* Apr. 15, 1985, 5.
45. Committee on Public Works, U.S. Senate, Subcommittee on Air and Water Pollution, 89th Congress, 2nd Sess., Air Pollution, Washington D.C., 1966, 203—228.
46. **Byers, R. K. and Lord, E. E.,** Late effects of lead poisoning on mental development, *Am. J. Disabled Chil.,* 66, 471—481, 1943.
47. **Byers, R. K.,** Lead poisoning. Review of the literature and report on 45 cases, *Pediatrics,* 23, 583—603, 1959.
48. U.S. Environmental Protection Agency, Air Quality Criteria for Lead, EPA-600/8-77-017, Research Triangle Park, NC, 1977.
49. **Patterson, C. C.,** contaminated and natural lead environments of man, *Arch. Environ. Health,* 11, 344—360, 1965.
50. **Patterson, C. C.,** Natural Levels of Lead in Humans, Carolina Environmental Essay Series, III, The University of North Carolina at Chapel Hill, 1982.

51. **Lovering, T. G., Ed.,** Lead in the environment, U.S. Geological Survey Professional Paper #957, Washington, D.C., 1976.
52. U.S. Senate, Committee on Public Works, Subcommittee on Air and Water Pollution, 89th Cong., 2nd Sess., Hearings, Air Pollution, 1966, Washington, D.C., 1966, 311—344.
53. **Sullivan, W.,** Warning is issued on lead poisoning, New York Times, p.71, Sept. 12, 1965.
54. **Budiansky, S.,** Lead: the debate goes on, but not over science, *Environ. Sci. Technol.*, 15, 243—246, 1981.
55. Letters, *Environ. Sci. Technol.*, 15, 722—724, 1981.
56. **Piomelli, S.,** The FEP (free erythrocyte porphyrins) test: a screening micromethod for lead poisoning, *Pediatrics*, 51, 254—259, 1973.
57. **Gillette, R.,** Lead in the air: industry weight on academy panel challenged, *Science*, 174, 800—2, 1971.
58. **Boffey, P. M.,** *The Brain Bank of America*, McGraw Hill, New York, 1975, 228—44.
59. **Goldsmith, J. R. and Needleman, H. L.,** Lead exposures of urban children: a handicap in school and life? unpublished manuscript, 1984.
60. **Wetstone, G. S., Ed.,** Meeting Record from Resolution of Scientific Issues and the Judicial Process: Ethyl Corporation v. EPA, Oct. 21, 1977, Washington, D.C., The Environmental Law Institute, 1981, 145.
61. *Environment Reporter*, Nov. 19, 1976, 1043; Dec. 10, 1976, 1173; Dec. 31, 1976, 1253; Jan. 14, 1977, 1361—1362, 1376; Feb. 4, 1977, 1486—1487; Mar. 4, 1977, 1701; Mar. 25, 1977, 1809—1810; Apr. 22, 1977, 1962; June 17, 1977, 274; July 8, 1977, 409—410; Sept. 2, 1977, 686—687; Oct. 14, 1977, 929—930; Dec. 16, 1977, 1235—1236, 1243—1244; Jan. 13, 1978, 1392, 1394; Feb. 24, 1978, 1651; Mar. 31, 1978, 1880.
62. **Needleman, H. L. and Piomelli, S.,** *The Effects of Low Level Lead Exposure*, Natural Resources Defense Council in cooperation with American Lung Association, New York, 1978.
63. Ethyl Corporation v. EPA, 8 Environment Reporter Cases, 1785.
64. *Environment Reporter*, Aug. 3, 1979, p.899.
65. Lead Industries Association v. EPA, June 27, 1980, 14 Environment Reporter Cases, 1906, Dec. 8, 1980, 15 Environment Reporter Cases, 2097.
66. **Patterson, C. C.,** An alternative perspective — lead pollution in the human environment: origin, extent, and significance, in *Lead in the Human Environment*, National Research Council, National Academy of Sciences, Washington, D.C., 1980.
67. **Bollier, D. and Claybrook, J.,** *Freedom From Harm; The Civilizing Influence of Health, Safety, and Environmental Regulation*, Public Citizen, Washington, D.C., 1986, 105—108, 144, 148.
68. **Claybrook, J.,** *Retreat From Safety: Reagan's Attack on America's Health*, Pantheon Books, New York, 1984, 77, 85—86, 127—128.
69. *Environment Reporter*, Aug. 14, 1981, 483; Dec. 11, 1981, 975—976; Feb. 5, 1982, 1293—1294; Apr. 9, 1982, 1609; Apr. 23, 1982, 1715.
70. *Environmental Health Letter*, Aug. 15, 1981, 1; Mar. 1, 1982, 6.
71. *Environmental Health Letter*, Apr. 15, 1982, 2—3.
72. *Environment Reporter*, April 16, 1982, 1639.
73. *Environment Reporter*, May 21, 1982, 61; June 11, 1982, 165; Aug. 20, 1982, 525—526.
74. *Environmental Health Letter*, Aug. 15, 1982, 4—5.
75. *Inside EPA*, Feb. 3, 1984, 1, 8—9; Mar. 16, 1984, 12—13; Mar. 30, 1984, 13.
76. *Environmental Health Letter*, Mar. 1, 1984, 1—2.
77. *Environment Reporter*, Mar. 2, 1984, 1899; Mar. 16, 1984, 2046; Apr. 6, 1984, 2206—2207.
78. *Inside EPA*, March 30, 1984, 13; July 20, 1984, 3.
79. **Marshall, E.,** EPA faults classic lead poisoning Study, *Science*, 222, 906—907, 1983.
80. **Needleman, H. L.,** letter, *Science*, 223, 116, 1984.
81. **Landrigan, P. J. and Houk, V. N.,** letter, *Science*, 223, 116, 1984.
82. **Ernhart, C. B.,** letter, *Science*, 223, 116, 1984.
83. *Inside EPA*, May 4, 1984, 11.
84. **Annest, J. L., Pirkle, J. L., Makuc, D., Neese, J. W., Bayse, D. D., and Kovar, M. G.,** Chronlogical trend in blood lead levels between 1976 and 1980, *N. Engl. J. Med.*, 308, 1373—1377, 1983.
85. **Schwartz, J.,** The Relationship Between Gasoline Lead Emissions and Blood Poisoning in Americans, unpublished manuscript, no date.
86. **Schwartz, J.,** The link between lead in people and lead in gas, *EPA J.*, 11, 12, May 1985.
87. *Environment Reporter*, Aug. 10, 1984, 571; Sept. 7, 1984, 718.
88. *Inside EPA*, May 4, 1984, 14.
89. *Inside EPA*, Mar. 23, 1984, 11.
90. *Environmental Health Letter*, Feb. 15, 1985, 3—4.
91. *Environment Reporter*, June 29, 1984, 380—381.
92. *Inside EPA*, June 29, 1984, 13.

93. U.S. Department of Health and Human services, Centers for Disease Control, Preventing Lead Poisoning in Young Children Atlanta, GA, 1985.
94. *U.S. Environmental Protection Agency,* Office of Policy, Planning and Evaluation, Costs and Benefits of Reducing Lead in Gasoline, Draft Final Report, EPA-230-03-84-005, Washington, D.C., March, 1984, and Final Report 1985.
95. *Environment Reporter,* Aug. 3, 1984, 532; Aug. 10, 1984, 585ff.
96. *Inside EPA,* Aug. 3, 1984, 8.
97. **Schwartz, J. and Otto, D.,** Blood lead, hearing thresholds, and neurobehavioral development in children and youth, *Arch. Environ. Health,* 42, 153—160, 1987.
98. *Inside EPA,* Aug. 31, 1984, 9; Jan. 23, 1987, 2.
99. **Goldstein, B.,** Health and the lead phasedown; an interview with Bernard Goldstein, assistant administrator for research and development, *EPA J.,* 11, 9—12, May 1985.
100. **Pitcher, H. M.,** Office of Policy, Planning and Evaluation, EPA, Comments on Issues Raised in the Analysis of the Neuropsychological Effects of Low Level Lead Exposure, no date.
101. *Environment Reporter,* Jan. 13, 1978, 1392—1393.
102. *Environment Reporter,* Aug. 10, 1984, 571.
103. **Cole, J. F.,** The lead in gasoline issue and EPA's lack of scientific objectivity, *Environ. Forum,* 3, 41, Nov. 1984.
104. *Environment Reporter,* May 11, 1984, 41; June 29, 1984, 375.
105. *Environmental Health Letter,* Sept. 1, 1982, 4—5.
106. *Environment Reporter,* June 29, 1984, 380—381.
107. *Inside EPA,* June 29, 1984, 13.
108. *Environmental Health Letter,* Feb. 15, 1985, 3—4.
109. *Inside EPA,* May 4, 1984, 11.
110. **Anon.** Profile — OMB's Jim Joseph Tozzi, *Environ. Forum,* 1, 11—12, May 1982.
111. *Environmental Health Letter,* Aug. 1, 1984, 3.
112. *Inside EPA,* June 29, 1984, 13; July 20, 1984, 5.
113. *Environmental Health Letter,* July 1, 1984, 3—4.
114. *Environment Reporter,* July 27, 1984, 465.
115. *Environment Reporter,* Aug. 10, 1984, 571.
116. *Environment Reporter,* July 29, 1984, 375.
117. *Inside EPA,* July 20, 1984, 3; July 27, 1984, 1, 11; Aug. 3, 1984, 8; Jan. 4, 1985, 1, 6.
118. *Environmental Health Letter,* Mar. 1, 1985, 1.
119. U.S. Puts Off Plan to Outlaw Leaded Gasoline, *New York Times,* p.A16 Dec. 25, 1987.
120. *Environment Reporter,* June 1, 1984, 146.
121. Environmental Protection Agency, Draft document, Air quality criteria for lead, EPA 600/8-83-028, Office of Research and Development, Research Triangle Park, 1984.
122. U.S. Environmental Protection Agency, Review of the national ambient air quality standards for lead: assessment of scientific and technical information, Draft staff paper, Office of Air Quality Planning and Standards Research Triangle Park, 1986.
123. *Inside EPA,* May 31, 1985, 12; Aug. 16, 1985, 4; Nov. 14, 1986, 13; Nov. 21, 1986, 1—2, 4; Dec. 5, 1986, 2; Dec. 12, 1986, 3—4; Jan. 23, 1987, 2; Apr. 17, 1987, 15; Sept. 2, 1988, 4.
124. *Environmental Health Letter,* Oct. 1, 1986, 4; Nov. 15, 1986, 7—8; April 1, 1987, 2—3.
125. *Environment Reporter,* May 1, 1987, 9—10; Oct. 9, 1987, 1503; Dec. 18, 1987, 1923—1924; Dec. 18, 1987, 1924—1925.
126. *Environment Reporter,* Sept. 7, 1984, 718.
127. **Lynam, D. R., Piantanida, L. G., and Cole, J. F.,** *Environmental Lead,* Academic Press, London, 1981.
128. **Lynn, F.,** The interplay of science and values in assessing and regulating environmental risks, *Sci. Technol. Human Values,* 11, 40—50, 1986.

Index

INDEX

A

Adrenergic effects and neurotoxicity, 95—96
Agency for Toxic Substances and Disease Registry (ATSDR), 37—38
Airborne lead
    anthropogenic sources, 67—69
    atmospheric deposition, 73—78
    characteristics, 69—73
    monitoring, 48—49
    natural sources, 66—69
δ-Aminolevulinic acid dehydratase inhibition test, 58—59
Anthropogenic emissions worldwide, 67
Anthropometric effects of low-level exposure, 234—237
Aphrodisiacal uses, 16
Asymptomatic nephropathy, 176—177
Atmospheric deposition, 73—78
Attention, learning, and memory in developmental exposure, 141—148
Auditory alterations, 116—118, 239
Australian paint poisoning studies, 25—27, 272
Automobile batteries as source, 28—29, 161, 164

B

Baltimore battery casing disease, 28—29
Basal meningitis, in paint poisoning, 25—26
Batteries, automobile, as source, 28—29, 161, 164
Behavioral impairment, 138—150, 192—205, see also Cognitive impairment
Beverages as source, 29—30
Biokinetic/metabolic aspects of monitoring, 52—54
Biological monitoring, 52—60
Birth weight, 234—236
Blood
    biokinetic/metabolic aspects of exposure, 52
    monitoring techniques, 54—56
Blood-brain barrier and neurotoxicity, 128—131
Blood levels
    behavioral effects and developmental exposure, 148—149
    blood pressure elevation and, 228
    growth rates correlated to, 236—237
    in iron deficiency, 239
    neurodevelopmental deficit and, 196
    normal limits, 34—35, 268
    renal effects and, 173—174
Blood pressure, see Hypertension
Bone lead, 54, 57—58, 176, 184
Brain lead, 93—94, 210—220, see also Cerebral neuropathy
Breastfeeding and cosmetic toxicity, 28

C

Calcium transport, 132, 181—182, 225

Cardiovascular disease, 180—181, 224—229, see also Hypertension
Catecholamines in lead-induced hypertension, 183—184
Cation transport in lead-induced hypertension, 181—184
Centers for Disease Control (CDC) lead advisory, 269, 274—277
Cerebral neuropathy, 93—94, 164
    developmental aspects, 126—133
    low-level exposure and, 210—220
Ceruse, 10
Chelatable lead, 52—54, 56, 173—174
Children
    absorption of lead in, 34—35
    acute lead nephropathy in, 170—172
    behavioral testing procedures, 140—141
    dosage and blood levels, 138—140
    epidemiology of lead poisoning, 33—38
    importance of monitoring, 50—52
    IQ and low-level exposure: meta-analysis, 244—254
    of lead workers, 163
    low-level exposure, 192—205, 210—220, 234—239, 244—254
    neurotoxicity in, 91—92
Chronic tubulointerstitial nephritis, 173—179
Clean Air Science Advisory Committee, 275—276, 278—280
Cognitive impairment, 91—92, 192—205, 272
    IQ and low-level exposure: metaanalysis, 244—254
Colic outbreaks, 29—30
Congenital anomalies, 237
Construction trades, 162—163
Cosmetics and toxicity in nursing infants, 28
Cyclic nucleotide metabolism, visual alterations and, 111—112

D

Dark/light adaptation, 110—111
Dental lead, see Tooth lead
Depression disease, 28—29
Developing countries, 272—273
    occupational exposure, 163—164
Developmental exposure
    auditory effects, 116—118
    neural effects, 126—133, 191—205, see also Neurodevelopmental effects
    primate studies of behavioral impairment, 138—150
    visual effects, 107—116
Drinking water, 32, 51—52
Drug contamination, 13
Dry deposition, 76—78

Dustborne lead, 32—33, 79, 164

E

Economic aspects of litigation, 260—265
Electroencephalography (EEG), correlation with neuropsychologic measures, 210—220
Electron microscopic studies ocular toxicity, 108—109
Electroretinographic studies, 109—110
Environmental lead
    environment/human pathways, 78—81
    monitoring, 48—52
    source/environmental pathways, 65—78
Environmental Protection Agency, 269, 274—277
Epidemiologic studies, see also Monitoring
    blood pressure, 226—229
    growth and low-level exposure, 234—238
    IQ and low-level exposure: meta-analysis, 244—254
    lead poisoning in children, 33—38
    methodology, 47, 244—248
Exposure types and monitoring systems, 46—47

F

Fanconi syndrome, 170—172
Fetal death, 237—238
Fetal effects, 38—39, see also Developmental exposure
Folk remedies, 33
Foodborne lead, 29—30, 32, 49—50, 80—81
Free erythrocyte protein level studies, 238—239

G

Galena, 6
Gasoline, 30—31, 68
    evolution of public policy, 269—277
Gestational age and low-level exposure, 234—236
Glazes, 30
Gout, 177—179
Growth rates and blood levels, 236—237

H

Hair levels, 54, 58
Hearing loss, 116—118, 239
Heme-related exposure indicators, 59—60
History
    early uses in medicine, 3—18
    modern (1890s–present), 23—39
    occupational exposure, 156—158
    public policy and environmental science, 269—274
Hormonal mediation in lead-induced hypertension, 183—184
Hypertension
    epidemiologic studies, 226—229
    lead-induced nephropathy and, 179—184, 224—229, see also Nephropathy

I

Industrial exposure, 158—166
Industry response to public policy, 277—279
Ingestion and human uptake, 81
Inhalation and human uptake, 81
IQ, see also Cognitive impairment
    litigational applications in paint poisoning cases, 260—265
    low-level exposure and: meta-analysis, 244—254
Iron deficiency, 238—239

J

Japanese studies, breastfed infants and cosmetic poisoning, 28
Job-related exposure, 158—166

L

Laboratory aspects of monitoring, 47—48
Lead acetate, 10—13
Lead colic, 29—30
Lead industry, response to public policy, 277—279
Lead nephropathy, see Nephropathy
Lead oxides, historical uses, 6—10
Lead paint, see Paint poisoning
Lead sulfide, historical uses, 6
Learning, attention, and memory in developmental exposure, 141—148
Legal/economic issues, 260—265
Light/dark adaptation, 110—111
Light microscopic studies of ocular toxicity, 107—108
Litharge, historical uses, 6—10
Low birth weight, 234—236
Low-level exposure
    EEG correlation with neuropsychologic measures, 210—220
    growth and development and, 234—239
    hypertension and, 224—229
    IQ and: meta-analysis, 244—254
    neurodevelopmental deficits and, 192—205
    neurological effects, 239

M

Memory, attention, and learning in developmental exposure, 141—148
Meningitis
    basal and paint poisoning, 25—26
    serous and cosmetic poisoning, 28
Meta-analysis: IQ and low-level exposure, 244—248
Metabolic studies, growth and low-level exposure, 238—239
Metallic lead, historical uses, 5—6
Methodological issues in epidemiology, 244—248
Mineral elixirs, 15—18
Monitoring
    biological, 52—60

environmental, 48—52
future directions, 60—61
methods, 47—48
rationale for, 46
types, 46—47
Moonshine, 30
Mortality, fetal, 237—238
Mortality studies, lead nephropathy/hypertension, 181
Motor fuels, see Gasoline

N

Na,-K-ATPase activity and visual alterations, 112—114
National Academy of Sciences report, 270, 273
Natural lead emissions worldwide, 67
Nephropathy
  in adults, 173—179
  asymptomatic, 176—177
  in children, 170—172
  gout, 177—179
  hypertension and, 179—184
Neurodevelopmental effects, 192
  exposure and severity, 194—197
  modification and variability, 198—201
  nature of deficit, 193—194
  response variability, 198—201
  reversibility/persistence, 201—204
Neuropathy
  central, 93—94
  peripheral, 239
Neurotoxicity
  auditory system, 116—118
  blood-brain barrier and, 128—131
  developmental neurobiology of, 126—133
  in industrial exposure, 164
  localized effects, 92—94
  mechanisms, 97—98
  neurochemical aspects, 95—96
  neurophysiological aspects, 94—95
  research directions, 96—97
  second-messenger metabolism in, 131—133
  systemic, 91—92
  visual system, 107—116
NHANES II study, 36—38, 179—180, 238—239, 276

O

Occupational exposure
  case management, 165
  current sources, 158—164
  health effects, 164
  history, 156—158
  prevention, 165—166
  public policy evolution, 271—272
Ocular toxicity, 107—114
Ototoxicity, 26, 114, 116—118, 239

P

Paint poisoning, 50—51

Australian studies, 25—27, 272
  in construction and demolition workers, 162—163
  economic aspects of litigation, 260—265
  representative legal case, 262—265
  in United States, 27—28
Particulates
  composition of lead aerosols, 70
  size distributions, 71—73
Peripheral neuropathy, 164, 239
  in paint poisoning, 25
Persistence/reversibility of cognitive effects, 201—204
Pharmacologic uses
  contamination of nonlead drugs, 13
  historical aspects of exposure, 3—18
  lead acetate, 10—13
  lead oxides, 6—10
  lead sulfide (galena), 6
  metallic lead, 5—6
  mineral elixirs, 15—18
  reproductive system diseases, 14—15
  white lead, 10
Pituitary/thyroid function, 238
Plasma lead, 55—56, 239
Pottery glazes, 30
Precipitation, lead content, 75—76
Prevention, 38—39, 165—166
Primate studies of behavioral impairment, 138—150
Protein kinase C, 132, 225—226
Psychologic testing, correlation with EEG, 210—220
Public policy aspects, 267—280

Q

Quality assurance/quality control, in monitoring, 48, 50—52
Quantitative EEG studies of low-level exposure, 210—220
Queensland, Australia, paint poisoning studies, 25—27
Queensland nephritis, 172

R

Renal effects, 170—184, see also Nephropathy
Renin-angiotensin-aldosterone system, 183
Reproductive diseases, early uses of lead, 14—15
Response variability, 198—201
Retinal ganglion cell axon studies, 114
Retinal studies of ocular toxicity, 107—114
Reversibility/persistence, of cognitive effects, 201—204, 272
Rhodopsin and ocular toxicity, 110

S

Sampling, 47, 49—52, 50
Science and public policy: value choices, 267—280

Second messenger metabolism in neurotoxicity, 131—133
Sex differences in cognitive effects, 198
Smelter employees, 160—161
Social behavior and developmental exposure, 150
Socioeconomic status and cognitive effects, 198
Sodium transport in lead-induced hypertension, 182—183
Soilborne lead, 32—33, 79, 164, 271
Statistical issues in epidemiologic studies, 244—248
Stature and low-level exposure, 234
Stillbirth, 237—238
Sugar of lead (lead acetate), 10—13
Synaptogenesis and neurotoxicity, 126—128

T

Tap water, 32, 51—52
Teratogenicity, 38—39, 237, see also Developmental exposure
Thyroid/pituitary function, 238
Tooth lead, 52, 56—57, 197
Treatment of occupational exposure, 165

U

Uptake in humans, 81

Urine levels, 52, 56, 175—176, see also Nephropathy

V

Value choices in science and public policy, 267—280
Variability of response, 198—201
Vasoactive hormones in lead-induced hypertension, 183—184
Visual alterations and lead exposure, 107—116
Vitamin D metabolism, 238

W

Water-borne lead, 32, 51—52, 79—80
Wet deposition, 73—76
White lead, historical uses, 10
WHO quality standards for monitoring, 48

XYZ

Zinc protoporphyrin analysis, 59—60